MODERN CONSTRUCTION MANAGEMENT

Ognjen Grkinic

TAP

Toronto Academic Press

MODERN CONSTRUCTION MANAGEMENT

Ognjen Grkinic

Toronto Academic Press
4164 Lakeshore Road
Burlington ON L7L 1A4
Canada
www.tap-books.com

Email: orders@arclereducation.com

ISBN: 978-1-77956-339-2

ABOUT THE AUTHOR

Ognjen Grkinić is a physicist with a profound expertise in applied and computer physics, particularly in data science and plasma physics. He is currently pursuing his PhD at the Faculty of Physics, University of Belgrade, where his research focuses on the behavior of ionized gases and plasma within the Attosecond-Plasmas Lab. Ognjen's academic achievements include a Master's degree in applied and computer physics, where he developed advanced software for processing scientific data, earning a near-perfect GPA of 9.80/10.00. In addition to his research, Ognjen has contributed significantly to education, having taught physics and mathematics at both high school and primary school levels. His passion for teaching is matched by his meticulous approach to scientific inquiry, which is evident in the books and papers he has authored and edited in the field of physics and other natural sciences. Beyond academia, Ognjen Grkinić's diverse experiences in research, education, and professional work underscore his commitment to advancing the field of physics and contributing to the broader scientific community.

Table of Contents

7 QUALITY MANAGEMENT 233

8 HEALTH AND SAFETY MANAGEMENT 265

List of Figures

List of Tables

PREFACE

Modern Construction Management (MCM) encompasses construction techniques that utilize off-site components manufactured and assembled away from the construction site to be later assembled on-site. This method has been gaining traction globally as a response to the limitations of traditional construction practices, often referred to as prefabrication, off-site, or industrial construction. MCM aims to alleviate challenges related to cost, time, defects, and risks to health and safety by implementing standardized procedures, a controlled manufacturing environment, and a single project phase owner. These features contribute to enhanced predictability and efficiency throughout the project lifecycle.

As the global population continues to grow at an unprecedented rate, the demand for housing escalates, making the rapid aggregation of housing units a critical aspect of global development. This surge necessitates the shortest possible production periods to meet the rising demand, which is achievable through the adoption of modern construction methodologies. For instance, in countries like India, the improvement in housing infrastructure has not kept pace with government targets, leading to a significant shortfall. The high demand in the housing market has also driven up costs, making it increasingly difficult for essential construction workers and other low-wage workers to secure affordable housing.

The production of affordable private sector housing often falls outside government control. There has been a significant push towards fostering innovative and advanced construction techniques to accelerate housing accumulation rates. This focus on modernization and innovation is crucial to addressing the housing shortages faced by many rapidly urbanizing regions.

Concurrently, digital technologies are becoming increasingly prevalent in both personal and professional spheres, including the construction industry, despite its reputation as one of the least digitalized sectors. This trend is explored in depth, examining how various technologies have been integrated into construction management responsibilities.

The interplay between these construction management responsibilities and digital technologies forms the foundation of modern construction management practices. While some technologies are already in widespread use among construction management professionals, many are still in their pilot stages. These emerging technologies not only impact present practices but also offer insights into the future direction of the entire discipline, highlighting the ongoing evolution towards a more technologically integrated construction industry.

Contents/Structure of the Book:

This book offers a comprehensive overview of the construction industry, covering 10 key chapters.

In Chapter 1, the book focuses on the construction industry, providing insights into its evolution, current trends, and future prospects.

Chapter 2 focuses on Construction Management, detailing the principles, processes, and best practices involved in managing construction projects effectively.

Production planning and control are discussed in Chapter 3, highlighting the importance of efficient planning and control mechanisms in ensuring project success.

Site management is the focus of Chapter 4, which explores the various aspects of managing construction sites, including logistics, resource allocation, and safety protocols.

Time management in construction projects is addressed in Chapter 5, emphasizing the significance of scheduling, progress tracking, and timely completion.

Cost management is the subject of Chapter 6, discussing strategies for cost estimation, budgeting, and cost control throughout the project lifecycle.

Chapter 7 focuses on Quality management, highlighting the importance of quality assurance and quality control measures in construction projects.

Health and safety management are discussed in Chapter 8, emphasizing the need for stringent safety protocols and practices to protect workers and minimize risks.

Environmental management in construction is covered in Chapter 9, focusing on sustainability, environmental regulations, and green building practices.

Lastly, Chapter 10 explores Construction Labor and Labor Management, discussing the challenges and strategies for managing construction workforce effectively.

This book offers a comprehensive guide for construction professionals, students, and enthusiasts, providing valuable insights into the key aspects of modern construction practices.

Audience:

This book comprises a diverse range of professionals and students involved in the construction industry. Firstly, it targets construction managers and project managers who seek to enhance their understanding of contemporary management techniques and best practices. These professionals will benefit from the detailed exploration of project planning, execution, and control strategies that are crucial for successful project delivery.

Additionally, the book is aimed at architects, engineers, and quantity surveyors who are integral to the construction process. By understanding the management aspect, these professionals can better collaborate with construction managers, ensuring that projects meet both design and budgetary requirements.

Students pursuing degrees in construction management, civil engineering, and related fields are another primary audience. The book serves as an educational resource, providing theoretical knowledge and practical insights that are essential for their academic and professional development.

Lastly, the book is beneficial for contractors and subcontractors who need to navigate the complexities of modern construction projects. By gaining a deeper understanding of management principles, these professionals can improve their operational efficiency and contribute more effectively to the overall success of their projects.

Learning Features:

- Start-of-chapter learning objectives;
- Key terms in each chapter;
- Summary Points at the end of each chapter;
- Review Questions at the end of each chapter;
- Case Studies and Activities in each chapter for solving real problems;
- References at the end of each chapter.

CHAPTER

1 Construction Industry

LEARNING OBJECTIVES

After studying this chapter, you will be able to:

- Examine the basics of construction;
- Define the role of construction industry;
- Identify the various construction industry sectors.

KEY TERMS FROM THIS CHAPTER

Structural work	Plumbing
Gross domestic products (GDP)	Fast-track construction
Skyscrapers	Industrial revolution
Building information modeling (BIM)	Financial closure risk
Paymaster	Financial structure
Mortgage bankers	Cost engineers
Financial management	Construction project

INTRODUCTION

The planning, designing, constructing, and maintaining of infrastructure and structures are all included in the broad spectrum of activities that make up the construction industry. Because it involves the construction of residential buildings, business establishments, industrial facilities, and different types of infrastructure like roads, bridges, and utilities, this industry is essential to the development and growth of communities. The construction industry is essential to economic growth because it generates a sizable portion of the GDP and employs a sizable workforce that includes laborers, engineers, architects, and skilled tradespeople. The process of converting unfinished materials and architectural blueprints into tangible structures is the foundation of the construction sector. Architects and engineers work to ensure that projects are both functional and aesthetically pleasing while also adhering to safety standards and regulations. This process starts with extensive planning and design. The real building process starts after the design stage and calls for the coordinated efforts of suppliers, subcontractors, and contractors. Every project has a number of phases, which include clearing the site, building the foundation, constructing the structure, installing the systems (like plumbing and electrical), and adding finishing touches like landscaping and painting.

The diversity of the construction industry, which includes a wide range of project types and sizes, is one of its distinguishing characteristics. From modest home improvements to large-scale infrastructure projects, the sector necessitates a wide range of abilities. Another characteristic of the industry is its cyclical nature, which is frequently impacted by governmental regulations, technological developments, and economic situations. The demand for new infrastructure and buildings tends to increase during times of economic expansion, which boosts activity in the construction industry. On the other hand, financial slumps may result in lower spending on building initiatives, which may have an effect on employment and industry investment. The building industry has been greatly impacted by technological innovation, which has brought new techniques and materials that improve sustainability and efficiency. The implementation of Building Information Modeling (BIM) facilitates enhanced project planning and visualization, thereby mitigating errors and enhancing stakeholder collaboration. New developments in building materials, like high-performance concrete and energy-saving systems, help to build structures that are both more resilient and ecologically friendly. Furthermore, the use of robotics and automation in building processes is starting to change the way buildings are built, with the potential for increased precision and lower labor costs.

Notwithstanding its growth and significance, the construction sector still faces a number of difficulties. Given the inherent hazards of construction sites, safety is still a top priority. It is crucial to guarantee employee well-being by implementing strict safety procedures and providing adequate training. In addition, the industry is experiencing a skilled labor shortage brought on by an aging workforce and a dearth of newcomers. Industry stakeholders must work together to address these problems, and part of that effort will need to be funding education and training initiatives to draw and keep talent.

1.1. BASICS OF CONSTRUCTION

Construction refers to the process of creating objects, systems, or organizations, combining art and science. The term originates from the Latin word constructio, meaning to pile up, and Old French construction. To "construct" is an action of building, while "construction" refers to how something is built or its structure. In a broad sense, construction includes all the steps involved in creating buildings, infrastructure, industrial facilities, and related activities until they are completed. This often begins with planning, financing, and design, continuing until the asset is ready for use. Construction encompasses maintenance, repair, expansion, and eventual demolition or decommissioning of the asset.

1.1.1. History

The earliest huts and shelters were built by hand or using basic tools. As civilizations grew in the Bronze Age, skilled craftsmen such as bricklayers and carpenters emerged, with occasional use of slaves for construction. During the Middle Ages, artisan craftsmen formed guilds. The 19th century saw the introduction of steam-powered machinery, followed by diesel- and electric-powered vehicles like cranes and excavators. The history of construction showcases the evolution of human civilization, from simple shelters to intricate skyscrapers and advanced infrastructures. Here's a brief overview of the significant milestones in the history of construction.

1.1.1.1. Ancient Civilizations

- Prehistoric Era: Early humans constructed simple shelters using natural materials such as branches, leaves, and animal skins. These were primarily used for protection from the elements and predators.
- Mesopotamia and Egypt: The construction of large-scale structures began with civilizations like the Sumerians and Egyptians. The ziggurats of Mesopotamia and the pyramids of Egypt are notable examples. The use of mud bricks in Mesopotamia and stone in Egypt marked significant advancements in building techniques.
- Greece and Rome: The ancient Greeks and Romans made substantial contributions to construction technology. The Greeks introduced architectural orders (Doric, Ionic, Corinthian) and used marble extensively.

1.1.1.2. Medieval Period

- Byzantine and Islamic Architecture: The Byzantine Empire saw the construction of iconic buildings like the Hagia Sophia, which featured advanced dome construction. Islamic architecture flourished with the construction of mosques, madrasas, and palaces, characterized by intricate tile work and geometric designs.

- Gothic Architecture: Originating in the 12th century, Gothic architecture is marked by pointed arches, ribbed vaults, and flying buttresses. Cathedrals such as Notre-Dame de Paris exemplify the height and grandeur achievable with these techniques.

1.1.1.3. Renaissance to Industrial Revolution

- Renaissance Architecture: The Renaissance period revived classical Greek and Roman architectural principles. Architects like Filippo Brunelleschi and Andrea Palladio emphasized symmetry, proportion, and geometry, leading to the construction of iconic buildings such as St. Peter's Basilica in Rome.
- Baroque and Rococo: The Baroque era introduced dramatic, elaborate designs with extensive use of curves, domes, and ornate detailing. This style evolved into the lighter and more playful Rococo style in the 18th century.
- Industrial Revolution: The advent of the Industrial Revolution in the 18th century brought significant changes to construction. The development of iron and steel production led to the creation of stronger and more flexible materials, facilitating the construction of larger and taller buildings. The Crystal Palace in London and the Eiffel Tower in Paris are notable examples.

1.1.1.4. Modern Era

- Early 20th Century: The early 20th century saw the rise of modernist architecture, emphasizing function, simplicity, and the use of new materials like reinforced concrete and glass. Architects such as Le Corbusier, Frank Lloyd Wright, and Ludwig Mies van der Rohe were influential in this movement.
- Post-World War II: The post-war era witnessed rapid urbanization and the construction of mass housing projects. The use of prefabricated materials and standardized construction techniques became widespread. Skyscrapers began to dominate city skylines, with the Empire State Building and later the World Trade Center as iconic examples.
- Late 20th Century and Beyond: The latter part of the 20th century and the early 21st century have seen a focus on sustainability and environmental impact. Green building practices and the use of energy-efficient materials and technologies have become prominent. Iconic modern structures include the Burj Khalifa in Dubai, which currently holds the title of the world's tallest building.

1.1.1.5. Technological Advancements

The constant progress in technology has also left its mark on the history of construction. The construction industry has seen continuous transformation due to technology, ranging from the utilization of simple hand tools to the creation of advanced machinery and software. The future of construction is being shaped by innovations like 3D printing, Building Information Modeling (BIM), and using drones to inspect sites.

1.1.2. Risk in Construction Sector

The construction industry is thought of as a service sector. It creates a significant amount of jobs and boosts the expansion of other manufacturing industries, including those that produce chemicals, bitumen, cement, iron and steel, bricks, paints, tiles, and so forth. It goes without saying that there is risk in the construction industry because big money attracts big risks. Each of the following parties has a different fear of taking on risks in the course of doing business: the project owner, construction companies, consultants, bankers and financial institutions, vendors and suppliers, and even service providers. The risks' exact magnitude is sometimes unknown.

What needs to be determined is:

- The proportion of real versus perceived risks;
- The monetary quantification of risks;
- The real import and the impact of a type of risk.

Undetermined risks are more dangerous than evaluated risks. The result is that banks and financial institutions are hesitant to lend to construction industry operators, or they lend without reliable information. Both scenarios harm the industry's growth and the economy. It is crucial to strengthen current operating systems to instill confidence in financial systems.

Broadly speaking, Construction Projects face the following type of risks:

- Business risks;
- Financial risks;
- Technology risk;
- Project risk;
- Political risk.

The risks in a standalone project are big. They include:

- Completion Risk: This is the risk that the project may not be completed on time, or at all, due to various reasons such as cost overruns, technology failure, force majeure, etc.
- Price Risk: This is the risk that the price of the project's output might be volatile due to supply-demand factors. If new capacities are coming up or if there is likelihood of fall in demand of the project output, the price risk is high.
- Resource Risk: This risk includes the non-availability of raw materials for the project operation. It also includes the risk that the raw material prices might move adversely.
- Technology Risk: This is the risk that the technology used in the project is not sufficiently proven.
- Operating Risk: This is a risk that the project operational and maintenance costs would escalate. It also includes the risk that the project will have operational problems.
- Political Risk: This risk relates to matters such as increased taxes and royalties, revocations or changes to the concession, exchange controls on proceeds, forced government participation in shares and refusal of import licenses for essential equipment.
- Casualty Risk: This is the risk of physical damage to the project

equipment. It also includes liabilities to third parties on account of accidents at the project site.

- Environmental Risk: This risk refers to increased project costs for complying with new environmental standards. There could also be environmental protests from the local populace against the project.
- Permission Risk: This is the risk that official clearances for the project may not be forthcoming or subject to expensive conditions.
- Exchange Rate Risk: This is the risk that the currency of sale of the project produce would depreciate with reference to the currency of the project loans. Even though the debt being rated might be Rupee denominated, the presence of foreign currency liabilities can decrease the debt service coverage ratio of the bonds in case there is adverse exchange rate movement.
- Interest Rate Risk: This is the risk that the floating interest rate of the project loans would increase beyond the levels assumed for preparing projected cash flows.
- Insolvency Risk: This is the risk of insolvency of contractors, project sponsors, suppliers, purchasers of project output, insurers, or a syndicate bank.
- Project Development Risk: This is the risk that the project development might not take place in an orderly manner.
- Site Risk: This is the risk that the project site might have legal encumbrances. It also includes the risk that the site has technical problems.
- Financial Closure Risk: This is the risk that the project might not reach financial closure.

1.1.2.1. Possible Ways of Mitigation of Risks

Even in the modern era, the underlying assumptions hold true and the classification is still valid, but the systems that have been implemented and the demands that are placed on them either increase or decrease the intensity of the encounter. Therefore, an attempt must be made to identify these risks, quantify them, and develop products, procedures, or other means of mitigating them. If not mitigated, the following could be the possible repercussions (behavioral) outcome:

- Dilemma of the project owner to invest in the project.
- Dilemma of the construction company to complete the project in time and within the scheduled costs.
- Dilemma of the consultants to provide effective support services.
- Dilemma of the Banker to provide the required financial support to the venture.
- Dilemma of the vendor to meet his supply obligations in time.

In terms of reducing these risks, the components stay the same, with differing levels of complexity resulting from the kind and type of mitigation employed. Once the elements are defined, they can be further divided into the following major categories. To lessen the impact, a risk distribution system can be developed by implementing improved work procedures and passing laws limiting independent practice to recognized masters of the craft. By developing subsystems and systems to support and operationalize the procedures in line with the first two premises (e.g., commencement of formal training and HRD), having thus outlined the tenets and removed the activities charter, a crucial question remains for all parties involved: how to develop the methods for risk profiling and quantification. The creation of a database using mathematical principles and what some refer to as "colloquial wisdom" or experience holds the key to the solution.

An attempt, therefore, is to be made to develop following:

- Risk identification instruments.
- Risk management instruments.
- Risk mitigation instruments and eventually.
- Damage control, containment, and resolutions instruments.

Actuaries are specialists who analyze the likelihood and magnitude of risks as numerical values. Actuaries record the specifics of the incidents that result in failures, analyze the reasons as mathematical expressions, ascertain the frequency, and assess the degree of losses or damage. As a significant economic activity, construction has recently attracted attention from a number of sources, and work on these crucial but long-ignored aspects has started in earnest. To continuously improve the performance of the industry as a whole, a number of services and support systems, including institutional systems for performance surveillance, insurance, and non-insurance backed product design, and the evolution of techno-commercial grading systems, are being designed and put into practice.

1.1.3. Construction Processes

In certain cases, the owner serves as the project's designer, paymaster, and laborer. These are minor repairs or renovations, such as repainting or sealing leaks. The owner may hire one or more specialized companies to handle the project's detailed planning, design, construction, and handover because larger, more ambitious projects typically call for more multidisciplinary knowledge and labor.

A contractor, designer, construction manager, or other advisor may be chosen by the owner to oversee the project; these professionals are typically chosen for their experience in project delivery and construction management. They will assist the owner in defining the project brief, deciding on a budget and timeline, communicating with pertinent government agencies, and obtaining materials and the services of other experts (the supply chain, which includes subcontractors and suppliers of materials). All companies enter into contracts for the provision of services, and these agreements are accompanied by comprehensive plans designed to guarantee the lawful, punctual, economical, and secure completion of the assigned tasks.

Aspects of design, finance, and law overlap and interact. The design must be both financially feasible to construct and legally permissible to use, in addition to being structurally sound and suitable for the intended use and location. The financial structure needs to pay for amounts that are legally owed and be sufficient to build the specified design. Legal frameworks enforce financial and other construction processes while integrating design with other activities.

These processes also affect procurement strategies. Clients may, for example, appoint a business to design the project, after which a competitive process is undertaken to appoint a lead contractor to construct the asset (design–bid–build); they may appoint a business to lead both design and construction (design-build); or they may directly appoint a designer, contractor, and specialist subcontractors (construction management). Some forms of procurement emphasize collaborative relationships (partnering, alliancing) between the client, the contractor, and other stakeholders within a construction project, seeking to ameliorate often highly competitive and adversarial industry practices.

Construction or refurbishment work in a "live" environment (where residents or businesses remain living in or operating on the site) requires particular care, planning, and communication.

1.1.3.1. Planning

When applicable, a proposed construction project has to abide by zoning and building code requirements as well as local land-use planning policies. A project's potential effects on nearby properties and on the current infrastructure—transportation, social infrastructure, and utilities like water supply, sewage, electricity, telecommunications, etc.—will normally be evaluated (by the authority having jurisdiction, AHJ), usually the municipality where the project will be located. Geotechnical investigations, site surveys, and analysis are methods that can be used to collect data. Planning permission is typically required before construction can begin, and it might also be necessary to complete preparatory work to make sure that any necessary infrastructure has been upgraded before building work can start. In order to prevent damage-causing outages and other dangerous circumstances, surveys of the current utility lines will also be part of the preparatory works.

Malum in se considerations, or the wish to stop undeniably negative phenomena, are the source of some legal requirements, e.g., explosions or the failure of bridges. Additional legal requirements stem from malum prohibitum considerations, which are customary or expected factors like separating residential from business areas or businesses from one another. An attorney may argue that a rule is unneeded (the bridge design won't collapse) or that a custom is no longer necessary (the community's acceptance of live-work spaces has grown) in order to seek changes or exemptions from the law that governs the land where the building will be built.

A municipal building inspector typically visits the ongoing construction site on a regular basis to make sure that the building is being built in accordance with the approved plans and the local building code. Any alterations to a building or other asset that impact safety after construction is finished, such as those related to use, expansion, structural integrity, and fire protection, typically need municipality approval.

1.1.3.2. Finance

Creating an overall plan for the financial management of a construction project may involve mortgage bankers, accountants, and cost engineers, depending on the nature of the project. Since the owner's equity in the property is the most obvious source of funding for a building project, the mortgage banker's presence is highly likely, even in relatively small projects. Accountants study the projected cash flow for the project and keep an eye on the payouts as they go along. Expertise is applied by professionals such as cost engineers, estimators, and quantity surveyors to connect the work and materials needed to an accurate valuation.

Before the project begins, financial planning guarantees that sufficient safeguards and backup plans are in place. It also guarantees that the plan will be correctly carried out throughout the project's duration. Financial issues that could have been avoided can affect construction projects. Underbids occur when contractors request insufficient funds to finish a project. Cash flow problems occur when the current funding level is insufficient to cover labor and material costs; these issues can occur even in situations where the overall budget is sufficient, posing a short-term challenge. Government project cost overruns have happened when the contractor found change orders or project modifications that raised costs. These changes are not competitive with other firms because they were already disqualified after the first bid. Another issue that is becoming more and more important in construction is fraud.

A building estimator's conceptual cost estimate is frequently the first step in large projects, which can involve extremely intricate financial plans. When a project is completed, its components may be sold, changing lenders or owners, but the logistical needs of having the appropriate materials and trades on hand for every stage of building construction remain the same. Major projects may also be implemented with the assistance of private finance initiatives (PFIs) or public-private partnerships (PPPs). In 2019, the great majority of major construction projects went over budget and took 20% longer than anticipated, according to McKinsey.

1.1.3.3. Legal

A construction project is a complicated web of legal requirements, construction contracts, and other documents that all involved parties need to carefully review. A contract is an agreement that outlines responsibilities and offers procedures for handling disagreements between two or more parties. For instance, delays in construction can be expensive, so contracts for such work clearly define expectations and provide guidelines for handling delays. Confusion and expensive disputes can result from poorly written contracts. Legal counsel aims to find unclear areas and other possible causes of concern in the contract structures at the beginning of a project and to offer solutions to avoid issues. They try to prevent and handle disagreements that come up during projects. The attorney helps each party to an obligation exchange that is appropriate for the project's reality.

1.1.3.4. Procurement

The most popular and tried-and-true method of construction procurement is design-bid-build. Under this arrangement, the project coordinator represents the client and is either an architect, engineer, or builder. They create the deliverables (models, drawings, etc.), design the works, and draft the specifications, oversee the contract administration, submit the work for tender, and oversee the project from start to finish. Concurrently, the main contractor and the client have direct contractual relationships, and the main contractor has direct contractual relationships with subcontractors. The agreement is upheld until the project is prepared for transfer of ownership.

1.1.3.5. Design-Build

With the advent of the design-build method in the latter half of the 20th century, clients could now hire a single company to handle both design and construction. In certain situations, locating the site, securing funding, and requesting all required statutory consents can also be included in the design-build package. Generally speaking, the client chooses a preferred supplier after inviting multiple Design and Build (D&B) contractors to submit bids in order to fulfill the project specifications. This is typically a partnership between a contractor and a design firm, sometimes including more than one of each. When states lack the necessary expertise or resources, especially for very large projects, design-build contracts are typically used by US transportation departments to advance projects.

1.1.3.6. Construction Management

The designer (engineer or architect), the construction manager, and individual trade contractors all sign separate contracts with the client under a construction management arrangement. The client assumes the contractual role, and the construction or project manager actively manages the various trade contracts, making sure that they all get completed efficiently and harmoniously. This method is frequently used to expedite procurement procedures, give the client more creative freedom in the design throughout the contract, permit the hiring of independent contractors, delineate individual contractual responsibilities throughout the contract, and give the client more control.

1.1.3.7. Design

Realizing designs through construction is a common practice in the industrialized world. The most typical case (i.e., for instance, in a design-bid-build project, the design group is hired by (i.e., under contract with) the landowner. Planning consultants, architectural consultants, mechanical engineers, electrical engineers, structural engineers, fire protection engineers, architects, and archaeologists may be on a design team, depending on the nature of the project. To assist in coordinating various disciplinary inputs to the overall design, a "lead designer" will typically be identified. The integration of formerly distinct disciplines—typically carried out by distinct firms—into multidisciplinary firms with specialists from all relevant fields or the formation of partnerships by firms to support design-build processes can both help with this. The need for design professionals versed in all stages of a project's life cycle and equipped with an understanding of the asset as a cutting-edge technological system necessitating close integration of numerous subsystems and their individual components, including sustainability, is driven by the growing complexity of construction projects. A new field called "building engineering" aims to address this new challenge in the construction industry.

The creation of sketches, engineering and architectural drawings, and specifications has historically been a part of design. Prior to the late 20th century, most drawings were created by hand. However, with the advent of computer-aided design (CAD) technologies, design productivity increased. The use of computer-generated models in building information modeling (BIM) processes, which can be used to create drawings and other visualizations, as well as capture non-geometric data about building systems and components, has become more common in the 21st century. For certain projects, design work may begin in parallel with the initial phases of on-site activity, or work on the foundations of a building may begin while interior space designs are still being finalized. A project may contain components that are intended to be built off-site (also see prefabrication and modular building), then they are delivered to the location prepared for assembly, erection, or installation.

1.1.3.8. On-Site Construction

Work on the project site can start as soon as contractors and other pertinent specialists are hired and the designs are sufficiently developed. A construction site usually consists of a secure perimeter to prevent unauthorized access, site access control points, welfare and office spaces for staff from the main contractor and other companies in the project team, and storage spaces for supplies, tools, and machinery. The first feature of the permanent structure, such as pile driving or the pouring of slabs or footings, can be considered the start of construction, according to the definition provided by the McGraw-Hill Dictionary of Architecture and Construction.

1.1.3.9. Commissioning and Handover

Commissioning is the process of verifying that all subsystems of a new building (or other assets) work as intended to achieve the owner's project requirements and as designed by the project's architects and engineers.

1.1.3.10. Defects Liability Period

A period after handover (or practical completion) during which the owner may identify any shortcomings in relation to the building specification ('defects'), with a view to the contractor correcting the defect.

1.1.3.11. Maintenance, Repair, and Improvement

Maintenance involves functional checks, servicing, repairing, or replacing necessary devices, equipment, machinery, building infrastructure, and supporting utilities in industrial, business, governmental, and residential installations.

1.1.3.12. Demolition

Demolition is the discipline of safely and efficiently tearing down buildings and other artificial structures. Demolition contrasts with deconstruction, which involves taking a building apart while carefully preserving valuable elements for reuse purposes (recycling – see also circular economy).

1.2. ROLE OF CONSTRUCTION INDUSTRY

The global development and expansion of economies and societies are significantly influenced by the construction industry. It is one of the biggest and most active industries, involving the construction and upkeep of commercial, industrial, residential, and infrastructure properties, which helps to shape our communities' physical environments. Its importance goes beyond physical space; it affects employment, social progress, economic prosperity, and technological advancement. Economically speaking, the construction sector significantly increases GDP (gross domestic product). It stimulates economic activity by generating tangible assets that support industry, trade, and transportation. Infrastructure projects lower transportation costs and promote commerce by improving connectivity and efficiency through the construction of roads, bridges, and railroads. Additionally, commercial and industrial buildings house businesses, encouraging innovation and entrepreneurship. Investments in construction have wide-ranging effects that boost demand in industries like manufacturing, raw materials, and services (Zhongfu et al., 2008).

Another crucial area where the construction industry has a big influence is employment. It employs a lot of people and offers a variety of jobs, from manual labor to highly specialized technical positions. Building projects ently result in employment opportunities for skilled craftspeople, such aslike carpenters, electricians, and plumbers, as well as architects, engineers, and project managers. Millions of people worldwide rely on these diverse job options to support their livelihoods and to lower unemployment rates. Furthermore, the industry supports workforce education and training by providing chances for career advancement and skill development. Technological innovation and advancement are also fueled by the construction industry. In order to increase productivity, safety, and sustainability, the industry is constantly adjusting to new materials, techniques, and technologies. Technological advancements like 3D printing, prefabrication, and Building Information Modeling (BIM) are transforming the design and construction of structures. With the use of these technologies, projects can be completed more quickly, with increased precision, decreased waste, and lower costs. The use of renewable materials, energy-efficient designs, and green buildings are all becoming more and more popular aspects of sustainable construction practices. This transition to sustainability helps the environment and satisfies the rising demand for environmentally friendly structures.

By supplying necessary infrastructure and amenities, the construction sector improves social conditions and people's quality of life. While housing projects take care of the basic need for shelter, healthcare, education, and community well-being are supported by schools, hospitals, and recreation centers. The livability of both urban and rural areas is improved by thoughtfully planned public spaces and infrastructure that foster social interaction, accessibility, and inclusivity. Additionally, building projects frequently result in urban regeneration and renewal, reviving abandoned areas into thriving communities. The sector is essential to resilience and recovery from disasters. The construction industry plays a crucial role in reconstructing and revitalizing areas devastated by natural calamities. This includes not just rebuilding damaged structures but also creating infrastructure and buildings that are resilient enough to survive future calamities. For communities affected by crises to recover and maintain long-term stability, such initiatives are essential.

1.2.1. Challenges in the Construction Industry

Despite being essential to both societal advancement and economic expansion, the construction sector faces numerous obstacles that may compromise the sustainability and successful completion of projects. The lack of skilled labor is one of the most urgent problems. Workers skilled in contemporary construction methods and technologies are becoming more and more necessary as the industry develops. However, this shortage has been made worse by an aging workforce and a lack of interest from younger generations to enter the field. This skilled labor shortage affects the caliber and safety of construction work in addition to causing delays in projects. One notable obstacle is the unpredictability of material costs. The cost of building materials like steel, wood, and concrete can change because of disruptions in the global supply chain, tariffs, and general economic conditions. Project managers may be forced to make concessions that lower the overall quality of the construction in order to avoid budget overruns caused by these erratic costs. Moreover, projects may be delayed by delays in the delivery of materials, resulting in longer schedules and higher expenses.

For the construction industry, technological integration offers a challenge as well as an opportunity. Even though technologies like automation, drones, and building information modeling (BIM) can increase accuracy and efficiency, putting them into practice will cost a lot of money and time. The financial strain of implementing new technologies and the learning curve that goes along with it can be particularly difficult for small and medium-sized construction companies. Also, since cyberattacks have the potential to disrupt business processes and jeopardize sensitive project data, maintaining cybersecurity in the increasingly digitalized construction industry is a growing concern. Another area that presents difficulties is regulatory compliance. There are numerous local, national, and international regulations pertaining to labor standards, safety, and environmental impact that construction projects must follow. To avoid legal problems and possible fines, navigating this complicated regulatory environment requires careful planning and monitoring. Regulations are always changing, which makes it harder to comply with the law and necessitates ongoing training and adaptation on the part of construction companies.

The building industry has made environmental sustainability a top priority due to growing awareness of climate change and the requirement for eco-friendly methods. However, because they require specialized knowledge and materials and come with higher upfront costs, implementing sustainable construction methods can be difficult. Innovative solutions and a readiness to prioritize long-term benefits over short-term cost savings are needed to strike a balance between the demand for sustainable building practices and financial restraints. Given the inherent hazards of the construction industry, safety is still a major concern. Personal protective equipment (PPE), frequent training, and strict adherence to safety procedures are all necessary to ensure worker safety. Construction sites continue to report high rates of accidents and injuries in spite of these precautions. In order to address safety concerns, it is necessary to promote a culture of safety in which employees feel empowered to put their own health and safety first, in addition to strictly enforcing adherence to safety laws.

Finally, coordination and project management are two issues that the construction sector must deal with. Numerous parties, including suppliers, contractors, subcontractors, and regulatory agencies, are frequently involved in large construction projects. For the project to be completed on schedule and successfully, these parties must effectively communicate and coordinate. Delays, cost overruns, and conflicts, however, can result from poor coordination, scope modifications, and misunderstandings. These problems can be avoided by using cutting-edge project management tools and encouraging teamwork, but doing so calls for proactive management.

1.2.2. The Economic Context of Building Construction

Similar to other economic goods, buildings can be purchased for a variety of amounts depending on the cost of production and the value they provide to the buyer. The building construction industry contributes significantly to the overall yearly value of the various national economies. For instance, in the United States in 1987, it accounted for roughly 10% of GDP, a figure that is comparable to that of the global economy. When compared to other products, the unit cost of buildings is relatively low, despite these high aggregate values. In 1987, the price of a new building in the US varied from roughly $0.50 to $2.50 per pound. Simple pre-engineered metal buildings have the lowest costs, and functionally complex buildings with lots of mechanical and electrical services, like hospitals and labs, have the highest costs. These unit costs are less than those of most other well-known consumer goods and are towards the low end of the manufacturer's scale, ranking alongside cheap foodstuffs. This price range serves as a crude barometer for the commodity's worth or usefulness to society. Food is necessary, but it is relatively simple to produce; on the other hand, aircraft, when produced at the high end of the scale, serve a useful purpose, but their intricate and costly mechanisms result in much higher unit prices, which are a reflection of the significant capital and research investments in addition to the labor and materials needed to produce them. Buildings are more valuable than food because they are more commonplace and necessary, but they can still provide the services that people need them to with relatively simple technology and low-cost materials. Because there hasn't been much of an incentive to invest in research given consumer expectations.

The cost of buildings is influenced by several specific factors within this broader economic context. The first are government building codes, which are performance-based and prescriptive standards that are implemented to safeguard public health and safety. The loads that buildings must support are described in structural requirements. These loads range from the regular, constant loads that gravity places on building contents to the more extreme, infrequent loadings caused by wind and earthquake forces. These are typically the maximum expected to occur with a 100-year frequency, and they are specified based on statistics. Safety factors for materials are designed to account for errors in quality control and unintentional overloading. Economic factors are also taken into account. For instance, while buildings must function well under normal gravity loads, they are not required by code to be resistant to direct wind exposure or the low-pressure effects of tornadoes because doing so would be extremely expensive.

Building setbacks from lot lines and restrictions on height and floor area are mandated by planning and zoning laws to guarantee that neighboring properties receive enough light and air. In addition, zoning laws define standards for the acceptable uses of buildings, parking lots, landscaping, and even how buildings should look. Another illustration would be the specifications for the building atmosphere, which include minimum (but not maximum) temperatures and air change rates to provide enough oxygen and lessen odors. Adequate stairways for emergency exit, emergency lighting, smoke detection and control systems, and fire-resistant building materials are all examples of life-safety requirements. Proper pipe sizes and an adequate number of plumbing fixtures are requirements for sanitation. Wire sizes, safety-related construction specifications, and outlet placement are examples of electrical requirements. There are market standards that go beyond government requirements and represent what users anticipate from buildings. Elevator systems are one example; while building codes do not mandate elevators, the number of elevators in office buildings in the United States is determined by a maximum waiting time of 30 seconds. Although it is not mandated by code, building atmosphere cooling is offered in regions and for certain types of buildings where market research has shown it to be economically advantageous.

Two dimensions of value are perceived for building systems and components. One is the purely functional dimension, which requires that the roof keep out rain and that the structure be able to withstand loads. The other is the psychological or aesthetic aspect: people believe that stone is more resilient than wood, and elevator systems that have a 30-second wait time are better than those that have a two-minute wait time. Many users are willing to pay more in exchange for these alleged disparities. The aesthetic factor plays a significant role in determining the value of buildings when symbolic structures like temples, cathedrals, and palaces are significant to society. For instance, the Parthenon in Athens and the Cathedral in Chartres demanded investments in their economies that could be roughly compared to the U.S. Apollo space program. However, the functional aspect of value predominates in the majority of buildings.

The building industry in industrialized countries is divided into numerous small enterprises due to its wide geographic distribution, highly variable demand, and relatively low level of technology. Because there is less centralization, research is typically discouraged and building components are kept straightforward and robust, adhering to tried-and-true formulas. There are several fairly well-defined markets within this diversity that are based on the types of buildings; these include high-rise buildings, long-span buildings, low-rise residential buildings, and low-rise commercial, institutional, and industrial buildings. Though the building industry is more centralized there, a somewhat similar pattern can be found elsewhere in Eastern Europe. A significantly smaller low-rise residential market exists as well, with high-rise structures housing the majority of newly constructed homes.

In developing countries, the major market is for low-rise residential buildings to house rapidly growing populations. Much of the construction is undertaken by local craftsmen using simple building products. Local timber is widely used, and masonry materials still include the ancient mud brick. More sophisticated long-span and high-rise technologies are found only in major cities.

1.2.3. Construction Industry Trends

Technological innovations, environmental concerns, and changing economic conditions are causing revolutionary changes in the construction industry. The planning, carrying out, and managing of construction projects are changing as a result of these trends, which will improve the built environment's resilience, sustainability, and efficiency. The integration of digital technologies is a prominent trend that is transforming every facet of the construction industry. Building information modeling (BIM), which makes it possible to create detailed digital representations of a building's functional and physical attributes, has emerged as a key component of contemporary construction methods. Better design visualization, increased stakeholder collaboration, and improved project management are all made possible by BIM. Drones have also become more popular for use in site surveys and inspections because they provide accurate data collection, increased safety, and quicker survey times.

Due to their substantial time and cost savings benefits, prefabrication, and modular construction are also becoming more and more popular. Using these techniques, building components are produced in a controlled factory setting and then assembled on-site. This method not only expedites the building process but also reduces waste and improves quality assurance. Large-scale projects and areas with a labor shortage benefit greatly from prefabrication's accuracy and repeatability. Greener practices in the construction industry are being driven by a critical trend called sustainability. Energy-efficient design, integration of renewable energy sources, and use of sustainable building materials are becoming more and more important. Building Research Establishment Environmental Assessment Method (BREEAM) and LEED (Leadership in Energy and Environmental Design) certifications are becoming industry standards for sustainable building. To lessen the environmental effect of construction operations, cutting-edge materials like low-carbon concrete, recycled steel, and cross-laminated timber (CLT) are being used.

Another important trend is the use of Internet of Things (IoT) technologies in smart buildings. These structures have integrated systems that keep an eye on and manage a number of operations, including lighting, security, HVAC (heating, ventilation, and air conditioning). Smart buildings with IoT capabilities improve operational effectiveness, occupant comfort, and energy efficiency. They also offer useful information for better building management and preventive maintenance. Infrastructure that is more adaptable and resilient is also becoming more prevalent in the construction sector. Buildings and infrastructure must be designed to withstand extreme weather and adapt to changing environments due to climate change and an increase in natural disasters. The use of sturdy materials, raised buildings in flood-prone areas, and designs that permit future adjustments are examples of resilient construction techniques.

The skilled labor shortage is the main focus of the evolving labor dynamics. The construction industry is allocating resources towards workforce development by means of training initiatives, apprenticeship schemes, and outreach to the younger demographic. Robotics and automation are augmenting human labor, especially for dangerous and repetitive tasks, to increase efficiency and security. Thanks to the widespread use of cutting-edge software tools that improve scheduling, planning, and resource allocation, project management is becoming increasingly sophisticated. Real-time data and analytics are made possible by these tools, which facilitate improved decision-making and effective project execution. Lean construction techniques and integrated project delivery (IPD) are being used to improve efficiency, cut waste, and encourage cooperation among project stakeholders.

1.2.4. Building Design and Construction

1.2.4.1. Design Programming

The architect of a building starts with the end user or owner, who considers factors like the building's perceived need, site selection, and estimated cost. A group of design experts, consisting of engineers and architects, receives these details from the user, or client, and uses them to create a set of construction documents that precisely outline the intended building and serve as a basis for its construction. Professionals licensed by the state to design buildings include structural, mechanical, and electrical engineers as well as architects. These individuals are required to formally attest that the buildings they create will adhere to all applicable laws and regulations. The principal professionals in design are architects, who oversee and coordinate the work of engineers and numerous other consultants specializing in vertical transportation, lighting, and acoustics.

The designers prepare their designs by consulting a variety of sources. Building science is the most fundamental of these, having been developed over the past 300 years. This covers the aspects of physical theory that have to do with construction, like the theories of light, electricity, and fluid flow, as well as the elastic theory of structures. A vast amount of data exists regarding the particular characteristics of building materials, which can be utilized in mathematical models to accurately predict the performance of construction projects. A substantial amount of research has also been done on human comfort standards related to building design, including temperature, lighting, and sound levels. The design team gathers particular information about the planned building site in addition to having a general understanding of building science. These consist of boundary and topographic surveys, studies of the subsurface for foundation and water-exclusion design, climate data, and other local components.

In tandem with gathering site data, the design team collaborates with the client to clarify the frequently nebulous concepts of building function into more specific and tangible language. These definitions are condensed into a building space program, which provides a written description of every necessary space, including floor area, equipment specifications, and functional performance standards. This agreement establishes the expected building size and performance between the client and the design team.

1.2.4.2. Design Development

The art of building design is the method by which the design team applies site information, building science, and the building space program. The process is intricate and involves choosing common building systems, then integrating and modifying them to create a structure that satisfies the client's requirements while staying inside the bounds of legal requirements and industry norms. These systems are now separated into several distinct sectors based on the kind of building that they are designed for. The systems for the building's foundation, structure, atmosphere, enclosure, space division, electrical distribution, water supply and drainage, and other functions are chosen during the design process. These systems allow for a great deal of variation in the final product, despite being composed of a restricted range of manufactured components. Once the systems and components have been selected, the design team prepares a set of contract documents, consisting of a written text and conventionalized drawings, to describe completely the desired building configuration in terms of the specified building systems and their expected performance. After the contract documents are finished, the building's final costs can usually be calculated with accuracy, and construction can start.

1.2.4.3. Construction

A dedicated construction team typically oversees a building's construction; this team is typically distinct from the design team, though some large organizations may integrate both responsibilities. Leading the construction team is a coordinating organization, also known as a general contractor, who enters into a contract with the building user to assume primary responsibility for building execution. Typically, the contract has a fixed price, but on larger projects where construction starts before the contract documents are finished and the building's scope is not entirely specified, cost-plus-fee contracts are employed. A team of specialty subcontractors working under contract to the general contractor completes the remaining work on the building; the general contractor may handle some of the actual construction work in addition to its coordination duties. A subcontractor is responsible for providing and installing one or more building systems, e.g., the electrical or structural framework. Subcontractors purchase the system components from manufacturers on their behalf. While the building is being constructed, the design team continues to represent the owner by ensuring that the systems and components meet the required performance and quality standards, and that the completed building complies with the contract documents.

1.2.5. Industry Scale and Characteristics

1.2.5.1. Economic Activity

The global construction industry produced an estimated $10.8 trillion in 2017, with projections of reaching $12.9 trillion by 2022 and around $14.8 trillion by 2030. Construction contributes more than 10% to the global GDP (6–9% in developed countries) and employs approximately 7% of the total global workforce, with over 273 million jobs in 2014. China has been the largest construction market since 2010, while the United States follows as the second largest market with a 2018 output of $1.581 trillion.

- In the United States in February 2020, around $1.4 trillion worth of construction work was in progress, according to the Census Bureau, of which just over $1.0 trillion was for the private sector (split roughly 55:45% between residential and nonresidential); the remainder was public sector, predominantly for state and local government.
- In Armenia, the construction sector experienced growth during the latter part of the 2000s. Based on the National Statistical Service, Armenia's construction sector generated approximately 20% of Armenia's GDP during the first and second quarters of 2007. In 2009, according to the World Bank, 30% of Armenia's economy was from the construction sector.
- Vietnam's national economy heavily depends on the construction sector. In recent years, Vietnam's construction industry has experienced some of the fastest growth in the Asia-Pacific area. The market's 2021 valuation was close to $60 billion. Vietnam's construction industry grew at a rate of 5.59% in the first half of 2022. Vietnam's construction sector generated over 589.7 billion Vietnamese dong, or more than 6% of the nation's GDP, in 2022. Vietnam's GDP is accounted for by the construction and industrial sectors at 38.26%. In addition, the sector has emerged as one of the most alluring for foreign direct investment (FDI) in recent times.

Construction is a major source of employment in most countries; high reliance on small businesses, and under-representation of women are common traits. For example:

- A little over 8% of all US workers were employed in the construction industry in 2020, accounting for about 11.4 million jobs. Another 1.8 million people worked in professional services related to architecture, engineering, and related fields. Of the over 8,43,000 organizations that employed construction workers, 8,38,000 were privately held companies. Around 64% of construction workers were employed by companies with fewer than 50 employees in March 2016. According to their percentage of all employment, women are notably underrepresented in the US labor force. In 2019, they made up 25.9% of workers in professional services and 10.3% of workers in the construction industry.
- In 2018, the construction industry in the UK contributed £117 billion, or 6% of the country's GDP, and it employed 2.44 million people in 2019, accounting for 0.66% of all jobs. These individuals worked for 3,43,000 "registered" construction companies or for "unregistered" companies, which were usually self-employed contractors. In 2019, just over 1 million small and medium-sized companies, primarily independent contractors, operated in the sector, making up roughly 18% of all businesses in the UK. In the UK, women made up 12.5% of the construction workforce.

According to McKinsey research, productivity growth per worker in construction has lagged behind many other industries across different countries including in the United States and in European countries. In the United States, construction productivity per worker has declined by half since the 1960s.

1.2.6. Safety

In the United States and the European Union, construction is the sector with the highest number of occupational fatalities, making it one of the riskiest jobs globally. Construction accounted for 1,061, or roughly 20%, of worker fatalities in the private sector in the United States in 2019. In the US, falls accounted for 366 out of 971 total fatalities in 2017, more than a third of construction-related deaths; in the UK, falls accounted for half of the average 36 fatalities annually over a five-year period to 2021. The risk of occupational injuries in the construction industry can be reduced by using the right safety gear, such as guardrails, hard hats, and harnesses, as well as by following protocols like climbing ladders securely and checking scaffolding. Trench cave-ins, transportation mishaps, and electrocution are some other common causes of death in the construction sector.

1.2.6.1. Sustainability

Sustainability is an aspect of "green building," defined by the United States Environmental Protection Agency (EPA) as "the practice of creating structures and using processes that are environmentally responsible and resource-efficient throughout a building's life-cycle from siting to design, construction, operation, maintenance, renovation, and deconstruction."

1.2.6.2. Decarbonizing Construction

If the construction sector is to successfully contribute to reaching the Paris Agreement's target of keeping the rise in global temperature to 1.5°C above pre-industrial levels, it may need to transform quickly and extensively. According to the World Green Building Council, there is a 40% reduction in embodied carbon emissions in buildings and infrastructure globally; however, this can only be accomplished through immediate change. Leaders in the industry have concluded that while the construction sector may face difficulties as a result of the net zero transformation, there is also potential. To meet the decarbonization targets, action is required from governments, standards organizations, the building industry, and the engineering profession.

Key areas to decarbonize the construction industry and the built environment as a whole were outlined in the 2021 report "Decarbonizing Construction: Building a New Net Zero Industry" by the National Engineering Policy Centre. This study outlined six overarching "system levers" that, when activated, will cause the construction sector to rapidly decarbonize. It also included about 20 recommendations for how engineers, the construction industry, and decision-makers should transform and decarbonize the sector.

These levels are:

- Setting and stipulating progressive targets for carbon reduction;
- Embedding quantitative whole-life carbon assessment into public procurement;
- Increasing design efficiency, materials reuse and retrofit of buildings;
- Improving whole-life carbon performance;
- Improving skills for net zero.

Taking a coordinated, systems-based approach to decarbonization within the construction industry and with other industries: Global progress toward decarbonization is being made, and this includes enhancements to sustainable procurement practices like the Danish Partnership for Green Public Procurement and the CO_2 performance ladder in the Netherlands. Additionally, there are currently examples of how the concepts of circular economy practices are being put into reality, such as the Brighton Waste House, ABN AMRO's sustainable pavilion, and Circl.

1.2.7. Long-Span Buildings

Long-span buildings create unobstructed, column-free spaces greater than 30 meters (100 feet) for a variety of functions. These include activities where visibility is important for large audiences (auditoriums and covered stadiums), where flexibility is important (exhibition halls and certain types of manufacturing facilities), and where large movable objects are housed (aircraft hangars). In the late 20th century, durable upper limits of span have been established for these types: the largest covered stadium has a span of 204 meters (670 feet), the largest exhibition hall has a span of 216 meters (710 feet). In these buildings, the structural system needed to achieve these spans is a major concern.

1.2.7.1. Structural Systems

Long-span building structural systems fall into two categories: funicular structures, which undergo either pure compression or pure tension, and bending systems, which experience both compressive and tensile forces. There has been a development interaction between bridges and long-span buildings because bridges are a common type of long-span structure. The girder, two-way grid, truss, two-way truss, and space truss are examples of bending structures. Their ideal depth-to-span ratios vary; for the one-way truss, they are between 1:5 and 1:15, and for the space truss, they are between 1:35 and 1:40. The parabolic arch, tunnel vault, and dome are examples of the funicular structures. They have a rise-to-span ratio of 1:10 to 1:2, and they act in pure compression. The cable-stayed roof, the bicycle wheel, and the warped tension surfaces act in pure tension. Along with other economic considerations, the materials and labor needed for assembly constitute a significant limitation in these general forms of long-span structures.

1.2.7.1.1. Timber Structures

One long-span material option is glue-laminated timber. Metal connectors can be used to prefabricate it into trusses with a maximum span of 45 meters (150 feet). However, its most affordable forms are ribbed domes, which can span up to 107 meters (350 feet), and multiple-arch vaults, which have pure compression shapes and span up to 93 meters (305 feet). Materials that corrode steel or concrete, like alumina, salt, and potash, are frequently stored in these industrial buildings. These kinds of wood buildings are typically only found close to forests because it is more expensive to transport wood to other locations.

1.2.7.1.2. Steel Structures

The primary material used in long-span structures is steel. Long-span buildings use bending components like plate girders and trusses, which were first designed for bridges. Although plate girders can span up to 60 meters (200 feet) and are deeper than typical rolled shapes, they are not very material-efficient. Instead, they are made from steel plates that are welded together. Trusses are made of hollowed-out beams into which stresses are directed into thin linear members formed of rolled shapes and fastened into sturdy triangular configurations via welding or bolting. The members of trusses operate in two ways: either in pure tension or pure compression. The forces acting on the top and bottom horizontal members are concentrated at the center of the span, while the forces acting on the vertical and diagonal members are concentrated at the supports. Trusses can span up to 190 meters (623 feet) and are incredibly effective at bending. Two-way grids can span square areas up to 91 meters (300 feet) in size using either plate girders or trusses; these structures are more cost-effective to construct but also more efficient.

For the longest spans, the most effective funicular forms are employed. For spans of up to 98.5 meters (323 feet), vaults composed of rows of parabolic arches—typically in truss form for greater rigidity—have been employed. Numerous sizable covered stadiums with a maximum span of 2,04,022 feet have opted for steel truss domes, specifically the Schwedler triangulated dome. Bridge building is the source of another structural system, which is cable-stayed roof construction. Steel cables that extend downward from masts above roof level support a flat roof structure in bending from above; spans of up to 72 meters (236 feet) have been constructed. A different example of a funicular structure is the bicycle-wheel roof, which is supported by columns and has two layers of radiating tension cables connecting a small inner tension ring to the outer compression ring, which is divided by tiny compression struts.

A taut surface of negative curvature, such as a saddle or trumpet shape, is formed by tension-cable networks, which use a mesh of cables stretched from masts or continuous ribs. The tension surface can also be formed by synthetic fabrics in place of the cable network. The air-supported membrane is another fabric structure that is supported by tension cables. A compression ring at the edge supports the assembly of cables and fabric, which is connected to a network of cables by continuous seams. To counteract the force of the wind outside, the building's interior air pressure is slightly raised. Even in large buildings with relatively small compressors, the increase can be maintained to as little as 1.5% of atmospheric pressure. In the event of an unintentional deflation, the cables support the fabric and stiffen it against flutter caused by uneven wind pressure.

1.2.7.1.3. Concrete Structures

Due to its inherent strength in compression, reinforced concrete is mostly utilized for funicular compression forms with long spans, such as shells, domes, and vaults. With spans up to roughly 90 meters (300 feet), thin parabolic shell vaults stiffened with ribs have been constructed. Concrete shells in more intricate designs, such as intersecting parabolic vaults and hyperbolic paraboloids, also known as saddle shapes, have been created. An illustration of the latter is the CNIT Exhibition Hall in Paris, which is made up of six double-shell parabolic vaults that intersect and are constructed to span a triangular area measuring 216 meters (708 feet) on one side and are only supported at the triangle's apexes. Usually with a parabolic section, reinforced concrete domes are constructed as thin shells or in ribbed form. These domes can extend up to 200 meters (660 feet) in diameter. The inverted dome, or dish, is another concrete funicular form; however, it is actually a composite structure. The structure is shaped like a dish with radial steel cables running inward and downward to a small steel tension ring at the center, supported by a concrete compression ring that is supported by columns around the perimeter. Encasing the cable network in a concrete dish that has been poured stiffens it against wind forces; structures of this type have been constructed with spans as high as 126 meters (420 feet).

1.2.7.2. Factors in the Built Environment

1.2.7.2.1. Acoustics

Acoustics play a role in long-span auditoriums since people want to hear music with the right tone and speakers with clarity. Regretfully, the acoustic specifications needed for music and speech quality frequently clash, making it challenging to create an auditorium that meets both needs. Rooms with short reverberation times, one to one and a half seconds, are good for speech intelligibility, while longer reverberation times, one to two and a half seconds, give musical performances more richness in tone. Therefore, while adding sound-absorbing materials to a hall enhances speech, it diminishes musical quality. As people are natural sound absorbers, the audience has a noticeable effect on the acoustics of an auditorium. To maintain this effect, auditorium seats are typically upholstered to resemble surrogate spectators with the same level of sound absorption. In auditoriums, curved surfaces are either avoided or covered with sound-absorbing material because they have a tendency to focus sound. Although they can help speakers in large halls, electronic sound-amplification systems are usually not good enough for music. The acoustics of other long-span structures, like covered stadiums and exhibition halls, are largely ignored.

1.2.7.2.2. Environmental Control Systems

Atmospheric systems in long-span buildings must handle the considerable heat and odor generation from population densities of less than one square meter (11 square feet) per person.

1.3. CONSTRUCTION INDUSTRY SECTORS

Broadly, there are three sectors of construction: buildings, infrastructure, and industrial:

- Building construction is usually further divided into residential and non-residential.
- Infrastructure, also called 'heavy civil' or 'heavy engineering,' includes large public works, dams, bridges, highways, railways, water or wastewater, and utility distribution.
- Industrial construction includes offshore construction (mainly of energy installations), mining and quarrying, refineries, and chemical processing, mills, and manufacturing plants.

Markets or sectors can also be used to categorize the industry. One US-based construction trade magazine that has gathered and published information on the size of design and construction contractors is Engineering News-Record (ENR). The data was divided into nine market segments in 2014: manufacturing, sewage/waste, telecom, hazardous waste, transportation, petroleum, buildings, power, industrial, and water. A tenth category was designated for other projects. ENR classified companies as heavy contractors based on information on water, sewage, transportation, and hazardous waste. Building construction, heavy and civil engineering construction, and specialty trade contractors are the three subsectors into which businesses that perform or engage in construction are categorized by the Standard Industrial Classification and the more recent North American Industry Classification System. Professional services firms are also divided into categories (e.g., project management, surveying, engineering, and architecture).

1.3.1. Building Construction

The process of adding structures to land, also referred to as real property sites, is called building construction. The owner of the property, who could be an individual or an organization, usually initiates or collaborates on projects; on rare occasions, the owner may have to forcibly purchase land for public use. The process of putting together structures that function as places for people to live, work, and participate in different activities is referred to as building construction. There are multiple stages to this intricate process, ranging from design and planning to building and finishing touches. There are four main types of building construction: residential, commercial, industrial, and institutional. Each has its own set of specifications and uses. Planning and design are the first stages of building construction, during which planners, engineers, and architects work together to produce blueprints that satisfy client requirements and comply with regional building codes and regulations. Environmental impact assessments, site selection, and feasibility studies are all included in this phase. The building's functionality, safety, and aesthetic appeal are guaranteed by meticulous structural calculations and architectural designs. Stakeholders take sustainability into account during this phase, attempting to include materials and energy-efficient systems that have the least negative impact on the environment.

Preconstruction starts after the design is complete. This includes choosing contractors, getting the required permits, and getting the site ready. Grading, clearing the ground, and erecting makeshift buildings for the construction workers are all part of site preparation. During this stage, utility connections for sewage, electricity, and water are also established. The smooth and timely progress of the construction process is contingent upon the implementation of effective project management and coordination among diverse teams. Laying the foundation is the first step in the actual construction phase and is essential to the building's stability and longevity. Simple slab foundations to intricate deep foundations can be found, depending on the size of the building and the state of the soil. Following the establishment of the foundation, the structural framework is built. The building's needs and design specifications will determine which material—steel, concrete, or wood—is used to construct this framework. The framework forms the fundamental structure of the building, supporting the roof, walls, and floors. The building is enclosed with external walls, windows, and roofing after the structural framework. This step is essential for providing insulation and shielding the interior from the elements. Interior systems like HVAC (heating, ventilation, and air conditioning) and plumbing are installed concurrently. To guarantee that these systems operate flawlessly and adhere to all safety regulations, exact coordination is necessary.

The emphasis moves to interior and exterior finishes once the building is enclosed and the systems are set up. Interior finishes enhance the building's usability and beauty and include painting, drywall installation, flooring, cabinetry, and fixtures. The building's external finishes, which improve its appearance and blend in with the surroundings, can include paving, painting, cladding, and landscaping. Inspection and commissioning constitute the last stage of building construction. This entails extensive inspections by building inspectors to guarantee adherence to all building rules and specifications. The building cannot be occupied until any flaws are fixed. Commissioning verifies that every system is up and running and satisfies performance requirements. After these inspections are finished, the building is turned over to the owner so they can occupy it.

1.3.2. Residential Construction

Home construction, also known as residential construction, is the process of building a house, apartment building, or other type of residential structure that is typically referred to as a home when taking into account the individuals who may live there now or in the future. Simple prehistoric shelters were the first homes, but building methods have advanced to create the enormous variety of living spaces that exist today. Homes with varying sizes, luxuries, and even defenses have been justified by varying degrees of wealth and power. There is a vast range of possible home structures due to the enormous diversity of architectural styles that have been created by cultural and environmental influences. Modern real estate trades, which often involve some degree of market force speculation, often control the price and availability of housing. While housing starts are used to indicate the amount of economic activity in the home construction section, it is actually measured in terms of distinct habitation units rather than distinct construction efforts. In the related concepts of housing tenure, affordable housing, and housing unit (aka dwelling), "housing" is also the selected term. Although there are many other trades, carpenters, masons, electricians, and plumbers are the four main trades involved in building a home.

The right to housing is not adequately supported in many economies, resulting in uneven access to homes across the globe. One of the objectives of Sustainable Development Goal 11 is to upgrade slums and provide basic services and adequate, safe, and affordable housing. The United Nations estimates that in order to meet demand, 96,000 new housing units must be constructed every day due to projected and actual global population growth. Modernizing and adapting existing structures to provide suitable housing is a crucial component of building new homes to fulfill this global demand. Residential construction can be done by general contractors, property developers, specialized housebuilders, individual landowners building on their own, public or social housing providers, and general contractors (e.g., housing associations, local government). When permitted by local planning or zoning laws, mixed-use developments may include both residential and non-residential buildings (e.g., retail, entertainment, workplaces, public structures, etc.).

The methods, materials, and technologies used in residential construction must adhere to the rules and guidelines set forth by the local building authority. The building materials used are typically determined by the materials that are easily accessible in the area (e.g., brick instead of stone instead of wood). The cost of building a house can vary significantly depending on a number of factors, including the site's characteristics, access routes, local laws, economies of scale (building custom-designed homes is typically more expensive), and the availability of skilled tradespeople. Because these structures are meant to house people rather than be places of business, residential construction differs from commercial construction. Contrary to popular belief, single-family homes do not only refer to single-family homes; they also include multi-family homes, apartments, condominiums, and even mobile homes. Every residential building project starts with the purchase, clearing, and grading of a plot of land. After adding the drainage system and other features, the building is constructed. After completion, the contractor moves on to their next project and someone moves into the completed home.

Since this kind of construction is currently very profitable, contractors must continue forward motion or risk losing out on upcoming projects. They must constantly search for new land, maintain a roster of competent laborers on hand, be aware of where to get materials at reasonable costs, and possess sufficient funds to cover all expenses until each home is completed. As with homeowners, certain contractors have a preference for working on a particular type of residential construction over others. The primary residential construction types are listed in subsections, along with the benefits and drawbacks of each.

1.3.2.1. Sinle Family Homes

Single-family homes are often found in neighborhoods or out in the country. These homes can have small yards, or enough green space to include a few farm animals, depending on where they are located. It is possible for people to create the home of their dreams when they construct a single-family home because they are not restricted to what their neighbors want. These homes always sell for more than a condo or townhome, which may not be helpful for a purchase, but it is certainly a plus when it comes time to sell. The only downside to one of these residential construction homes is that they can cost a lot to maintain, and the owner is responsible for all the costs.

1.3.2.2. Condominiums

Although they can be found in smaller towns as well, condominiums are most frequently found in urban areas. Due to their numerous amenities—such as lounge areas, pools, gyms, and other features that single-family homes lack—many homeowners choose to own one of these residential construction homes. HOA fees, or homeowner's association fees, are typically associated with these kinds of homes and pay for upkeep of the pool, the hallways, and any other building or external grounds maintenance. Certain repairs are not covered by the current HOAs; therefore, owners bear some of the financial burden for those repairs.

1.3.2.3. Townhouses

Townhouses are a combination of single-family homes and condos, where they offer some privacy and not a ton of maintenance. A few townhomes will have HOAs, but they are still more affordable than most single-family homes. The only downside to owning a townhome is that there are not all the extra amenities like a condo has.

1.3.2.4. Multi-Family Homes

Although the number of multi-family homes built annually is relatively small, this does not preclude contractors from building them upon request. Some contractors will build these residential construction units in order to make a little extra cash for their own investments, as they are frequently used as investments. Many choose to rent these units rather than buy them because, in the event that they are rented out, the owner is in charge of maintenance. These homes' greatest feature is that multiple generations can live there and still have their own private space.

Remarkably, a lot of outbuildings, including garages, are included in the definition of residential construction. Thus, new sheds, garages, and any other building a homeowner desires on their property are being built by residential contractors. Naturally, since many of these buildings will have plumbing, electrical, and other fixtures that require installation by a qualified specialist in that field, contractors may find themselves having to use subcontractors for these projects.

The majority of residential construction structures are of types 1A, V-A, and V-B. Every one of these kinds describes the materials the building is constructed of and the level of fire protection it has. Naturally, contractors want to establish a reputation for providing their homeowners with the safest possible residential buildings. To ensure that everyone has access to the best homes possible, contractors must stay abreast of the continuous advancements in the materials used in residential construction. This implies that contractors are lifelong learners who modify their operations accordingly.

1.3.3. Non-Residential Construction

Non-residential building construction can be commissioned by a variety of public and private entities, including retailers, hoteliers, property developers, local governments, educational and religious institutions, transportation companies, and other private businesses, depending on the type of building. In these industries, general contractors handle the majority of construction. The construction and remodeling of buildings not meant for residential use is referred to as non-residential construction. Numerous types of projects are included in this sector, such as infrastructure projects, institutional buildings, commercial buildings, and industrial facilities. The construction industry relies heavily on non-residential construction to power economic expansion, supply basic services, and mold communities'-built environments. Buildings for offices, stores, lodging, and entertainment are all included in the category of commercial construction. Because they create areas where companies can operate and where customers can access goods and services, these projects are essential to economic activity. Office buildings, for example, serve as the operational hub for numerous businesses and organizations, promoting economic productivity and streamlining business operations. Retail locations, like malls and shopping centers, are essential to business because they provide places where goods and services can be sold. In addition to drawing tourists and stimulating local economies, hotels, and entertainment venues also serve the tourism and hospitality sectors.

Building factories, warehouses, distribution centers, and other facilities essential to manufacturing and logistics is all part of the industrial construction process. This industry is essential to the manufacturing and distribution of goods, which support economic and industrial activity. The hub of industrial production is a factory or manufacturing plant, where raw materials are turned into final goods. In the supply chain, warehouses and distribution centers are essential because they make sure that products are moved and stored effectively. The construction of these facilities promotes the expansion of industry and the creation of jobs, which greatly boosts economic growth.

Buildings for public services such as healthcare, education, and government are examples of institutional construction. As places for instruction, learning, and research, schools, universities, and research institutes are crucial for innovation in education. Because they provide communities with access to medical services and treatment, healthcare facilities—such as hospitals, clinics, and specialty medical centers—are essential to maintaining public health. Courthouses, city halls, and public safety facilities are examples of government buildings that facilitate the operation of public administration and the provision of necessary services to the populace. Institutional construction promotes societal development and well-being by guaranteeing that communities have access to essential services and facilities.

Despite being frequently regarded as a separate category, infrastructure projects constitute a substantial portion of non-residential construction. Building highways, bridges, ports, airports, and utility systems are all included in this. Connectivity, mobility, and the supply of utilities like water, power, and communication networks all depend on the development of infrastructure. Through the provision of necessary services, these projects improve economic activity, ease the flow of people and goods, and raise living standards. The non-residential construction industry is distinguished by its emphasis on innovation and sustainability. Sustainable materials, energy-efficient designs, and green building techniques are becoming more and more common in non-residential projects. These procedures have a positive effect on the environment, save operating expenses, and give residents better spaces. Furthermore, the design, construction, and operation of non-residential buildings are changing due to technological advancements like prefabrication, smart building systems, and Building Information Modeling (BIM).

1.3.4. Infrastructure Construction

The services and facilities required for a nation, city, or other area's economy, households, and businesses to operate are collectively referred to as its infrastructure. Public and private physical structures like roads, railroads, bridges, tunnels, water supplies, sewage systems, electrical grids, and telecommunications (including broadband and Internet access) are all considered infrastructure. Infrastructure is generally understood to be the tangible parts of interconnected systems that offer goods and services necessary to support, maintain, or improve societal living standards and the environment. Modern infrastructure talks usually center on sustainable development and green infrastructure, especially in light of the enormous societal changes required to mitigate and adapt to climate change. In recognition of this significance, the international community has established policies through the Sustainable Development Goals, particularly Goal 9: Industry, Innovation, and Infrastructure, that are centered on sustainable infrastructure.

Different forms of infrastructure can be categorized into two groups: hard infrastructure and soft infrastructure. The actual networks required for the operation of a contemporary industrial society or industry are known as hard infrastructure. This covers railways, bridges, and roads. Any institution that upholds a nation's standards for social, cultural, environmental, health, and economic aspects is considered soft infrastructure. This includes official statistics, parks and leisure centers, emergency services, law enforcement, and educational initiatives. The design, building, and upkeep of the natural and man-made environment falls under the purview of civil engineering. This includes public works like highways, bridges, canals, dams, tunnels, airports, water and sewage systems, pipelines, and railroads. Civil engineering contractors are businesses that focus on this industry and may have specializations in specific infrastructure types. Some general contractors are also skilled in civil engineering.

Infrastructure construction refers to the process of designing, planning, and building physical structures and facilities that are necessary for the functioning of a society, economy, or industry. This can include building highways, bridges, tunnels, airports, railways, water treatment plants, power plants, and other essential facilities.

Infrastructure construction projects are often complex, involving multiple stakeholders and requiring significant financial resources and technical expertise. They may involve the use of various materials, such as concrete, steel, and asphalt, and require specialized equipment and machinery (Tabei & Sayed, 2019).

Infrastructure construction is critical for the economic growth and development of a country, as it creates jobs, improves connectivity, and facilitates the movement of goods and people. It also plays a crucial role in improving public safety, enhancing the quality of life, and ensuring the sustainability of communities.

1.3.4.1. Importance

Building infrastructure is essential to the development of society because it supports social cohesion, economic expansion, and general quality of life. It is impossible to exaggerate the significance of infrastructure construction, which includes the creation of utilities, public buildings, transportation networks, and communication systems, among other necessary facilities and systems. These fundamental components provide the required framework for a variety of services and activities, making them essential to the smooth operation of a modern society.

Building infrastructure first and foremost promotes economic growth. The smooth movement of goods and people is made possible by efficient transportation networks, which include roads, bridges, railroads, and airports. This lowers costs and boosts productivity. Reliable utilities are necessary for communities to prosper and for businesses to run efficiently. Examples of these utilities include electricity grids, sewage systems, and water supplies. Additionally, building infrastructure projects create jobs, boosting the economy and increasing GDP growth. Long-term economic benefits come from the jobs created during the construction phase and the subsequent upkeep and operation of infrastructure facilities.

Apart from the financial benefits, building infrastructure has a big influence on social welfare. Public health is fundamentally dependent on having access to clean water, dependable energy, and effective waste management systems. Adequate infrastructure lowers the risk of illness, boosts sanitation, and improves people's quality of life in general. As components of social infrastructure, healthcare, and education facilities are essential for the development of human capital. Modern amenities in schools, colleges, hospitals, and clinics guarantee that communities have access to necessary services, promoting a population that is healthier and better educated. Additionally, essential to improving connectivity and communication is infrastructure. The digital divide is closed by the development of telecommunication networks, which include internet and mobile services and give people access to information and communication technologies. For business operations, personal communication, and access to government services, this connectivity is essential. Strong communication infrastructure is essential in the digital age to foster innovation, entrepreneurship, and the expansion of the knowledge economy, all of which advance society.

An additional crucial component of building contemporary infrastructure is environmental sustainability. In order to reduce their negative effects on the environment, sustainable infrastructure projects use efficient resource management, renewable energy sources, and green building techniques. Infrastructure development helps to mitigate climate change, conserve natural resources, and lower carbon footprints by putting sustainability first. Not only do projects like smart grids, green buildings, and public transportation systems help the environment, but they also improve community health. Building infrastructure also improves preparedness for disasters and resilience. Well-built infrastructure is resilient to natural calamities like hurricanes, floods, and earthquakes, which lessens community vulnerability and guarantees service continuity in an emergency. Resilient infrastructure investment is essential for reducing the risk of disaster, facilitating a quicker recovery, and minimizing financial losses.

1.3.5. Industrial Construction

Along with mining and quarrying, refineries, breweries, distilleries, and other processing plants; power plants; steel mills; warehouses; and factories, industrial construction also includes offshore construction (primarily of energy installations: oil and gas platforms, wind power). Power plants, warehouses, and other extremely specialized buildings are all part of the industrial construction business. The mechanical and structural elements of these facilities are also included in their design, installation, and upkeep. Any company that manufactures products is referred to as industrial. Special transportation needs, like round-the-clock shipping and receiving, easy access to major highways, and nearby air or train transportation, are common in industrial buildings. Floor plans that facilitate easy foot traffic and concentrate on coordinating manufacturing and distribution are key components of the building design for an industrial facility. Because they must house heavy machinery that must adhere to industry-specific regulations, industrial buildings are constructed with functionality rather than style in mind. Permitting and occupancy requirements are typically met by industrial projects in order to appease local communities. While larger industrial projects need to comply with federal, state, and local building codes, commercial projects operate similarly with regard to permits.

1.3.5.1. History of Industrial Buildings

To elaborate, an industrial building is any large facility, such as a factory, that is used for the production or storage of goods, services, or raw materials for commercial gain. The first American textile mill was established by Samuel Slater in 1790, which is when the country's industrial architecture began. The American Industrial Revolution is widely regarded as having begun with the establishment of that textile mill. Slater provided evidence of the economic advantages of employing industrial structures to significantly boost output. The industrial facilities market is estimated to be worth $32 billion, and they have completely changed the way we live today.

1.3.5.2. Types of Industrial Buildings

1. Warehouse and Distribution: Goods are transported and stored in warehouses and distribution centers. These structures are typically one story and can have 5,000 square feet or hundreds of thousands of square feet. Because racking and storage systems are housed inside, ceilings are typically at least 60 feet high. These facilities also have big garage doors, loading docks, and parking spaces big enough to handle the semi-trailers that are used for distribution.

2. Manufacturing: The capital-intensive machinery needed to produce goods and materials is housed in manufacturing or heavy industrial buildings. Industries that use manufacturing facilities include those in the oil, mining, and shipbuilding sectors. In order to supply sufficient voltage to operate the large machinery housed inside, manufacturing buildings are equipped with a three-phase electric power supply. To create clean air and remove dangerous chemicals and fumes from the building, they also have large ductwork, high-capacity ventilation, and exhaust systems. For the machinery to operate properly and for functional purposes like floor drains and storage tanks, they will also have pressurized air or water lines.

1.3.5.3. How Is Industrial Construction Different From Commercial Construction?

Commercial and industrial construction differs somewhat, despite their similarities. There is a big difference in site planning. When it comes to lot size, traffic patterns, building locations, and other aspects, commercial and industrial construction differs substantially. The main priority in commercial construction, similar to retail establishments, is customer access. Specialized conveyance requirements, such as 24-hour shipping, are the focus of industrial construction. Additional significant distinctions are in the infrastructure and building design. Smooth foot traffic for customers is emphasized in a commercial setting such as a retail store. The main focus of industrial facilities is functionality in relation to the kind of manufacturing they facilitate. In an industrial setting, heavy machinery is the focus of the supporting infrastructure. In order to meet various set goals, managing either of these construction types requires experienced project managers and artisans on the job site.

1.3.5.4. Contact Stevens for Your Next Industrial Construction Project

Starting a new industrial construction project will necessitate anticipating potential issues and knowing how to resolve them. An experienced industrial construction company will be able to identify these possible issues and provide quick solutions. Stevens is prepared to assist you in beginning your upcoming industrial construction project. We maintain in-house construction crews to provide the high-quality construction services for which we are renowned, and we have over 75 years of operational experience. We are pleased to provide our reputable services to clients in the oil and gas, government, energy and power, manufacturing and industrial, ferrous and non-ferrous, and oil and gas sectors. Regardless of the nature of the project, STEVENS will satisfy your building requirements and provide top-notch industrial contractor services.

CASE STUDY

MODELING COVID-19 IMPACTS AND RESPONSE STRATEGIES IN THE CONSTRUCTION INDUSTRY:

COVID-19 has caused various issues and problems for the modern world's healthcare, economic, and social systems. The construction industry, as well as other industries, has been negatively impacted by the pandemic. COVID-19 not only affects the construction industry in terms of construction projects, but also the workforce and construction organizations.

COVID-19 not only affects construction projects, but also causes negative impacts on the construction workers in the field as well as in the office. The study categorizes the impacts into organizational, economic, psychological, individual, and moderating factors. The results show that construction workers experience unexpected work suspension and have suffered psychologically and emotionally from homesickness due to the travel ban. Construction organizations also experienced negative impacts due to COVID-19. In addition to that, the impacts that affect construction organizations in Ghana include decreased work rate, delays in payments, and increased material cost due to border closure.

Pandemic Response

Policymakers are developing response strategies to address the pandemic's impacts. For example, policymakers in Sri Lanka and Australia have developed response strategies to overcome the pandemic's impacts. At the organizational level, the response strategies include keeping standard operating procedures, establishing successful relationships with suppliers, and working in shifts. There are 11 strategies that can be used to develop pandemic resiliency among construction organizations; these strategies include portfolio diversification, collaborative contracting methods, industrialized construction, circular economy, remote working, integrated design management using building information modeling (BIM), staffing and skills training, reversible building design, augmented reality, automation, three-dimensional printing, and lean construction.

COVID-19 in the Malaysian Construction Industry

COVID-19 has emerged in society, devastatingly impacting many industries, including the construction industry. The construction industry contributes significantly to Malaysia's economic growth, with an annual total return of MYR204 billion on construction projects. Like many other nations, the Malaysian government enforced movement restrictions during the outbreak due to the constant and rapid growth of COVID-19 cases. As a result, construction production declined by 13.1% yearly to MYR31.4 billion in the third quarter of 2020. Material supply shortages also occurred in many areas due to travel restrictions or post-travel quarantines. Another serious issue is human resources, as many foreign workers cannot re-enter Malaysia or leave their home countries due to travel restrictions. Construction organizations and contractors were forced to deal with labor scarcity resulting from restrictions on the admittance of foreign workers. In addition, contractors had to cope with contract delays or increased expenses, as well as other changes that needed to be made. Material shortages are frequently caused by changes and delays in material acquisition, site operations, and supply chain issues.

Researchers in Malaysia are conducting research to investigate and mitigate the pandemic's impact on the Malaysian construction industry. The research identified five critical pandemic impacts, including reduced construction productivity, reduced foreign investment in the construction industry, reduced demand for construction-related works, disruption in the supply chain, and a reduced number of public projects. The most significant problem faced by workers is challenges in adopting new norms on site, followed by workforce shortage, planning and schedule disruption, workforce health, and workforce management issues. While the most effective strategy is increasing the use of communication technology, followed by redefining risk and safety management on-site and adjusting working spaces based on the standard operating procedure. The consequences of the lockdown implemented by the Malaysian government (i.e., the Movement Control Order or MCO) towards construction project success. The result shows that MCO has negatively impacted project success in terms of regulation compliance, safety, additional time for project delivery, increase in development cost, limited human resources supplies, and limited resource availability on-site. The study concluded that there are four critical response strategies for small-medium enterprises, including forming a special task force to provide support in maneuvering COVID-19, providing infrastructure investment budgets to local governments, developing employee assistance programs that fit all types of working groups, and diversifying existing supply chain. Large enterprises have two distinct critical response strategies, including providing help in digitizing existing construction projects and mandating COVID-19 as force majeure.

Knowledge Gap

This subsection synthesizes the knowledge gaps that exist in the current literature to support the rationale for conducting the study. Although prior works have studied COVID-19's impacts and response strategies in the construction industry, much remains unknown, as COVID-19 is a relatively new topic. In addition, previous works lack insights into the relationships between COVID-19 impacts and response strategies. Prior works only focused on identifying the impacts and response strategies independently, without connecting them. Developing appropriate response strategies is critical for addressing the targeted impacts. The current study leverages the knowledge gap by identifying (i) the underlying impacts of COVID-19 on the Malaysian construction industry; (ii) the construction industry response strategies to COVID-19; and (iii) the relationships between the underlying impacts and response strategies.

CLASS ACTIVITY

CONSTRUCTION SITE SIMULATION

- **Objective:** To give students a hands-on experience of managing a construction project, understanding the challenges, and practicing problem-solving skills.
- **Materials Needed:** Paper, Pens/pencils, site layout diagram, project scenario, etc.
- **Assessment:** Distribute the construction site layout diagram to each group. This diagram should represent the layout of the construction site, including different areas such as excavation, foundation laying, framing, etc. Assign specific tasks to each group based on the scenario provided. For example, one group might be responsible for scheduling and coordinating subcontractors, while another might oversee quality control. As the simulation progresses, introduce unexpected events or challenges (e.g., inclement weather, supply chain disruptions) that the groups must address in real-time.
- **Reflection:** Conclude the activity by asking students to individually reflect on what they learned from the simulation. Prompt them to consider how they can apply these lessons to real-life construction projects in the future.

SUMMARY

- The construction industry encompasses a wide range of activities involved in the planning, design, building, and maintenance of structures and infrastructure.
- Technological innovation has significantly impacted the construction industry, introducing new methods and materials that enhance efficiency and sustainability.
- Construction is a general term meaning the art and science of forming objects, systems, or organizations. It comes from the Latin word constructio (from com – "together" and struere "to pile up") and Old French construction.
- The construction sector is viewed as a service industry. It generates substantial employment and provides growth impetus to other manufacturing sectors like cement, bitumen, iron and steel, chemicals, bricks, paints, tiles, etc.
- Financial planning ensures adequate safeguards and contingency plans are in place before the project is started, and ensures that the plan is properly executed over the life of the project. Construction projects can suffer from preventable financial problems.
- A construction project is a complex net of construction contracts and other legal obligations, each of which all parties must carefully consider. A contract is the exchange of a set of obligations between two or more parties and provides structures to manage issues.
- Design and construction. In some cases, the design-build package can also include finding the site, arranging funding, and applying for all necessary statutory consents.
- In a construction management arrangement, the client enters into separate contracts with the designer (architect or engineer), a construction manager, and individual trade contractors.
- Commissioning is the process of verifying that all subsystems of a new building (or other assets) work as intended to achieve the owner's project requirements and as designed by the project's architects and engineers.
- The construction industry plays a pivotal role in the development and growth of economies and societies worldwide. As one of the largest and most dynamic sectors, it encompasses the building and maintenance of infrastructure, residential, commercial, and industrial properties, thereby shaping the physical landscape of our communities.
- The construction industry is also a driver of technological advancement and innovation. The sector continuously adapts to new materials, methods, and technologies that improve efficiency, safety, and sustainability.
- The construction industry, while pivotal to economic growth and societal development, faces a myriad of challenges that can hinder project execution and sustainability. One of the most pressing issues is the shortage of skilled labor.

- Environmental sustainability has become a critical focus in the construction industry, driven by increasing awareness of climate change and the need for eco-friendly practices. However, implementing sustainable construction methods can be challenging due to higher upfront costs and the necessity for specialized knowledge and materials.
- The construction industry is undergoing transformative changes driven by technological advancements, environmental concerns, and evolving economic conditions. These trends are reshaping how construction projects are planned, executed, and managed, leading to increased efficiency, sustainability, and resilience in the built environment.

REVIEW QUESTIONS

1. What are the primary risks associated with the construction sector?
2. What challenges does the construction industry face?
3. How does the economic context impact building construction practices?
4. Which types of considerations are involved in building design and construction?
5. Define the key aspects of residential construction.
6. What are the complexities and challenges of infrastructure construction?

REFERENCES

1. Abdel-Wahab, M., & Vogl, B. (2011). Trends of productivity growth in the construction industry across Europe, US and Japan. Construction Management and Economics, 29(6), 635–644. https://doi.org/10.1080/01446193.2011.573568.
2. Asheim, B. T., & Isaksen, A. (1997). Location, agglomeration and innovation: Towards regional innovation systems in Norway? European Planning Studies, 5(3), 299–330. https://doi.org/10.1080/09654319708720402.
3. Chau, K. W., & Walker, A. (1988). The measurement of total factor productivity of Hong Kong construction industry. Construction Management and Economics, 3, 309–224. https://doi.org/10.1080/01446198800000019.
4. Chen, M., & Yang, W. G. (2011). Total factor productivity change measurement of construction industry based on scale economy. Journal of Engineering Management. https://doi.org/10.3969/j.issn.1674-8859.2011.05.007.
5. Dai, H., Chen, S., & Li, Y. (2020). Characteristics of regional energy consumption of China's construction industry from the perspective of life cycle. China Population, Resources and Environment, 30(7), 9. https://doi.org/10.12062/cpre.20191124.
6. Dubois, A., & Gadde, L. E. (2002). The construction industry as a loosely coupled system: Implications for productivity and innovation. Construction Management and Economics, 20(7), 621–631. https://doi.org/10.1080/01446190210163543.

7. Fang, J., & Li, Z. (2009). An empirical analysis on economies of scale and scope in China large contractors. Journal of Quantitative and Technical Economics, 26(2), 47–55.

8. Gao, J. Y. (2003). A positive analysis on total factor productivity of Chinese construction industry. Journal of Shandong University of Architecture and Engineering, 18(78), 82. https://doi.org/10.3969/j.issn.1673-7644.2003.04.020.

9. Goh, M. B. H. (2003). Total factor productivity growth accounting in the construction industry of Singapore. Construction Management and Economics, 10, 707–718. https://doi.org/10.1080/0144619032000056126.

10. Leontief, R. W. (1941). Production and distribution theories: The formative period by George J. Stigler. American Economic Review, 31(4), 855–858. https://doi.org/10.2307/1803252.

11. Li, B., Wang, X., & Liu, B. (2011). Systematic evaluation of the competitive advantage of Chinese regional construction industry based on principal component and DEA-DA. China Civil Engineering Journal, 2, 143–150.

12. Li, Z., & Li, Y. (2009). Performance evaluation of infrastructure investment in China based on DEA: An empirical analysis from 2003 to 2007. Journal of Systems Management, 18(3), 7.

13. Li, Z., Zhang, X., & Liu, G. (2008). An empirical study on the migration of total factor productivity of the Chinese construction industry in 1996–2005. China Civil Engineering Journal, 41, 58.

14. Nasereddin, M. (2015). Causes and effects of delays in Jordanian construction industry. American Journal of Engineering Research (AJER), 5(5), 87–94. https://doi.org/10.13140/RG.2.1.1011.6963.

15. Park, Y. S., Egilmez, G., & Kucukvar, M. (2015). A novel life cycle-based principal component analysis framework for eco-efficiency analysis: Case of the United States manufacturing and transportation Nexus. Journal of Cleaner Production, 92, 327–342. https://doi.org/10.1016/j.jclepro.2014.12.057.

16. Sambasivan, M., & Soon, Y. W. (2007). Causes and effects of delays in Malaysian construction industry. International Journal of Project Management, 25(5), 517–526. https://doi.org/10.1016/j.ijproman.2006.11.007.

17. Tabei, S. M. A., Bagherpour, M., & Mahmoudi, A. (2019). Application of fuzzy modelling to predict construction projects cash flow. Periodica Polytechnica Civil Engineering. https://doi.org/10.3311/ppci.13402.

18. Wang, X., Zhou, S., & Liu, B. (2011). An empirical study on regional construction efficiency in China based on three stage DEA model. Construction Economics, 11(12), 8–12.

19. Wang, Y., & Zhang, G. (1999). Research on the production efficiency and the affecting factors in the construction industry of Hong Kong. Journal of Shenzhen University, 16(4), 53–60. https://doi.org/10.3969/j.issn.1000-2618.1999.04.010.

20. Xing, M., Luo, F., & Fang, Y. (2021). Research on the sustainability promotion mechanisms of industries in China's resource-based cities—From an ecological perspective. Journal of Cleaner Production, 315, 128114. https://doi.org/10.1016/j.jclepro.2021.12811.

21. Zhang, X., Li, Z., & Wang, L. (2008). An empirical study on the migration of total factor productivity of Chinese large scale construction contractors from the year 1996–2005. Journal of Systems Management, 17(4), 423–427.

CHAPTER

2 Construction Management

LEARNING OBJECTIVES

After studying this chapter, you will be able to:

- Give an overview of construction management system;
- Discuss the project risk management;
- Explain the risk identification

KEY TERMS FROM THIS CHAPTER

Risk breakdown structure (RBS)
Cost accounting
Critical path method (CPM)
Business enterprise
Constraint
Budget
Construction management
Risk responses
Cost management
Field operations

INTRODUCTION

From project conception to completion, planning, coordinating, and carrying out construction projects are the main focuses of the specialized field of construction management. It combines several fields, such as project management, engineering, and architecture, to guarantee that projects are completed on schedule, within budget, and to the necessary quality standards. Overseeing every facet of the construction process, such as resource allocation, project planning, risk management, and stakeholder communication, is the main responsibility of a construction manager.

Project planning is one of the core components of construction management. This entails establishing goals, specifying the project's parameters, and creating thorough budgets and schedules. Planning a project well guarantees that resources are distributed effectively and that all activities are in line with the overarching objectives. Construction managers plan and track the progress of their projects using a variety of instruments and methods, including Gantt charts and the Critical Path Method (CPM).

Allocating resources is yet another crucial aspect of construction management. This entails overseeing the personnel, supplies, and machinery required for the undertaking. To prevent delays and cost overruns, construction managers need to make sure that the appropriate resources are available when they're needed. To keep a constant supply of goods and services, they must also communicate with suppliers and subcontractors.

An essential component of construction management is risk management. Because construction projects are complex and involve many different stakeholders, they are by nature risky. Construction managers identify possible risks and create plans to reduce them, such as defects in the design, site conditions, and regulatory changes. This proactive approach guarantees the project's smooth progress and helps to minimize disruptions.

Stakeholder management and effective communication are critical to the success of building projects. For clients, architects, engineers, and contractors, among other project stakeholders, construction managers serve as the primary point of contact. To make sure that everyone is in agreement and that any problems are dealt with right away, they facilitate prompt and clear communication. Working together is essential to settling disputes and coming to wise decisions.

A key factor in the effective completion of building projects is the diverse discipline of construction management. It calls for careful preparation, effective resource management, proactive risk mitigation, and clear communication. Construction managers can guarantee that projects are finished on schedule, within budget, and to the highest quality standards by incorporating these components, which will satisfy the expectations of all parties involved.

2.1. OVERVIEW OF CONSTRUCTION MANAGEMENT SYSTEM

Construction management is a professional service that utilizes specialized project management techniques to supervise the planning, design, and construction of a project from start to finish. The goal of construction management is to oversee a project's schedule, budget, and quality, which are often referred to as the "triple constraints." Professional construction managers are typically hired for large-scale, high-budget projects such as commercial real estate, transportation infrastructure, industrial facilities, and military infrastructure, known as capital projects.

Source: https://www.letsbuild.com/wp-content/uploads/2018/09/onstruction-time-waste-1280x720.jpg.

Every step of the building process for any built environment is the focus of the professional field of construction management. Single-family residential building sites are typically excluded, but commercial or multi-unit residential building sites are usually included. A general contractor or another organization may be in charge of site management; either way, the staff or company reports to the developer and is in charge of managing the project's entirety from beginning to end.

While many professionals in this field hold an associate's or bachelor's degree, in certain circumstances a person with extensive experience can become certified as a construction manager even though they have not attended college. Furthermore, for specific projects, some regions might need a master's degree in construction management.

Advanced mathematics, design and construction laws, architecture, estimating, accounting, and business are among the subject areas covered. Being one of the biggest industries in the US, construction has a bright future for future professionals. There are jobs available in the government and the private sector.

Construction management staff are in charge of managing every stage of a project and resolving any differences between the actual implementation of the blueprint and the original design. They have to keep labor and material costs under control and are also responsible for the project's overall cost. They frequently collaborate in both an office and a hardhat setting concurrently, working with one or more project managers, architects, and superintendents specific to the site. Apart from supervising the project's physical construction stages, construction management also needs to be knowledgeable about design and construction laws, including US building codes and OSHA regulations, and act as a point of contact for all matters pertaining to construction. Delivering a project's phases on schedule and within budget, all the way to the finished building or buildings, is equally important.

2.1.1. Need for Project Management

There is typically only one chance for the contractor to set the price (the bid) in construction contracts. From then on, the project manager's capacity to save costs through improved daily operations planning and decision-making will decide profits. A project's field operations must be closely managed if it is to be completed within the allocated budget and time frame. Construction operations planning, scheduling, and control are crucial due to factors like technical complexity, timely completion of the project, resource constraints, and significant costs. Regretfully, once the construction process starts, it is not a self-policing system and needs professional supervision to ensure that things go according to plan.

It is important to remember that projects are unique endeavors with a set time frame, involving varying and non-standardized work. Construction projects are particularly susceptible to unforeseen events that can significantly impact costs and timelines. In such unpredictable conditions, planning plays a crucial role in determining the success or failure of the project. The higher the level of uncertainty, the more essential it becomes to have thorough project planning and skilled management.

The general contractor manages construction operations under the majority of fixed-sum, competitively bid contracts that only require construction services. In these kinds of situations, self-interest is the primary driving force, as the contractor is bound by contract to complete the project by a given date and for a predetermined amount. Using a project management system is the most reliable way for the contractor to meet both its own goals and the owner's as well. Under other types of contracts, project control is primarily focused on serving the owner's best interests. The primary goal of field management in design-construct, construction management, and many cost-plus contracts is to offer the owner expert management and advisory services in order to best meet the owner's goals.

Source: https://img.etimg.com/thumb/width-1200,height-900,imgsize-509283,resizemode-1,msid-74948012/wealth/personal-finance-news/realty-developers-body-naredco-urges-govt-to-lift-lockdown-on-construction-activity.jpg.

2.1.2. Project Management Characteristics

In its most common form, 'management' refers to the planning, organizing, directing, and controlling of a business. Business management is an ongoing internal activity that involves a company's personnel, finances, property, and resources. On the other hand, construction project management is focused on overseeing a specific project, with different phases being carried out by various organizations. This means that managing a construction project is more about coordinating and regulating all the necessary elements to complete the job, rather than managing the internal affairs of a single company. Project managers in construction often work closely with other organizations, and their authority is often defined by contractual agreements or power of agency, making it less direct than that of a traditional business manager. Project management relies heavily on collaboration between employees of different employers.

2.1.3. Time and Cost Management

Effective project time and cost management relies on the development of schedules and the utilization of an information system to compare expected performance with actual progress. This system constantly evaluates job progress against the planned performance, allowing the project manager to quickly address any deviations that may arise. In construction projects with volatile production conditions, costs and time can easily spiral out of control, highlighting the importance of prompt detection and correction of any issues.

Timely cost and time control information is crucial, giving the project manager the opportunity to assess alternatives and implement corrective actions before problems escalate.

In essence, management focuses on controlling costs by ensuring efficient completion of safe and high-quality projects, which benefits both the contractor in terms of construction savings and the owner in terms of utilization. However, in reality, time and cost management are often treated as distinct but interconnected processes. This separation is evident in the varying job breakdown structures used for time and cost control. The unique nature of these procedures necessitates dividing the project into two sets of elements: project components for time management and work classifications for cost management.

Practically speaking, it is impossible to maintain strict control over every detail when working on a field project. As such, it is imperative to acknowledge that the time and cost management techniques covered in this book are not perfect processes, providing managers with far from absolute control but with results that are reasonably accurate. Project management techniques don't provide a magic bullet for issues with construction.

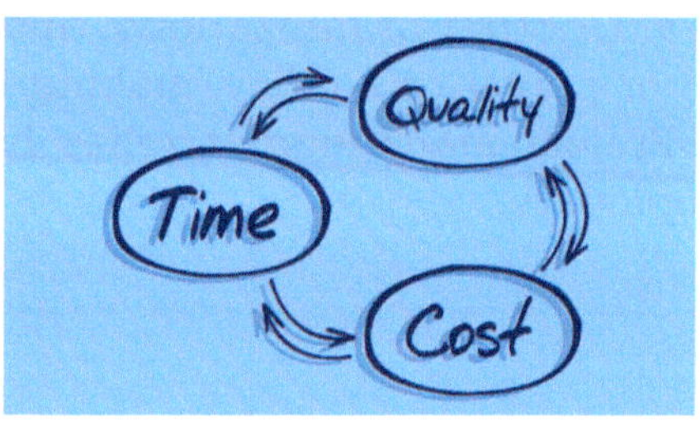

Source: https://www.projectmanager.com/wp-content/uploads/2018/03/181119_Blog_Feature_Triple_Constraint.jpg.

They don't offer any miracle cures, and the quality of the management information they produce is equal to that of the input data. However, a passably solid foundation is created for well-informed decision-making.

2.1.4. CPM Procedure

The critical path method (CPM), also known as critical path scheduling, is the most commonly used scheduling technique. This method determines the project's minimum completion time as well as the potential start and end times for each activity. In fact, a lot of managers and texts consider critical path scheduling to be the only workable and realistic scheduling technique. Critical path scheduling computer programs and algorithms are widely available and effectively manage projects with thousands of activities.

The critical path is the sequence of predecessor/successor activities that will take the longest time to complete. Its duration is the total of the durations of the activities along that path. In essence, the critical path is the longest route through the project's network of activities and represents the minimum time needed to finish the project. Delays on the critical path would result in the project taking longer to complete.

There could be multiple critical paths within the project activities, meaning that delaying any activities along these paths could result in a delay in completing the entire project. For instance, in a project with two activities running concurrently, each taking three days to complete, both activities are critical for the project to be finished in three days.

Critical path scheduling is a formal assumption that a project has been broken down into discrete activities with predetermined durations and clear predecessor relationships. A predecessor relationship indicates that in the schedule, one activity has to come before another. In the most basic form of critical path scheduling, no resource constraints are recognized other than those implied by precedence relationships.

In the implementation of critical path scheduling, construction planners often use a precedence relation to represent resource constraints. Constraints limit a manager's options and are typically due to limited availability of resources such as equipment, material, space, or labor. For instance, when two activities require the same equipment, one activity is arbitrarily assumed to come before the other to ensure they do not overlap. Additionally, most critical path scheduling algorithms have restrictions on the relationships between activities and network configurations used. This means that the construction plan is represented as a network plan with activities as nodes, each with a unique number or designation. Two additional nodes are added to symbolize the project's start and completion.

The computerized representation of the project schedule typically includes a detailed list of activities, their durations, necessary resources, and predecessor activities. Utilizing graphical network representations instead of a simple list can aid in visualizing the plan and ensuring that all mathematical criteria are satisfied. Inputting data into a computer program can be done by filling in blanks on a screen menu, importing an existing data file, or directly typing in the information with appropriate identifiers.

In an activity-on-branch network, dummy activities can be added to give distinct activity labels and ensure the proper order of activities. These dummy activities are considered to have zero duration and are typically represented by a dashed line in the network diagram. Several cases in which dummy activities are useful are illustrated in Figure 2.1(a). The elimination of activity C would mean that both activities B and D would be identified as being between nodes 1 and 3. However, if a dummy activity X is introduced, as shown in part Figure 2.1(b) of the figure, the unique designations for activity B (node 1 to 2) and D (node 1 to 3) will be preserved. Furthermore, if the problem in part Figure 2.1(a) is changed so that activity E cannot start until both C and D are completed but that F can start after D alone is completed, the order in the new sequence can be indicated by the addition of a dummy activity Y, as shown in part Figure 2.1(c). In general, dummy activities may be necessary to meet the requirements of specific computer scheduling algorithms, but it is important to limit the number of such dummy link insertions to the extent possible.

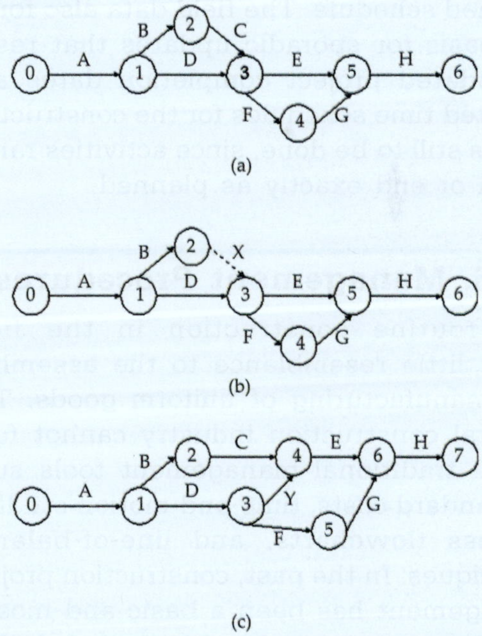

Figure 2.1. Dummy activities in a project network.

Source: https://www.cmu.edu/cee/projects/PMbook/10_Fundamental_Scheduling_Procedures.html.

2.1.5. Time Monitoring and Control

The project will move forward in line with an approved job plan when field operations start. Field progress is measured during construction and reported on a regular basis to track the progress of the work. To ensure that the work is completed on time, these data are examined, and time-control strategies are implemented as needed.

For time-control purposes, progress measurement is an approximate process that starts with figuring out the time status of each individual task. Typically, progress is determined by recording completed tasks and projecting the amount of time needed to finish ongoing tasks. These data provide the manager with an instantaneous indication of the time status of each job activity when compared to the most recent planned schedule. The field data also forms the basis for sporadic updates that result in updated project completion dates and updated time schedules for the construction that is still to be done, since activities rarely begin or end exactly as planned.

2.1.6. Management Procedures

The routine construction in the field bears little resemblance to the assembly-line manufacturing of uniform goods. The general construction industry cannot fully utilize traditional management tools such as standard costs, time-and-motion studies, process flowcharts, and line-of-balance techniques. In the past, construction project management has been a basic and mostly intuitive process, made easier by helpful but insufficient manufacturing adaptations—the bar chart being one such example.

In recent years, advancements in scientific management have led to the development and implementation of new concepts in the construction industry. These principles have been applied to improve the management of construction costs, schedules, resources, and project finances, by treating the construction process as a cohesive system. From the beginning to the end of construction, comprehensive management techniques are utilized to ensure successful project completion.

Field project management begins when construction commences, with the development of a detailed budget and timeline. These serve as the baseline for the project and are used to guide the construction process. Throughout the project, monitoring systems are implemented to track costs and progress, with regular updates provided to compare against the original plan. Any deviations are quickly identified for management action, allowing for adjustments to be made to accurately forecast the project's completion.

The process described is commonly referred to as a management-by-exception approach. When implemented in a project, it focuses on quickly and clearly identifying any deviations from the established plan or norm. Reports that point out exceptions from the standard allow the manager to promptly identify areas of the project that need attention. If a task is progressing as planned, no action is necessary; however, there are often many issues that do require intervention. This book highlights the usefulness of management-by-exception tools and emphasizes their application.

The field management system must take into account not only time and money, but also job resource management and project financial control. In this sense, resources include labor, building supplies, machinery, and subcontractors. The main steps in resource management are anticipating project needs, planning and coordinating the timely delivery of resources, and modifying demands as needed. The project manager is accountable for both the contract's terms and the total cash flow produced by the construction work when it comes to project financial control.

Source: https://www.build-review.com/wp-content/uploads/2019/12/A-guide-to-construction-safety-for-homebuilders.jpg.

2.1.7. Estimating the Project

A comprehensive cost estimate is created once the project design is complete. This estimate is used by the contractor in the bidding process and for subsequent cost control. Similar estimates are prepared for the owner's cost control during construction in cost-plus and construction management contracts. A thorough quantity takeoff, which is an accumulation of all necessary elementary work classifications, forms the basis of the final estimate. The quantity of work required determines how much labor, building supplies, and equipment will cost. The general contractor receives bids from subcontractors, and those bids are used to determine the subcontract amounts. As needed, taxes, overhead, and surety bonds are added.

Labor and construction equipment costs are the hardest to predict and manage of all the expenses associated with building a project. Production rates are the foundation upon which such costs are estimated. The number of man-hours or equipment time needed to complete a unit amount of a particular work type can be used to express a production rate. The amount of man-hours needed to erect one ton of structural steel is an illustration of this. The number of units of a work type that can be completed in a given amount of time, such as an hour, is another way to express production rates. The quantity of excavation in cubic yards that a power shovel can complete in an hour is an illustration of this type of production rate. Production rates are frequently converted to costs per unit of work for easy and quick application. The company's cost accounting system is the source of production rates and unit costs. The project control budget is created after the cost estimate is finished. During field operations, the actual production costs are measured against this cost schedule.

2.1.8. Project Cost Accounting

Cost accounting for projects involves gathering information on actual production rates and unit costs as projects progress. This data is essential for monitoring project costs and estimating future work. Labor and equipment costs, which can be unpredictable, are closely monitored during construction and are key components of the cost accounting system. Labor and equipment work hours, hourly expense rates, and the amount of work completed are all analyzed to produce regular cost reports. These reports compare budgeted and actual costs for each aspect of the project, helping to control costs and predict the final project cost. Additionally, they provide valuable data for future cost estimates and record production rates for reference.

Unlike financial accounting, cost accounting is not exclusively done in terms of expenses. The amount of work, the number of labor hours, and the equipment usage must all be ascertained in order to calculate production rates and unit costs. As a result, precise measurement and reporting of work quantities completed as well as the corresponding number of man-hours and equipment hours used in the field are essential components of a cost accounting system.

Source: https://www.michaelpage.com/sites/michaelpage.com/files/rising_costs_in_the_construction_industry.jpg.

2.1.9. Resource Management

Subcontractors, labor, materials, and construction equipment are all considered resources. These resources are crucial for the job's progress and need to be carefully managed during construction. Future resource requirements are scheduled, and proactive measures are taken to ensure that sufficient job support will be available when needed. Timely material deliveries require attention to purchasing, expediting, quality control, shop drawing review, and approval. Construction equipment and labor teams need to be planned and scheduled. Subcontractors should be informed in advance of when their services are needed, kept updated on the overall job schedule, and have their work coordinated with the project's overall effort.

2.1.10. Project Financial Control

A construction project is managed as an independent endeavor that requires resources from various sources. Each project is managed independently for income, expenses, profit or loss, and general financial accounting purposes. The project manager's responsibility for overseeing financial matters related to the project is significant. Financial considerations range from project cash flow to routine contract administration issues. Financial matters under management control procedures include monthly pay requests, an estimated schedule of payments by the owner, project cash forecasts, contract modifications, and payments to material dealers and subcontractors.

Source: https://assets.dmagstatic.com/wp-content/uploads/2018/12/dallas-construction.jpg.

2.1.11. Automating Project Management Tasks

Large amounts of numerical data must be created, manipulated, summarized, and interpreted by the project manager in order for the contractor to manage and control the project's time, cost, resources, and finances during the field construction phase. Typically, computer software is utilized by project managers to generate and apply information in an optimal manner. It can perform a wide range of data processing functions and automate repetitive tasks. Project managers have to respond swiftly to shifting circumstances and make decisions with the assurance that the information they are acting upon is sufficient, accurate, and up to date. Making this information available and providing information for the assessment of alternative courses of action can be greatly aided by modern management software. The project manager can focus more of their time on problem-solving and creating more lucrative strategies since the majority of the labor-intensive data manipulation has been automated.

However, this volume does not attempt to discuss individual hardware configurations or the operation of currently used project management software due to the constant emergence of new machines and software support. Instead, focus is on information sources, the management value of computer-generated data, and the ways in which this data is used to manage a construction project.

2.1.12. Manual Methods

The previous conversation about using computer applications for job management does not suggest that manual methods are irrelevant. Project managers may still use manual methods for small parts of a project, to do quick calculations, analyze the impact of changes, or focus on a specific area of the work.

The project manager needs to comprehend the computational processes that are an integral part of the employed techniques, even in cases where the calculations are automated. The manager will have an intuitive feel and understanding of a project that cannot be acquired in any other way due to their close familiarity with the inner workings of the procedures. This book covers the manual calculation of the various types of construction management tools in detail because manual methods are valuable in and of themselves and because applying project control methods correctly requires a thorough understanding of the computational methods involved with the computer generation of management data.

2.2. PROJECT RISK MANAGEMENT

The processes involved in risk management planning, identification, analysis, responses, monitoring, and control on a project are all included in project risk management; the majority of these processes are updated as the project progresses. Enhancing the likelihood and significance of favorable events and reducing the likelihood and significance of unfavorable events for the project are the goals of project risk management. An overview of the Project Risk Management processes is given in Figure 2.2, and a process flow diagram showing those processes' inputs, outputs, and other relevant Knowledge Area processes is shown in Figure 2.3.

The Project Risk Management processes include the following:

- Risk Management Planning – deciding how to approach, plan, and execute the risk management activities for a project.
- Risk Identification – determining which risks might affect the project and documenting their characteristics.
- Qualitative Risk Analysis – prioritizing risks for subsequent further analysis or action by assessing and combining their probability of occurrence and impact.
- Quantitative Risk Analysis – numerically analyzing the effect on overall project objectives of identified risks.
- Risk Response Planning – developing options and actions to enhance opportunities, and to reduce threats to project objectives.
- Risk Monitoring and Control – tracking identified risks, monitoring residual risks, identifying new risks, executing risk response plans, and evaluating their effectiveness throughout the project life cycle.

Both these processes and the processes in the other Knowledge Areas interact with one another. Depending on the requirements of the project, each process may require the efforts of one or more individuals or groups of individuals. Every process happens at least once during a project, and if a project is divided into phases, it happens during one or more of those phases. The processes may overlap and interact in ways not covered here, even though they are shown here as distinct elements with well-defined interfaces.

Project risk is defined as an unforeseen circumstance or event that, if it materializes, could have a favorable or unfavorable impact on at least one project goal, such as scope, budget, time, or quality (i.e., where the goal of the project is to deliver within the predetermined budget; where the goal of the project is to deliver according to the predetermined schedule, etc.). If a risk materializes, it could have one or more causes as well as one or more effects. One reason could be that the project design team is understaffed or that work requires an environmental permit. The possibility exists that the permitting body will take longer than anticipated to issue a permit or that the assigned and available design staff will not be sufficient for the project. The project's budget, timeline, or performance could all be impacted if either of these uncertain events materializes. Risk conditions may include elements of the project's or organization's surroundings that could raise project risk, like subpar project management techniques, a deficiency of integrated management systems, the execution of several projects at once, or reliance on uncontrollable external parties.

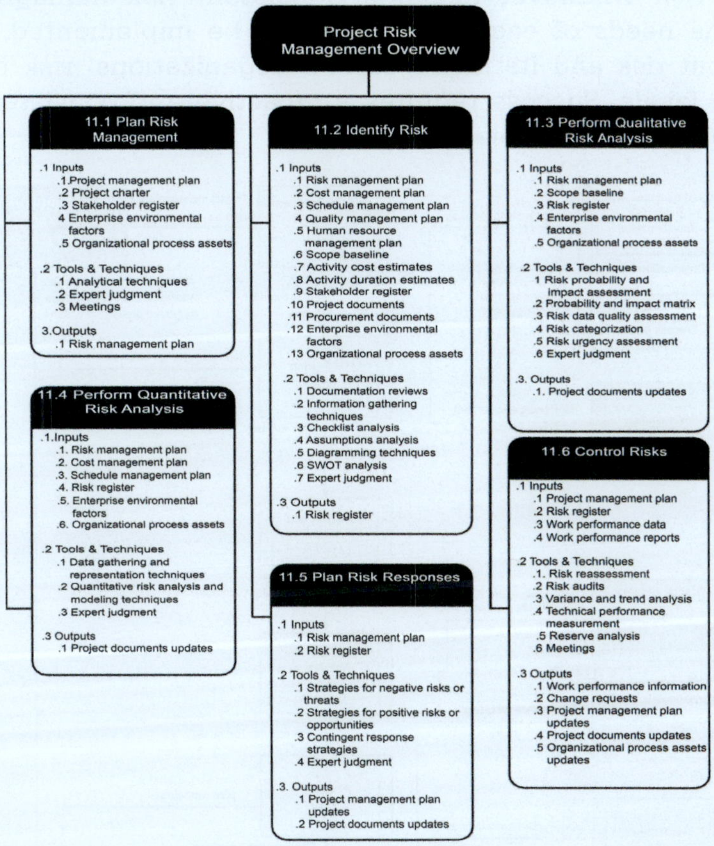

Figure 2.2. Project risk management overview.

Source: http://pmworkplace.com/PMP/A_Guide_to_the_Project_Management_Body_of_Knowledge_PMBOKGuide/images/9342239-00_0.jpg.

The uncertainty inherent in every project is the source of project risk. Risks that have been identified and analyzed are referred to as known risks, and using the processes, it might be possible to plan for them. Since it is impossible to manage unknown risks proactively, the project team would be wise to set aside general contingency against them as well as against any known risks for which developing a proactive response might not be feasible or cost-effective.

Risk is seen by organizations as opportunities to increase project success chances or as threats to project success. If the potential reward from taking the risk is greater than the risk itself, then risks that pose a threat to the project may be approved. For instance, committing to an expedited timeline that might incur delays is a risk taken in order to meet an earlier completion date. Opportunities that present risks, like the potential for increased productivity from hiring more workers, can be seized in order to further the goals of the project.

Individuals and organizations possess attitudes towards risk that impact their perception of risk accuracy and response strategies. It is important to clearly define attitudes towards risk whenever feasible. Consistent risk management approaches tailored to meet the needs of each project should be implemented, with transparent communication about risk and its management. Organizations' risk responses indicate their risk tolerance levels. Success requires a proactive and consistent commitment to managing risk throughout the project lifecycle.

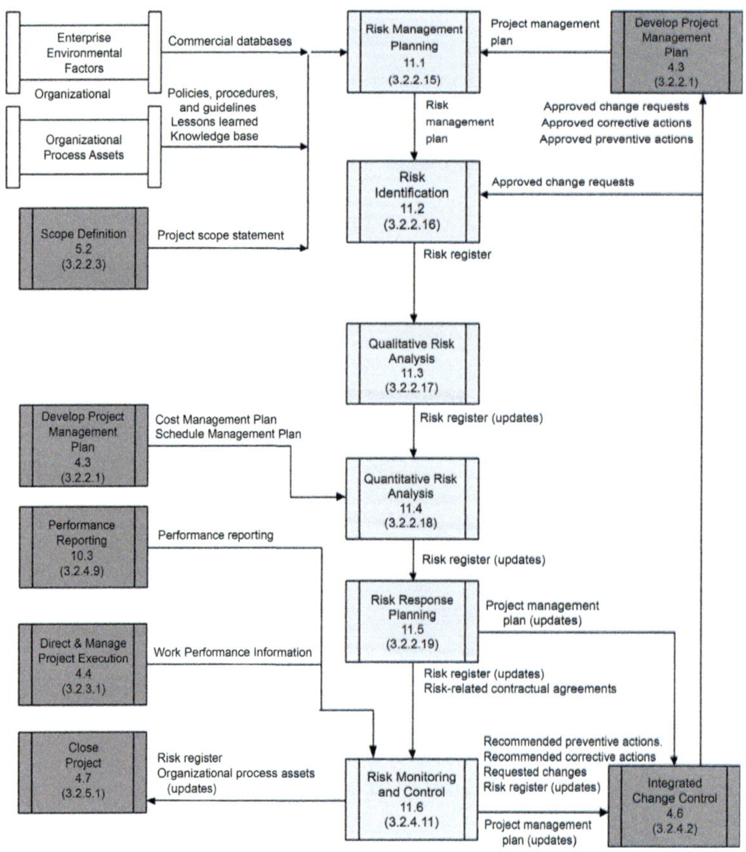

Figure 2.3. Project risk management process flow diagram.

Source: http://pmworkplace.com/PMP/A_Guide_to_the_Project_Management_Body_of_Knowledge_PMBOKGuide/images/9342241-00_0.jpg.

2.2.1. Risk Management Planning

Thorough and detailed planning increases the likelihood of success for the five remaining risk management processes. Risk Management Planning involves determining the approach and execution of risk management activities for a project. It is essential to plan out these processes to ensure that the level, type, and visibility of risk management align with the project's significance to the organization. Adequate resources and time must be allocated for risk management activities, and a common understanding of risk evaluation should be established. Risk Management Planning should be done early in the project planning phase to effectively carry out the remaining processes (Figure 2.4).

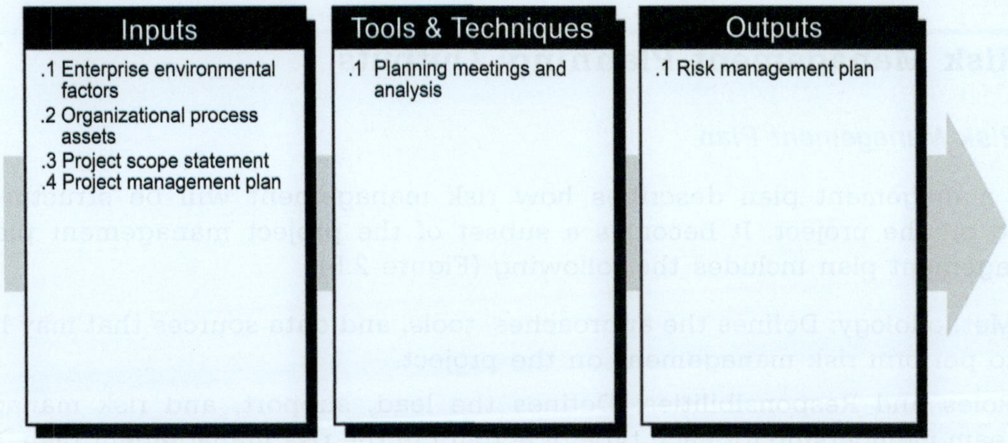

Figure 2.4. Risk management planning: Inputs, tools and techniques, and outputs.

Source: http://pmworkplace.com/PMP/A_Guide_to_the_Project_Management_Body_of_Knowledge_PMBOKGuide/images/9342242-00_0.jpg.

2.2.2. Risk Management Planning: Inputs

1. Enterprise Environmental Factors: The attitudes toward risk and the risk tolerance of organizations and people involved in the project will influence the project management plan. Risk attitudes and tolerances may be expressed in policy statements or revealed in actions.
2. Organizational Process Assets: Organizations may have predefined approaches to risk management such as risk categories, common definitions of concepts and terms, standard templates, roles and responsibilities, and authority levels for decision-making.

2.2.3. Risk Management Planning: Tools and Techniques

2.2.3.1. Planning Meetings and Analysis

To create the risk management plan, project teams meet for planning. Participants in these meetings may comprise the project manager, specific members of the project team and stakeholders, any individual within the organization tasked with overseeing the risk planning and execution processes, and additional individuals as required. These meetings establish the fundamental strategies for carrying out the risk management tasks.

The project budget and schedule will incorporate risk cost elements and schedule activities, respectively. Risk duties are going to be delegated. The probability and impact matrix, as well as general organizational templates for risk categories and definitions of terms like levels of risk, probability by type of risk, and impact by type of objectives, will be customized for each project. The risk management plan will include a summary of these activities' results.

2.2.4. Risk Management Planning: Outputs

2.2.4.1. Risk Management Plan

The risk management plan describes how risk management will be structured and performed on the project. It becomes a subset of the project management plan. The risk management plan includes the following (Figure 2.5):

- Methodology: Defines the approaches, tools, and data sources that may be used to perform risk management on the project.
- Roles and Responsibilities: Defines the lead, support, and risk management team membership for each type of activity in the risk management plan, assigns people to these roles, and clarifies their responsibilities.
- Budgeting: Assigns resources and estimates costs needed for risk management for inclusion in the project cost baseline.
- Timing: Defines when and how often the risk management process will be performed throughout the project life cycle, and establishes risk management activities to be included in the project schedule.
- Risk Categories: The structure provided ensures a thorough process of systematically identifying risks at a consistent level of detail, enhancing the effectiveness and quality of Risk Identification. Organizations can utilize a pre-established categorization of common risks, such as a risk breakdown structure (RBS) or a simple listing of project aspects. It is recommended to regularly review and adjust risk categories during the Risk Management Planning process before applying them to the Risk Identification process. Categories based on previous projects may require customization or expansion to suit the specific needs of the current project.
- Definitions of Risk Probability and Impact: The quality and credibility of the Qualitative Risk Analysis process require that different levels of the risks' probabilities and impacts be defined. General definitions of probability levels and impact levels are tailored to the individual project during the Risk Management Planning process for use in the Qualitative Risk Analysis process.

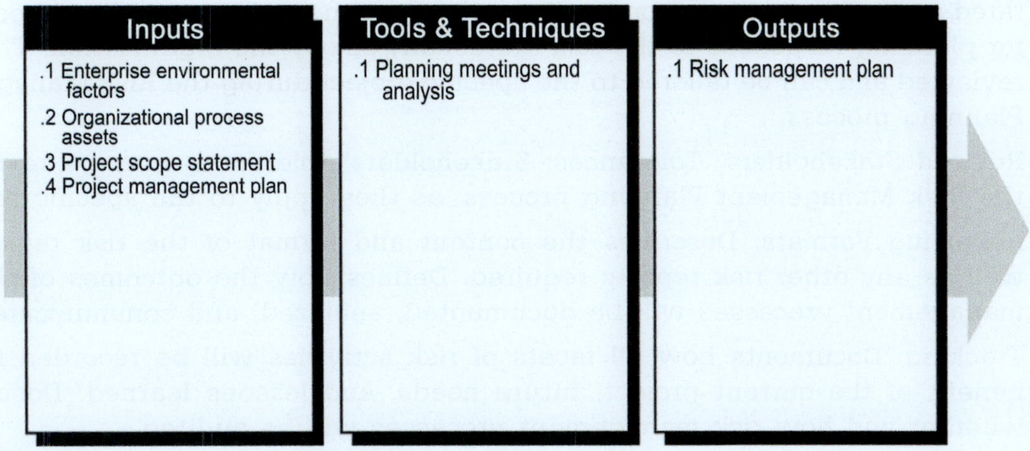

Figure 2.5. Example of a risk breakdown structure (RBS).

Source: http://pmworkplace.com/PMP/A_Guide_to_the_Project_Management_Body_of_Knowledge_PMBOKGuide/images/9342244-00_0.jpg.

A relative scale representing probability values from 'very unlikely' to 'almost certain' could be used. Alternatively, assigned numerical probabilities on a general scale (e.g., 0.1, 0.3, 0.5, 0.7, 0.9) can be used. Another approach to calibrating probability involves developing descriptions of the state of the project that relate to the risk under consideration (e.g., the degree of maturity of the project design).

The impact scale measures the importance of potential impact, whether it is negative (threats) or positive (opportunities), on each project goal in the event of a risk. These scales are tailored to the specific objective at risk, the scale and scope of the project, the organization's strategies and financial health, and the organization's level of vulnerability to certain impacts. Impact scales are categorized into relative descriptors like 'very low,' 'low,' 'moderate,' 'high,' and 'very high,' representing escalating levels of impact as defined by the organization. Alternatively, numerical scales assign values to these impacts.

These values may be linear (e.g., 0.1, 0.3, 0.5, 0.7, 0.9) or nonlinear (e.g., 0.05, 0.1, 0.2, 0.4, 0.8). Using nonlinear scales can help organizations prioritize avoiding high-impact threats or taking advantage of high-impact opportunities with low probability. When utilizing nonlinear scales, it is essential to understand the numbers' meaning, their relationships, how they were determined, and their impact on project objectives. Figure 2.6 demonstrates the drawbacks of using specific definitions to evaluate risk impacts on four project objectives, showcasing both relative and numeric (nonlinear) methods.

The figure aims to compare these two approaches within the same image rather than separately:

- Probability and Impact Matrix: Risks are prioritized according to their potential implications for meeting the project's objectives. The typical approach to prioritizing risks is to use a look-up table or a Probability and Impact Matrix. The specific combinations of probability and impact that lead to a risk being rated as 'high,' 'moderate,' or 'low' importance-with the corresponding importance for planning responses to the risk are usually set by the organization. They are reviewed and can be tailored to the specific project during the Risk Management Planning process.

- Revised Stakeholders' Tolerances: Stakeholders' tolerances may be revised in the Risk Management Planning process, as they apply to the specific project.
- Reporting Formats: Describes the content and format of the risk register as well as any other risk reports required. Defines how the outcomes of the risk management processes will be documented, analyzed, and communicated.
- Tracking: Documents how all facets of risk activities will be recorded for the benefit of the current project, future needs, and lessons learned. Documents whether and how risk management processes will be audited.

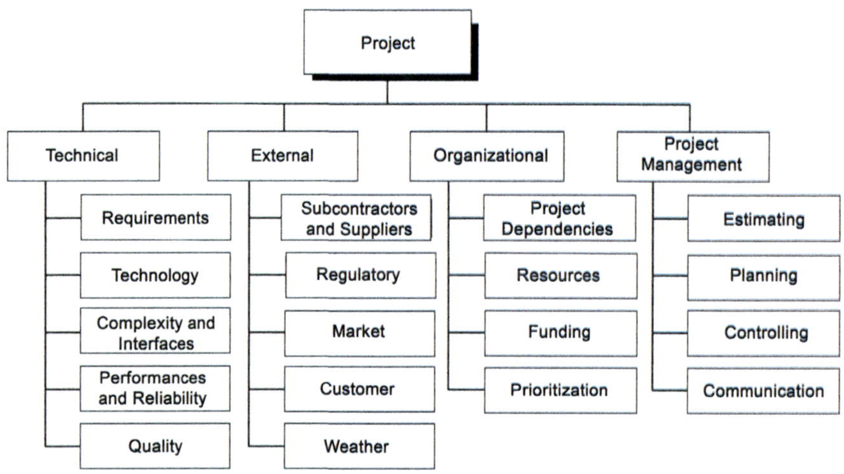

The Risk Breakdown Structure (RBS) lists the categories and sub-categories within which risks may aris for a typical project. Different RBSs will be appropriate for different types of projects and different types organizations. One benefit of this approach is to remind participants in a risk identification exercise of th many sources from which project risk may arise.

Figure 2.6. Definition of impact scales for four project objectives.

Source: http://pmworkplace.com/PMP/A_Guide_to_the_Project_Management_Body_of_Knowledge_PMBOKGuide/images/9342245-00_0.jpg.

2.3. RISK IDENTIFICATION

Risk identification identifies the risks that could have an impact on the project and records their attributes. When appropriate, the following individuals may participate in risk identification activities: the project manager, members of the project team, the risk management team (if designated), outside subject matter experts, customers, end users, other project managers, stakeholders, and experts in risk management. All project staff members should be encouraged to identify risks, even though these individuals are frequently essential in the risk identification process.

As a project moves through its life cycle, new risks might become apparent, making risk identification an iterative process. The number of iterations and the participants in each cycle will differ depending on the circumstances. In order for the project team to grow and preserve a sense of ownership and accountability for the risks and related risk response actions, they should be included in the process. Stakeholders not affiliated with the project team might offer more unbiased data. Qualitative risk analysis is typically the result of the risk identification process. Alternatively, when carried out by a skilled risk manager, it can lead straight to the Quantitative Risk Analysis procedure. Sometimes just identifying a risk might indicate how to address it; these should be noted for later review and incorporation into the risk response planning procedure (Figure 2.7).

Defined Conditions for Impact Scales of a Risk on Major Project Objectives
(Examples are shown for negative impacts only)

	Relative or numerical scales are shown				
Project Objective	Very low /0.05	Low /0.10	Moderate /0.20	High /0.40	Very high /0.80
Cost	Insignificant cost increase	< 10% cost increase	10 – 20% cost increase	20 – 40% cost increase	> 40% cost increase
Time	Insignificant time increase	< 5% time increase	5 – 10% time increase	10 – 20% time increase	> 20% time increase
Scope	Scope decrease barely noticeable	Minor areas of scope affected	Major areas of scope affected	Scope reduction unacceptable to sponsor	Project end item is effectively useless
Quality	Quality degradation barely noticeable	Only very demanding applications are affected	Quality reduction requires sponsor approval	Quality reduction unacceptable to sponsor	Project end item is effectively useless

This table presents examples of risk impact definitions for four different project objectives. They should be tailored in the Risk Management Planning process to the individual project and to the organization's risk thresholds. Impact definitions can be developed for opportunities in a similar way.

Figure 2.7. Risk identification: Inputs, tools and techniques, and outputs.

Source: http://pmworkplace.com/PMP/A_Guide_to_the_Project_Management_Body_of_Knowledge_PMBOKGuide/images/9342246-00_0.jpg.

2.3.1. Qualitative Risk Analysis

Methods for prioritizing identified risks for further action, such as Quantitative Risk Analysis or Risk Response Planning, are included in Qualitative Risk Analysis. By focusing on high-priority risks, organizations can effectively improve project performance. The priority of identified risks is assessed in Qualitative Risk Analysis by considering factors such as probability of occurrence, impact on project objectives, time frame, and risk tolerance within the project constraints of cost, schedule, scope, and quality.

Definitions of the levels of probability and impact, and expert interviewing, can help to correct biases that are often present in the data used in this process. The time criticality of risk-related actions may magnify the importance of a risk. An evaluation of the quality of the available information on project risks also helps understand the assessment of the risk's importance to the project.

Qualitative Risk Analysis is usually a rapid and cost-effective means of establishing priorities for Risk Response Planning, and lays the foundation for Quantitative Risk Analysis, if this is required. Qualitative Risk Analysis should be revisited during the project's life cycle to stay current with changes in the project risks (Figure 2.8).

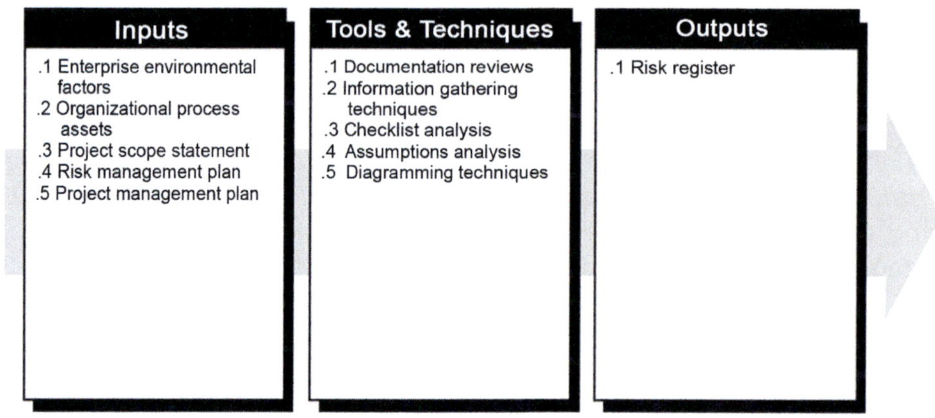

Figure 2.8. Qualitative risk analysis: Inputs, tools and techniques, and outputs.

Source: http://pmworkplace.com/PMP/A_Guide_to_the_Project_Management_Body_of_Knowledge_PMBOKGuide/images/9342250-00_0.jpg.

2.3.2. Quantitative Risk Analysis

Quantitative Risk Analysis evaluates risks that have been identified as significant during Qualitative Risk Analysis, assessing their potential impact on the project's objectives. This process assigns a numerical rating to each risk event and provides a structured method for decision-making under conditions of uncertainty.

This process uses techniques such as Monte Carlo simulation and decision tree analysis to:

- Quantify the possible outcomes for the project and their probabilities;

- Assess the probability of achieving specific project objectives;
- Identify risks requiring the most attention by quantifying their relative contribution to overall project risk;
- Identify realistic and achievable cost, schedule, or scope targets, given the project risks;
- Determine the best project management decision when some conditions or outcomes are uncertain.

Though skilled risk managers occasionally carry out quantitative risk analysis immediately after risk identification, it usually follows the qualitative risk analysis procedure. Effective risk responses may sometimes be developed without the need for quantitative risk analysis. Which method(s) to use on a given project will depend on a number of factors, including time, money, and the requirement for qualitative or quantitative statements regarding risk and impacts. After risk response planning, quantitative risk analysis should be conducted again as part of risk monitoring and control to see if the overall project risk has been sufficiently reduced. Trends may point to the necessity of taking certain risk management measures. It is a component of the process of risk response planning (Figure 2.9).

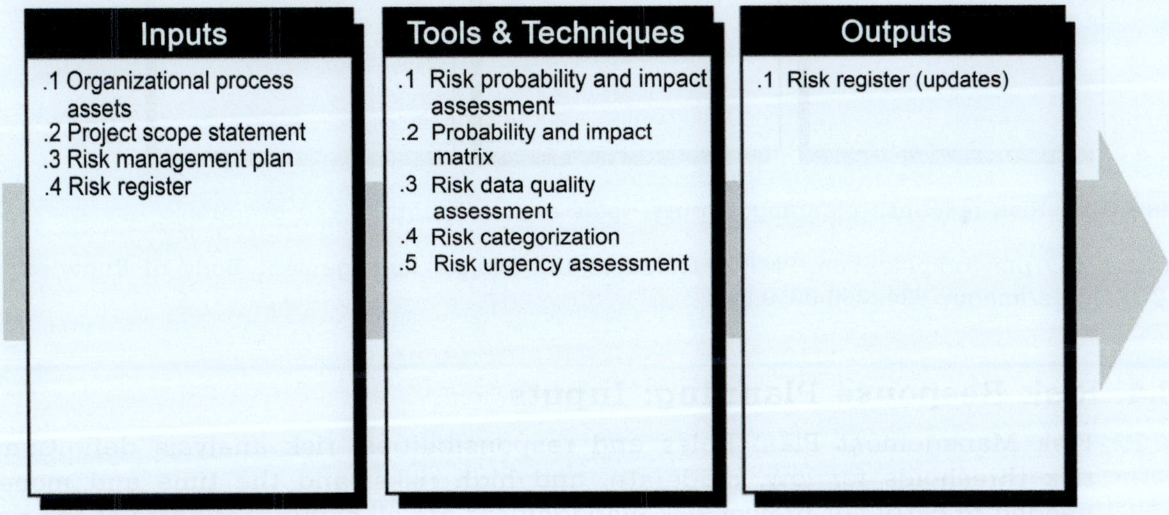

Figure 2.9. Quantitative risk analysis: Inputs, tools and techniques, and outputs.

Source: http://pmworkplace.com/PMP/A_Guide_to_the_Project_Management_Body_of_Knowledge_PMBOKGuide/images/9342254-00_0.jpg.

2.3.3. Risk Response Planning

The process of creating options and choosing how to proceed in order to maximize opportunities and lessen risks to the project's goals is known as risk response planning. It adheres to both the quantitative and qualitative risk analysis procedures. It involves designating and assigning one or more people to assume accountability for each approved and funded risk response (referred to as the "risk response owner").

Risk Response Planning prioritizes the risks and adds resources and activities to the project management plan, schedule, and budget as needed. Planned risk responses need to be reasonable given the risk's importance, timely, cost-effective in addressing the challenge, realistic in the context of the project, approved by all stakeholders, and owned by an accountable individual.

It's frequently necessary to choose the best risk response from a range of possibilities. Commonly used methods for organizing responses to the risks are presented in the section on risk response planning. Responses are addressed for each risk, which includes opportunities and threats that may have an impact on the project's success (Figure 2.10).

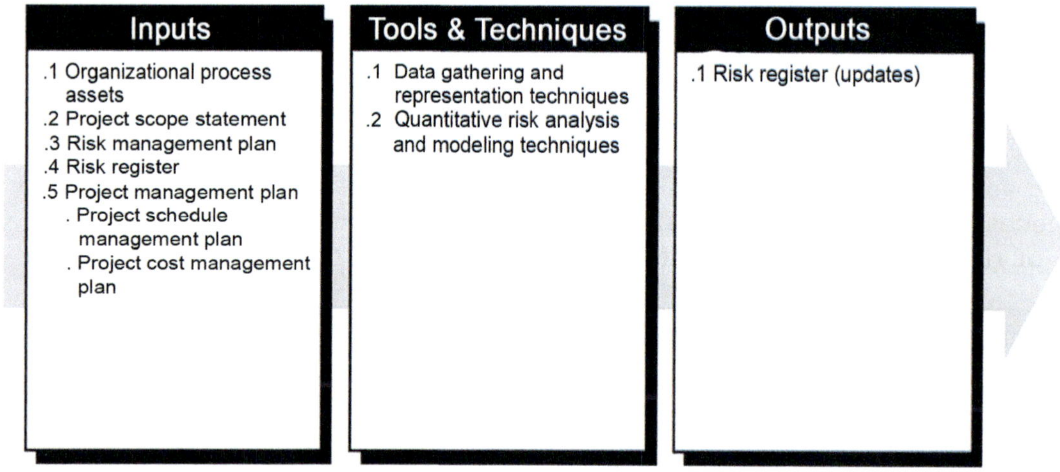

Figure 2.10. Risk response planning: Inputs, tools and techniques, and outputs.

Source: http://pmworkplace.com/PMP/A_Guide_to_the_Project_Management_Body_of_Knowledge_PMBOKGuide/images/9342260-00_0.jpg.

2.3.4. Risk Response Planning: Inputs

1. Risk Management Plan: Roles and responsibilities, risk analysis definitions, risk thresholds for low, moderate, and high risks, and the time and money needed to carry out project risk management are all significant parts of the risk management plan. Risk thresholds for low, moderate, and high risks can be used as crucial inputs to the risk response planning process. Other components of the risk management plan include staff assignment, scheduling, and budgeting for risk response planning.
2. Risk Register: The risk identification process creates the risk register initially, and the Qualitative and Quantitative Risk Analysis processes update it. When creating risk responses, the Risk Response Planning process might need to consult previously identified risks, their underlying causes, lists of possible reactions, risk owners, symptoms, and warning indicators.

The relative rating or priority list of project risks, a list of risks that need immediate attention, a list of risks requiring further investigation and action, patterns in the findings of qualitative risk analysis, the causes of the risks, the risks categorized, and a watchlist of low priority risks are all significant inputs to risk response planning. The process of Quantitative Risk Analysis involves additional updates to the risk register.

2.3.5. Risk Response Planning: Tools and Techniques

There exist multiple strategies for responding to risks. For every risk, the most likely successful strategy or combination of strategies should be chosen. To determine the best answers, risk analysis techniques like decision tree analysis can be applied. Subsequently, targeted initiatives are formulated to execute that plan. One may choose between primary and backup strategies. If an accepted risk materializes or if the chosen course of action proves to be ineffective, a backup plan can be created and put into action. A contingency reserve is frequently set aside for expenses or time. Eventually, it will be possible to create backup plans and determine the circumstances under which they should be implemented.

2.3.5.1. Strategies for Negative Risks or Threats

Three strategies typically deal with threats or risks that may have negative impacts on project objectives if they occur. These strategies are to avoid, transfer, or mitigate:

- Avoid: Risk avoidance entails modifying the project management plan to remove the threat that an unfavorable risk poses, to shield the project's goals from the risk's effects, or to loosen the scope or schedule of the goal that is in danger. By getting information, enhancing communication, acquiring experience, or clarifying requirements, some risks that surface early in the project can be avoided.
- Transfer: Risk transference involves moving the negative effects of a threat and responsibility for managing it to a third party, without completely eliminating the risk. This approach is most effective in addressing financial risks. Typically, a risk premium is paid to the party assuming the risk. Different tools like insurance, performance bonds, warranties, and guarantees can be used for risk transference. Contracts may also be utilized to transfer liability for specific risks to another party. Depending on the contract type, cost risks may be passed on to the buyer with a cost-type contract, while a fixed-price contract could transfer risk to the seller if the project's design is secure.
- Mitigate: Risk mitigation implies a reduction in the probability and/or impact of an adverse risk event to an acceptable threshold. Taking early action to reduce the probability and/or impact of a risk occurring on the project is often more effective than trying to repair the damage after the risk has occurred. Adopting less complex processes, conducting more tests, or choosing a more stable supplier are examples of mitigation actions. Mitigation may require prototype development to reduce the risk of scaling up from a bench-scale model of a process or product. Where it is not possible to reduce probability, a mitigation response might address

the risk impact by targeting linkages that determine the severity. For example, designing redundancy into a subsystem may reduce the impact from a failure of the original component.

2.3.5.2. Strategies for Positive Risks or Opportunities

Three responses are suggested to deal with risks with potentially positive impacts on project objectives. These strategies are to exploit, share, or enhance.

- Exploit: This strategy may be selected for risks with positive impacts where the organization wishes to ensure that the opportunity is realized. This strategy seeks to eliminate the uncertainty associated with a particular upside risk by making the opportunity definitely happen. Directly exploiting responses include assigning more talented resources to the project to reduce the time to completion, or to provide better quality than originally planned.
- Share: Sharing a positive risk involves allocating ownership to a third party who is best able to capture the opportunity for the benefit of the project. Examples of sharing actions include forming risk-sharing partnerships, teams, special-purpose companies, or joint ventures, which can be established with the express purpose of managing opportunities.
- Enhance: This strategy modifies the 'size' of an opportunity by increasing probability and/or positive impacts, and by identifying and maximizing key drivers of these positive-impact risks. Seeking to facilitate or strengthen the cause of the opportunity, and proactively targeting and reinforcing its trigger conditions, might increase probability. Impact drivers can also be targeted, seeking to increase the project's susceptibility to the opportunity.

2.3.5.3. Strategy for Both Threats and Opportunities

Acceptance: A method used because it is rarely feasible to completely remove risk from a project. This tactic suggests that the project team is either unable to come up with another workable response strategy or has chosen not to alter the project management plan in order to address a risk. It can be applied to opportunities or threats. This tactic can take an active or passive form. The project team is left to handle opportunities and threats as they arise with passive acceptance, which calls for no action on their part. Establishing a contingency reserve, which includes time, money, or resources to handle known threats or opportunities—or occasionally even potentially unknown ones—is the most popular active acceptance strategy.

2.3.5.4. Contingent Response Strategy

Some responses are designed for use only if certain events occur. For some risks, it is appropriate for the project team to make a response plan that will only be executed under certain predefined conditions, if it is believed that there will be sufficient warning to implement the plan. Events that trigger the contingency response, such as missing intermediate milestones or gaining higher priority with a supplier, should be defined and tracked.

2.3.6. Risk Response Planning: Outputs

2.3.6.1. Risk Register (Updates)

In Risk Identification, the risk register is created, and it is updated in Qualitative and Quantitative Risk Analysis. The process of Risk Response Planning involves selecting, deciding upon, and adding appropriate responses to the risk register. The level of detail in the risk register should match the priority ranking and the intended course of action. The high and moderate risks are frequently covered in great detail. Hazards deemed to be low priority are added to a "watchlist" for sporadic inspection.

Components of the risk register at this point can include:

- Identified risks, their descriptions, area(s) of the project (e.g., WBS element) affected, their causes (e.g., RBS element), and how they may affect project objectives.
- Risk owners and assigned responsibilities.
- Outputs from the qualitative and quantitative risk analysis processes, including prioritized lists of project risks and probabilistic analysis of the project.
- Agreed-upon response strategies.
- Specific actions to implement the chosen response strategy.
- Symptoms and warning signs of risks' occurrence.
- Budget and schedule activities required to implement the chosen responses.
- Contingency reserves of time and cost designed to provide for stakeholders' risk tolerances.
- Contingency plans and triggers that call for their execution.
- Fallback plans for use as a reaction to a risk that has occurred, and the primary response proves to be inadequate.
- Residual risks that are expected to remain after planned responses have been taken, as well as those that have been deliberately accepted.
- Secondary risks that arise as a direct outcome of implementing a risk response.
- Contingency reserves that are calculated based on the quantitative analysis of the project and the organization's risk thresholds.

2.3.6.2. Project Management Plan (Updates)

The project management plan is updated as response activities are added after review and disposition through the Integrated Change Control process. Integrated change control is applied in the Direct and Manage Project Execution process to ensure that agreed-upon actions are implemented and monitored as part of the ongoing project. Risk response strategies, once agreed to, must be fed back into the appropriate processes in other Knowledge Areas, including the project's budget and schedule.

2.3.6.3. Risk-Related Contractual Agreements

Contractual agreements, such as agreements for insurance, services, and other items as appropriate, can be prepared to specify each party's responsibility for specific risks, should they occur.

2.3.7. Risk Monitoring and Control

Planned risk responses that are included in the project management plan are executed during the life cycle of the project, but the project work should be continuously monitored for new and changing risks.

The process of identifying, analyzing, and planning for newly emerging risks, tracking identified risks and those on the watch list, reanalyzing current risks, keeping an eye on trigger conditions for backup plans, monitoring residual risks, and reviewing and assessing the efficacy of risk responses is known as risk monitoring and control. Techniques like variance and trend analysis, which depend on performance data produced during project execution, are used in the risk monitoring and control process. For the duration of the project, risk monitoring and control, along with the other risk management procedures, are continual processes.

Other purposes of Risk Monitoring and Control are to determine if:

- Project assumptions are still valid;
- Risk, as assessed, has changed from its prior state, with analysis of trends;
- Proper risk management policies and procedures are being followed;
- Contingency reserves of cost or schedule should be modified in line with the risks of the project.

Among the steps in risk monitoring and control are selecting different approaches, putting a backup plan into action, making necessary corrections, and adjusting the project management schedule. Periodically, the risk response owner provides the project manager with updates on the plan's efficacy, any unexpected consequences, and any mid-course adjustments required to properly manage the risk. For the benefit of upcoming projects, risk monitoring and control also entails updating organizational process assets, such as databases of project lessons learned and risk management templates (Figure 2.11).

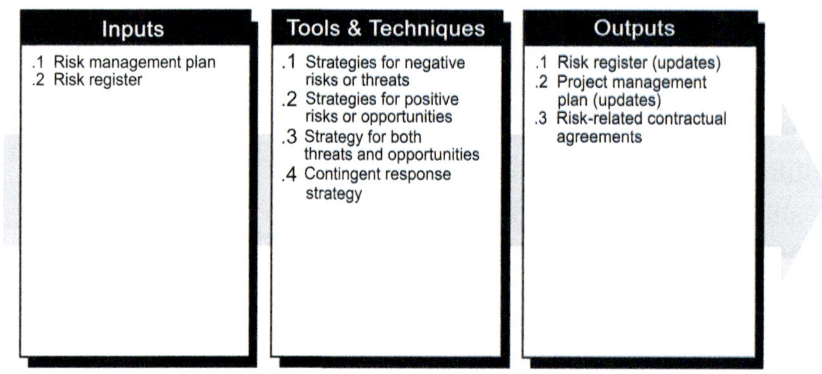

Figure 2.11. Risk monitoring and control: Inputs, tools and techniques, and outputs.

Source: http://pmworkplace.com/PMP/A_Guide_to_the_Project_Management_Body_of_Knowledge_PMBOKGuide/images/9342265-00_0.jpg.

CASE STUDY

CONSTRUCTION MANAGEMENT OF THE SYDNEY OPERA HOUSE:

The Sydney Opera House, one of the most iconic structures in the world, serves as a compelling case study in construction management. Designed by Danish architect Jørn Utzon, the project faced numerous challenges from inception to completion, offering valuable lessons in various aspects of construction management.

- Project Overview:
  - **Location:** Sydney, Australia;
  - **Construction Period:** 1959–1973;
  - **Cost:** Originally estimated at $7 million, ultimately costing $102 million.
- Key Challenges and Management Strategies:
- **Complex Design and Engineering:**
  - **Challenge:** The innovative, sail-like design of the Opera House posed significant engineering challenges, especially in the construction of the roof shells.
  - **Management Strategy:** Utzon and his team adopted a trial-and-error approach, involving extensive research and development. The use of computer-aided design (CAD) for structural analysis was pioneering at the time, helping to solve complex geometrical problems.
- **Budget and Schedule Overruns:**
  - **Challenge:** The project experienced severe budget overruns and delays, with costs escalating from $7 million to $102 million.
  - **Management Strategy:** Detailed cost management and continuous budget reassessments were necessary. While the initial phases suffered from poor financial oversight, later stages saw tighter budget controls and more rigorous financial planning.
  - **Resource Allocation:**
  - **Challenge:** Efficiently managing labor, materials, and equipment was crucial, given the project's scale and complexity.
  - **Management Strategy:** The project required meticulous resource scheduling and allocation. Coordination among multiple subcontractors and suppliers was critical to ensure timely delivery and efficient use of materials and labor.

- **Stakeholder Management:**
  - **Challenge:** The project involved multiple stakeholders, including the New South Wales government, local community, and international experts.
  - **Management Strategy:** Effective communication channels were established to manage stakeholder expectations and involvement. Regular updates and transparent communication helped in maintaining stakeholder support, despite the project's challenges.
- **Risk Management:**
  - **Challenge:** Numerous risks, including design feasibility, financial viability, and public scrutiny, needed careful management.
  - **Management Strategy:** Proactive risk identification and mitigation strategies were implemented. This included continuous design modifications, financial audits, and public relations efforts to maintain the project's credibility.

Outcome and Legacy: Despite the significant challenges and controversies, the Sydney Opera House was completed in 1973 and has since become a UNESCO World Heritage site. The project underscored the importance of flexibility, innovation, and effective stakeholder management in construction management. It highlighted how complex projects could benefit from adaptive management strategies and the integration of new technologies.

The Sydney Opera House remains a testament to the potential of visionary architecture and the critical role of construction management in bringing such visions to fruition.

CLASS ACTIVITY

For a class activity on construction management, divide students into small groups and assign each group a different aspect of construction management to research: project planning, resource allocation, risk management, and stakeholder communication. Each group will prepare a brief presentation (5 minutes) highlighting key points, challenges, and best practices related to their topic. After the presentations, conduct a group discussion to compare the different aspects and how they interrelate. Conclude with a collaborative exercise where all groups work together to create a mini project plan for a hypothetical construction project, integrating the concepts learned from each presentation.

SUMMARY

- Construction management is a specialized field that focuses on the planning, coordination, and execution of construction projects from inception to completion.
- Construction management (CM) is a professional service that uses specialized project management techniques to oversee the planning, design, and construction of a project, from its beginning to its end.
- In its most common context, the term "management" relates to the planning, organizing, directing, and controlling of a business enterprise.
- Project time and cost management are based on time and cost schedules developed for the project and an information system that will provide data for comparing expected with actual performance.
- When field operations begin, the order in which the project proceeds is in accordance with an approved job plan.
- Field project management starts with the onset of construction, at which point a comprehensive construction budget and detailed time schedule of operations are prepared.
- Project cost accounting is the process of obtaining actual production rates and unit costs from ongoing projects. This system provides the basic information for project cost control and for estimating new work.
- The management and control of project time, cost, resources, and finance by the contractor during the field construction process require that the project manager originate, manipulate, summarize, and interpret large volumes of numerical data.
- Project Risk Management includes the processes concerned with conducting risk management planning, identification, analysis, responses, and monitoring and control on a project; most of these processes are updated throughout the project.
- Risk Identification is an iterative process because new risks may become known as the project progresses through its life cycle. The frequency of iteration and who participates in each cycle will vary from case to case.
- Quantitative Risk Analysis is performed on risks that have been prioritized by the Qualitative Risk Analysis process as potentially and substantially impacting the project's competing demands.
- Risk Response Planning is the process of developing options and determining actions to enhance opportunities and reduce threats to the project's objectives.

REVIEW QUESTIONS

1. What are the primary responsibilities of a construction manager in a construction project?
2. How does the Critical Path Method (CPM) assist in project planning and scheduling?

3. What is the importance of risk management in construction management, and what are common strategies to mitigate risks?
4. How do construction managers ensure that projects stay within budget?
5. What role does effective communication play in the success of construction projects?
6. Describe the process and importance of resource allocation in construction management.
7. What are the key components of a project plan in construction management?
8. How do construction managers handle changes in project scope and ensure they do not disrupt project timelines?
9. What are some common challenges faced in construction management, and how can they be addressed?
10. How does the use of technology and software tools improve the efficiency and effectiveness of construction management?

REFERENCES

1. Bielefeldt, A. (2013). Pedagogies to achieve sustainability learning outcomes in civil and environmental engineering students. Sustainability, 5(10), 4479–4501.
2. Bielefeldt, A. R. (2011). Incorporating a sustainability module into first-year courses for civil and environmental engineering students. Journal of Professional Issues in Engineering Education and Practice, 137(2), 78–85.
3. Bourke, K. (2001). Whole life costing–An overview. In Construction Industry Research and Information Association, Service Life Planning/Whole Life Costing, Report of A Joint Workshop With The Ric's and Rias Held at The Scottish Executive, Victoria Quay, Edinburgh, 7 February, Members' Report E1102.
4. Chadwick, A., & Morfett, J. (2002). Hydraulics in civil and environmental engineering (Vol. 1, pp. 11–30). CRC Press.
5. Construction Best Practice Programme. (2001). Themes–Whole Life Costing. Retrieved from: http://www.cbpp.org.uk/ (accessed on 1 July 2024).
6. Construction Industry Research and Information Association. (2001a). Asset management. CIRIA News, 1, 6.
7. Construction Industry Research and Information Association. (2001b). Service life planning/whole life costing, Report of a joint workshop with the RICS and RIAS held at the Scottish Executive, Victoria Quay, Edinburgh, 7 February, Members' Report E1102.
8. Ottosen, L. M., Christensen, I. V., Rörig-Dalgaard, I., Jensen, P. E., & Hansen, H. K. (2008). Utilization of electromigration in civil and environmental engineering—Processes, transport rates and matrix changes. Journal of Environmental Science and Health, Part A, 43(8), 795–809.

CHAPTER

3 Construction Planning and Control

LEARNING OBJECTIVES

After studying this chapter, you will be able to:

- Define basic concepts in the development of construction plans;
- Defining work tasks;
- Explain estimating activity durations;
- List the estimating resource requirements for work activities;
- Describe construction control;
- Discuss the control measures for civil construction under modern concepts.

KEY TERMS FROM THIS CHAPTER

Construction planning	Construction control
Expenditure	Task precedence's
Budget information	Work breakdown
Tedious process	Convention
Construction monitoring	Pavement
Reinforcement	Structural integrity
Trench preparation	Concrete preparation

INTRODUCTION

Construction planning and control are essential components of successful construction project management. These processes ensure that projects are completed on time, within budget, and to the required quality standards, while managing resources efficiently and mitigating risks.

Construction planning is the foundational phase of any construction project, involving the detailed mapping out of tasks, resources, timelines, and budgets required to complete the project. It begins with defining project objectives and scope, which includes understanding the client's requirements, regulatory constraints, and technical specifications. This phase also involves the development of a project schedule, which breaks down the project into smaller, manageable tasks or activities, and assigns start and end dates to each.

One key aspect of construction planning is resource allocation, which includes labor, materials, and equipment. Effective planning ensures that the right resources are available at the right time, preventing delays and cost overruns. Construction planning also encompasses procurement planning, where the timing of purchasing materials and services is scheduled to align with project timelines.

Risk management is another critical element, identifying potential risks such as weather delays, labor shortages, or supply chain disruptions, and devising mitigation strategies to minimize their impact. Effective planning also involves stakeholder management, ensuring clear communication and coordination among all parties involved, from contractors and suppliers to regulatory bodies and the client.

Construction control, on the other hand, is the process of monitoring and adjusting the project plan to ensure it stays on track. It involves ongoing oversight of project progress, comparing actual performance with planned objectives, and making necessary adjustments to address any deviations. Key performance indicators (KPIs) such as schedule adherence, cost control, and quality standards are closely monitored.

Construction control includes regular progress reports, site inspections, and quality control checks to ensure compliance with project specifications and standards. It also involves managing change orders and variations, which are inevitable in construction projects. Effective control ensures that changes are managed in a way that minimizes their impact on project timelines and budgets.

Financial control is a crucial aspect of construction control, involving budget tracking and cost management to prevent cost overruns and ensure financial accountability. Additionally, safety management is integral to construction control, ensuring that safety protocols are adhered to, thereby minimizing the risk of accidents and ensuring a safe working environment.

3.1. BASIC CONCEPTS IN THE DEVELOPMENT OF CONSTRUCTION PLANS

One of the most important and difficult tasks in the management and completion of construction projects is construction planning. It entails selecting the appropriate technology, defining the tasks to be completed, estimating the time and resources needed for each task, and determining any interactions between the various work tasks. A solid construction plan serves as the foundation for creating the work schedule and budget. Even if the plan is not written down or otherwise officially documented, creating the construction plan is an essential part of managing the construction process. Aside from these technical aspects of construction planning, organizational decisions regarding the relationships between project participants and even the inclusion of certain organizations in a project may also need to be made. For instance, during construction planning, it's common to decide how much of a project will use subcontractors.

Forming a construction plan is a highly challenging task. As Sherlock Holmes noted:

Most people, if you describe a train of events to them, will tell you what the result would be. They can put those events together in their minds and argue from them that something will come to pass. There are few people, however, who, if you told them a result, would be able to evolve from their own inner consciousness what the steps were which led up to that result. This power is what I mean when I talk of reasoning backward (Doyle, 1993).

A planner starts with a result, just like a detective does (e.g., a facility design) and must combine the actions necessary to achieve this outcome. The creation of necessary tasks, consideration of these tasks' consequences, and selection of alternative methods of completing tasks are all crucial components of construction planning. But unlike a detective cracking a single case, construction planners also have to make a normative decision about which of several alternative plans to choose. Furthermore, a planner must envision the finished facility as it is outlined in the plans and specifications, while a detective must deal with an observable outcome. As shown in Figure 3.1, when creating a construction plan, it is typical to place a strong emphasis on either schedule control or cost control. Certain projects are mainly broken down into cost categories and related expenses. Construction planning is cost- or expense-oriented in these situations. Costs incurred directly in carrying out an activity and costs incurred indirectly in completing the project are distinguished within the expenditure categories. For instance, borrowing costs are typically classified as indirect costs when they relate to project financing and overhead. For other projects, the planning process places a strong emphasis on the scheduling of work activities over time. In this instance, the planner makes sure that the right priorities are upheld among the tasks and that the available resources are scheduled as effectively as possible. Conventional scheduling techniques place a strong emphasis on maintaining task priorities, which leads to critical path scheduling techniques, or making effective use of resources over time, which leads to job shop scheduling techniques. Lastly, the majority of complex projects necessitate taking scheduling and cost into account over time, which means that both aspects must be taken into account in planning, monitoring, and record keeping. The integration of budget and schedule data is crucial in these situations.

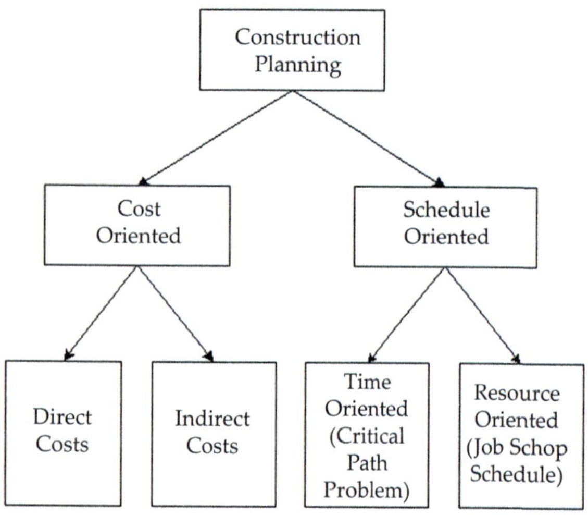

Figure 3.1. Alternative emphases in construction planning.

Source: https://www.cmu.edu/cee/projects/PMbook/images/fig9_1.gif.

We'll talk about the functional requirements for construction planning, including budgeting, work breakdown, and technology selection. Planning for construction is not a task that has to be limited to the time following the award of a construction contract. It ought to be a crucial task when designing the facility. Re-planning is also necessary in the event that issues emerge during construction.

3.1.1. Choice of Technology and Construction Method

Similar to the creation of suitable alternatives for facility design, the selection of suitable technology and construction techniques is frequently an unstructured process but is essential to the project's success. For instance, choosing to pump concrete or move it in buckets will have an immediate impact on how much it costs and how long it takes to complete construction-related tasks. When choosing between these two options, one should take into account the relative costs, equipment availability, and dependability of each transport method. Unfortunately, the precise effects of various approaches depend on a number of factors, such as worker experience and expertise or the specific subsurface conditions at a site, about which information may be hazy during the planning phase.

It might be required to create several construction plans based on different approaches or presumptions in order to choose between alternative technologies and methods. The effects of the alternative approaches on cost, time, and reliability can be examined once the complete plan is available. In bidding competitions, where multiple alternative designs may be submitted or value engineering for alternative construction methods may be allowed, this examination of multiple alternatives is frequently made clear. In this situation, prospective builders might want to draft construction plans for every alternate design utilizing the recommended method in addition to draft plans for alternate construction techniques that would be suggested during the value engineering process. Using a formal computer-based simulation technique or the planner's imagination to simulate the construction process is a useful method for creating a construction plan (Paulson et al., 1983). Comparisons between various plans or issues with the current plan can be found by looking at the outcome. When a specific piece of equipment is chosen for an operation, for instance, the issue of whether or not there is enough access space for the equipment arises right away. Within a computer-aided design (CAD) system, three-dimensional geometric models can be useful for both simulating operational space requirements and identifying potential interferences. Similarly, by including more resources in the construction plan, issues with resource availability that were discovered during the simulation of the building process may be successfully avoided.

- Example 1: A Roadway Rehabilitation: A Pittsburgh, Pennsylvania roadway rehabilitation project can be used as an example to highlight the significance of careful construction planning and the impact of technology selection. This project involved replacing both the highway's pavement and the decks on overpass bridges. When the highway surface beneath the overpass bridges was being replaced, the original construction plan called for working outward from each end of the bridges. As a result, there were significant issues with equipment and concrete truck access to the overpass bridges. On the other hand, the highway work could be scheduled so that at specific times each overpass bridge could be accessed from below. Lower expenses and faster completion were achieved by pumping concrete from the highway below to the deck of the overpass bridge.
- Example 2: Laser Leveling: Utilizing laser leveling equipment to increase the efficiency of excavation and grading is an example of a technological decision. According to Paulson (1985), these systems precisely determine the relative height of mobile equipment by erecting laser surveying equipment on the site. To metating laser beam revolving laser light is flashed across the construction site on a level plane, and the location of the light's reflection on receptors on mobile equipment, like graders, is precisely observed. The height at which the laser shines anywhere on the construction site provides an accurate indication of the height of a receptor on a piece of mobile equipment because laser light does not disperse appreciably. A blade, bucket from an excavator, or any other piece of equipment can then be measured using the receptor height. When coupled with electro-hydraulic control systems installed on mobile machinery like bulldozers, graders, and scrapers, these systems enable accurate and automated control of the height of the excavation and grading blades. The automation of blade heights has improved product quality as indicated by the desired amount of excavation or the degree to which a final grade achieves the desired angle, and has, in some cases, reduced costs by over 80%. These systems also make it possible to use less experienced operators and smaller machines. However, in order to use these semi-automated systems, equipment for laser surveying must be purchased, and equipment must be modified to allow electronic feedback control units. Nevertheless, in many cases, laser leveling seems to be a great technological option.

3.2. DEFINING WORK TASKS

Determining the numerous tasks that need to be completed is a parallel step in the planning process that happens concurrently with the selection of technology and overall approach. The framework needed to schedule construction activities is represented by these work tasks, which also estimate the resources needed for each task individually and any necessary sequences or precedences between tasks. In construction plans, the terms "tasks" and "activities" are frequently used synonymously to refer to distinct, well-defined work items. A project would be referred to as a "job" and an activity as an "operation" in job shop or manufacturing terminology, but the meaning of the terms is the same. According to Baker (1974), the scheduling problem is figuring out the right combination of activity start times, resource allocations, and completion dates that will enable the project to be finished on time and effectively. Planning for construction is an essential step before scheduling. The planning process usually involves defining work tasks, technology, and construction method in parallel or iteratively.

Although it can be a time-consuming and tedious process to define suitable work tasks, this information is essential for the application of formal scheduling procedures. Construction projects may require thousands of distinct work tasks, so this definition stage can be costly and time-consuming as well. Thankfully, a lot of work can be duplicated in various areas of the building, and building plans from previous projects can serve as broad templates for new ones. For instance, the tasks required to construct a building floor might be repeated for each floor of the building with only minor modifications. Additionally, most tasks have standard definitions and nomenclature. As a result, not every aspect of the project needs to be approached from scratch by the individual planner defining work tasks. The amount of work required can be decreased by repeating tasks in different locations or by replicating tasks from previous projects, but there aren't many computer tools available to help define activities. A plethora of computer programs are available for scheduling. However, there will probably always be a reliance on the expertise, judgment, and experience of the construction planner for the crucial task of defining activities.

More formally, any division of project tasks is referred to as an activity. A project's defined set of activities should be comprehensive or all-inclusive in order to include all required work tasks in one or more activities. Every design feature in the proposed facility will usually have one or more related project activities. An activity needs time and resources to be completed, including labor and equipment, as the following section explains. The duration of an activity is the amount of time needed to complete it. Activities' start and finish serve as milestones or signposts that show how far the project has come. There are times when it's helpful to designate non-durational activities to commemorate significant occasions. Receiving equipment at the construction site, for instance, could be considered an activity because it would depend on other activities and the project manager might want official notification of its arrival. In a similar vein, the project plan would specifically note receiving regulatory approvals.

In building project plans, the amount of work involved in any given activity can vary greatly. In fact, it's typical practice to start with fairly broad definitions of activities and then further break tasks down as the plan gets more precise. As a result, as the plan is being created, the definition of the activities changes. This process leads to a naturally occurring hierarchy of activities, wherein broad, abstract functional activities are repeatedly broken down into ever-more-specific subtasks. Sub-activities related to installing forms, installing reinforcing steel, pouring concrete, finishing the concrete, removing forms, and other tasks, for instance, would be part of the process of placing concrete on site. Subtasks can be defined even more precisely, such as the removal and cleaning of forms following the placement of concrete.

Even further, the sub-task "clean concrete forms" could be subdivided into the various operations:

- Transport forms from on-site storage and unload onto the cleaning station.
- Position forms on the cleaning station.
- Wash forms with water.
- Clean concrete debris from the form's surface.
- Coat the form surface with an oil release agent for the next use.
- Unload the form from the cleaning station and transport to the storage location.

The process of programming or designing a robot to perform this activity requires a detailed task breakdown of the activity "clean concrete forms," which is something that is typically not done in standard construction planning. This is because the various specific tasks involved in this activity must be well defined for a robot to be implemented (Skibniewski, 1985).

Generally speaking, it is beneficial to create an explicit work activity hierarchy in order to streamline the creation and presentation of a schedule. For instance, "site clearance" may have been defined in the original plan as a single task. Subsequently, this one task could be further broken down into "grading," "relocating utilities," "removing vegetation," etc. These tasks, nevertheless, might still be classified as components of the larger activity of "site clearance." The creation of summary charts and reports, which combine specific operations into aggregate or "super"-activities, is made easier by this hierarchical structure.

Formally speaking, a hierarchical approach to work task definition breaks down the work activity into its constituent pieces, which are represented as a tree. While branches in the tree lead to smaller components and work activities, higher levels in the tree represent decision nodes or summary activities. Numerous relationships between the different nodes can be specified or enforced, such as the precedence relationships between the various tasks as outlined below. Decisions about technology can be broken down into choices made at specific nodes within the tree. For instance, decisions about plumbing technology may be made independently of decisions about other functional tasks.

Of course, each construction plan can have a multitude of distinct activity hierarchies defined for it. For instance, lower-level activities may be divided into the necessary construction operations after upper-level activities are connected to facility components, such as foundation elements. Alternatively, lower work divisions might represent the application of these operations to particular facility components, while upper-level divisions might represent general types of activities like electrical work. A third option is for the initial divisions to symbolize various spatial locations within the proposed facility. The planner's convenience and the intended structure for condensing work information determine which hierarchy to use. Multiple hierarchies can be stored in computerized databases, allowing for the creation of various aggregations or views of the WBS.

It is up to convention or judgment to determine how many and what kind of activities to include in a construction plan. Depending on the planner's choices and the size of the project, construction plans can span from fewer than 100 to thousands of specified tasks. Overly refined subdivided activities lead to unmanageable network sizes and high planning costs. If reasonably accurate work breakdown estimates and the necessary resources cannot be determined at the detailed level, then subdivision is not beneficial. However, it is hard to create realistic schedules and resource requirements details during a project if the activities specified are too coarse. Better control and more realistic scheduling are made possible by more thorough task definitions. It is useful to define separate work tasks for:

- those activities which involve different resources; or
- those activities which do not require continuous performance.

For instance, dividing the task of "preparing and checking shop drawings" into separate tasks is necessary because different people are responsible for each task and there could be a delay between preparation and checking.

In actuality, the appropriate degree of detail will depend on the project's size, significance, and difficulty as well as the particular scheduling and accounting practices that are used. But more often than not, schedules are created with too little information rather than too much. Remember that the definition of tasks will form the foundation for scheduling, communicating the construction plan, and overseeing the construction process. Task completion is frequently the prerequisite for receiving progress payments from the owner. Therefore, it can be beneficial to have more thorough task definitions. However, more thorough task breakdowns are only useful to the degree that each activity's resource requirements, duration, and activity relationships are realistically estimated. Giving thorough work task breakdowns without making an equal effort to estimate resource requirements realistically is not helpful. It is reasonable to anticipate that the number of work tasks will rise as more robust computer-based scheduling and monitoring systems are implemented. This will also make it easier to define and modify tasks.

- Example 3: Task Definition for a Road Building Project: Let's say we want to create a plan for building a road with two culverts as an exstruction planning (Aras,begin, weWe begin by categorizing categorize project activities into three groups, as indicated by Figure 3.2: general, roadway, and structures. The primary categories of design elements to be built serve as the basis for this division. Earthwork and pavement constitute a further sub-division of the roadway work. We define watering, compaction, and paving sub-activities associated with pavement within these subdivisions, and we identify clearing, excavation, filling, and finishing (including seeding and sodding) associated with earthwork. Lastly, we observe that because the roadway segment is rather lengthy, distinct physical segments of the roadway path can be assigned to distinct activities. Each paving and earthwork task is broken down into tasks unique to each of the two road segments in Figure 3.2. We define the sub-divisions of structural excavation, concreting, and reinforcing for the construction of the culvert. To be more precise, structural excavation is separated into two parts: the actual excavation and the necessary backfill and compaction. Comparably, the steps involved in concreting are putting in concrete forms, pouring concrete, removing the forms, and curing the concrete. Detailed activities for reinforcing each of the two culverts are defined as the last stage of the structural planning process. The definition of general work activities includes setup, general oversight, and cleanup. This planning has resulted in the definition of more than 30 distinct, in-depth activities.

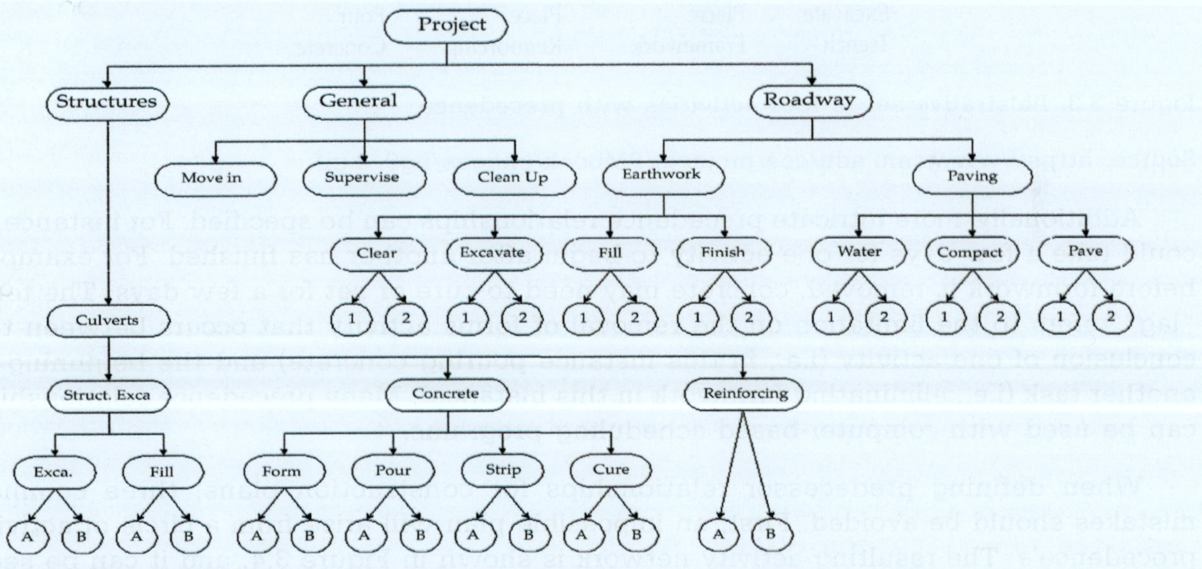

Figure 3.2. Illustrative hierarchical activity divisions for a roadway project.

Source: https://www.cmu.edu/cee/projects/PMbook/images/fig9_2.gif.

The planner may decide to define extra activities for this project at their discretion. For instance, ordering supplies or painting lanes could be included as distinct tasks. Additionally, a planner might specify a different hierarchy of work breakdowns than the one depicted in Figure 3.2. For culverts, for instance, installing reinforcing might have been a sub-activity under concreting. Separating the placement of reinforcement could serve as a means of highlighting the various materials and resources needed for this particular task. It is also possible that the separation of the roadway into distinct segments and culverts was implemented earlier in the hierarchy. With all of these possible variations, the most crucial thing is to make sure that the final plan includes all required activities.

3.2.1. Defining Precedence Relationships Among Activities

The relationships between the work activities can be specified once they have been defined. Activities that have precedence relationships indicate that they must occur in a specific order. Because of regulations, other technical requirements, and the need for structural integrity, there are many inherent sequences for construction activities. Design drawings, for instance, cannot be reviewed before they are created. Diagrammatically, a network or graph with arrows representing the activities can be used to show precedence relationships. In the activity network, the arrows in Figure 3.3 are referred to as branches or links, and the circles that indicate the start or finish of each arrow are referred to as nodes or events. In this diagram, nodes stand in for milestone events, and links for specific activities.

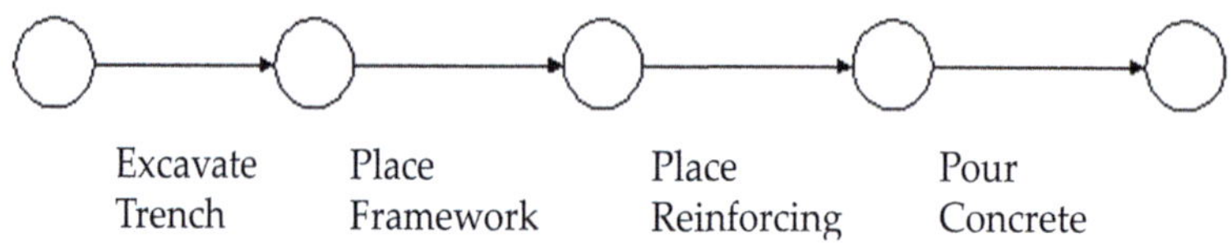

Figure 3.3. Illustrative set of four activities with precedence.

Source: https://www.cmu.edu/cee/projects/PMbook/images/fig9_3.gif

Additionally, more intricate precedence relationships can be specified. For instance, it could take a few days for one activity to begin after another has finished. For example, before formwork is removed, concrete may need to cure or set for a few days. The term "lag" refers to the limitation on the removal of forms activity that occurs between the conclusion of one activity (i.e., in this instance pouring concrete) and the beginning of another task (i.e., eliminating formwork in this instance). Many precedence relationships can be used with computer-based scheduling programs.

When defining predecessor relationships for construction plans, three common mistakes should be avoided. First, an impossible plan will arise from a circle of activity precedence's. The resulting activity network is shown in Figure 3.4, and it can be seen that the project can never be started or finished if, for instance, activity A comes before activity B, activity B comes before activity C, and so on. Fortunately, any such mistakes in the construction plan's logic can be detected by formal scheduling techniques and reliable computer scheduling software.

More subtle is the forgetting of an essential precedence relationship. Assume, for instance, that drywall installation ought to be completed before floor finishing. If this relationship between precedence and scheduling is ignored, both activities might be scheduled at the same time. Making changes on the fly could lead to higher expenses or poor-quality work when the project is finished. Regrettably, there aren't many methods available for identifying precedence omissions except consulting with experienced managers or comparing to similar projects. Another, albeit rarely employed, method for verifying precedents is to physically or virtually simulate the building process and look for any issues.

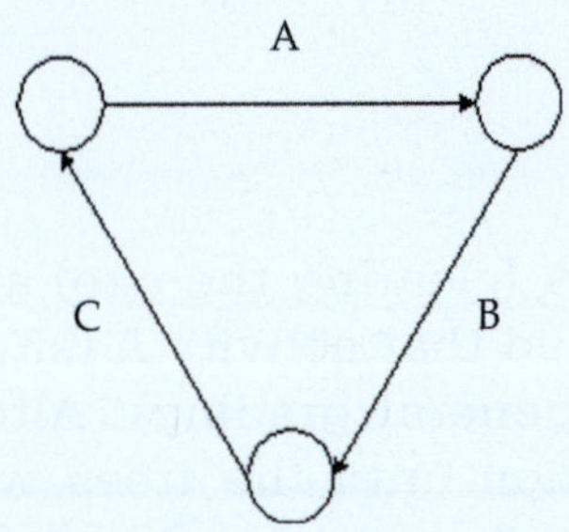

Figure 3.4. Example of an impossible work plan.

Source: https://www.cmu.edu/cee/projects/PMbook/images/fig9_4.gif.

Finally, it is important to realize that different types of precedence relationships can be defined and that each has different implications for the schedule of activities:

- Some activities have a necessary technical or physical relationship that cannot be superseded. For example, concrete pours cannot proceed before formwork and reinforcement are in place.
- Some activities have a necessary precedence relationship over a continuous space rather than as discrete work task relationships. For example,. formwork may be placed in the first part of an excavation trench even as the excavation equipment continues to work further along in the trench. Formwork placement cannot proceed further than the excavation, but the two activities can be started and stopped independently within this constraint.
- Some "precedence relationships" are not technically necessary but are imposed due to implicit decisions within the construction plan. For example, two activities may require the same piece of equipment so a precedence relationship might be defined between the two to ensure that they are not scheduled for the same time period. Which activity is scheduled first is arbitrary. As a second example, reversing the sequence of two activities may be technically possible but more expensive. In this case, the precedence relationship is not physically necessary but only applied to reduce costs as perceived at the time of scheduling.

It is important to understand that different types of precedence relationships have quite different implications for the flexibility and cost of making changes to the construction plan when scheduling changes as work progresses. Regretfully, a lot of formal scheduling systems are unable to display this kind of flexibility. The manager is therefore responsible for making these choices and ensuring that the schedules are practical and efficient. It should come as no surprise that a manager in such circumstances would give little priority to creating or updating the official, computer-based construction plan given all of their other duties as a project manager. Formal construction plans, however, might be necessary for effective project management in complex situations.

- Example 4: Precedence Definition for Site Preparation and Foundation Work: Suppose that a site preparation and concrete slab foundation construction project consists of nine different activities:

A. Site clearing (of brush and minor debris);

B. Removal of trees;

C. General excavation;

D. Grading general area;

E. Excavation for utility trenches;

F. Placing formwork and reinforcement for concrete;

G. Installing sewer lines;

H. Installing other utilities;

I. Pouring concrete.

Since they are independent of each other, activities A (clearing the site) and B (tree removal) do not have any preceding activities. It is assumed that activity A (site clearing) comes before activities C (general excavation) and D (general grading). Alternatively, the planner may have wanted to postpone any excavation until the trees were taken out, in order to prioritize B (tree removal) over C (general excavation) and D (general grading). Since they require further excavation and trench preparation, Activities E (trench excavation) and F (concrete preparation) cannot start until general excavation and tree removal are finished. Activity E (trench excavation) comes before activities G (install lines) and H (install utilities), which are installation in the utility trenches and cannot be attempted until the trenches are prepared.

Table 3.1. Precedence Relations for a Nine-Activity Project Example

Activity	Description	Predecessors
A	Site clearing	–
B	Removal of trees	–
C	General excavation	A
D	Grading general area	A
E	Excavation for utility trenches	B, C
F	Placing formwork and reinforcement for concrete	B, C
G	Installing sewer lines	D, E
H	Installing other utilities	D, E
I	Pouring concrete	F, G

Source: https://www.cmu.edu/cee/projects/PMbook/09_Construction_Planning.html.

Additionally, activity D (general grading) comes before activities G (install sewers) and H (install utilities) because we believe that grading must be finished before utilities are installed to prevent equipment conflicts. Lastly, activities F and G come before activity I, which is pouring concrete, because it cannot start until the sewer line is installed and the formwork and reinforcement are ready. It is possible that additional utilities will be routed over the slab foundation, so activity H (install utilities) need not come before activity I (pour concrete). Table 3.1 displays the immediate precedences that are the outcome of our planning.

Using this data, the next challenge is to identify all of the precedence relationships between the activities and to depict them in a network diagram. Figure 3.5 depicts one network representation of these nine activities, where the activities are represented as links or branches connecting nodes. The milestones of potential beginning and starting times are represented by the nodes. We refer to this illustration as an activity-on-branch diagram. It should be noted that node 0 defines the beginning of an activity, and node 5 indicates the conclusion of all activities.

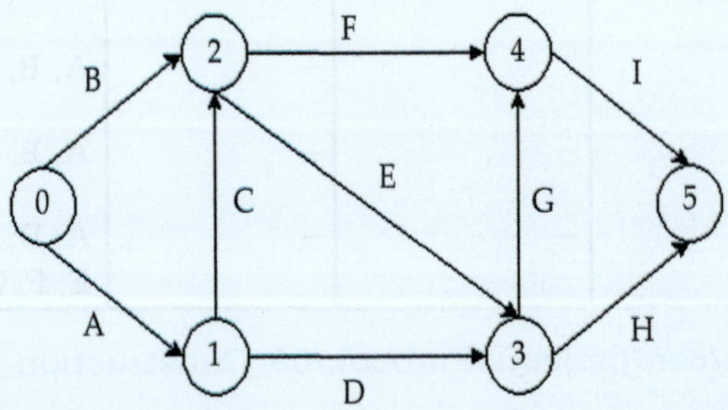

Figure 3.5. Activity-on-branch representation of a nine-activity project.

Source: https://www.cmu.edu/cee/projects/PMbook/images/fig9_5.gif.

As shown in Figure 3.6, the nine activities could also be represented as nodes and the predecessor relationships as branches or links. An activity-on-node diagram is the end product. To commemorate these significant turning points, new activity nodes signifying the start and finish of construction have been added to Figure 3.6.

When visualizing the different activities and their relationships for a project, these network representations of activities can be very useful. The representation of activities as branches or nodes mostly depends on organizational policy and individual preference.

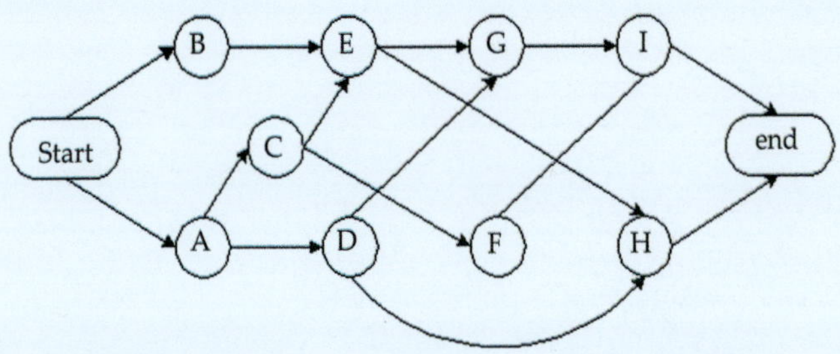

Figure 3.6. Activity-on-node representation of a nine-activity project.

Source: https://www.cmu.edu/cee/projects/PMbook/images/fig9_6.gif.

Table 3.2. All Activity Precedence Relationships for a Nine-Activity Project

Predecessor Activity	Direct Successor Activities	All Successor Activities	All Predecessor Activities
A	C, D	E, F, G, H, I	–
B	E, F	G, H, I	–
C	E, F	G, H, I	A
D	G, H	I	A
E	G, H	I	A, B, C
F	I	–	A, B, C
G	I	–	A, B, C, D, E
H	–	–	A, B, C, D, E
I	–	–	A, B, C, D, E, F, G

Source: https://www.cmu.edu/cee/projects/Pmbook/09_Construction_Planning.html.

It's also noteworthy that Table 3.1 only includes relationships between immediate predecessors. It is evident that there exist additional precedence relationships involving multiple activities. For instance, since the activity "grading general area" (Activity D) must come before activity G and must come after activity A, "installing sewer lines" (Activity G) cannot be started before "site clearing" (Activity A) is finished. Table 3.1, which only includes information about immediate predecessors, is an implicit precedence list. Every activity that came before activity G would be included in an explicit predecessor list. All of these predecessor relationships that the project plan suggests are displayed in Table 3.2. This table can be generated algorithmically by following every path back through the network from a specific activity (Christofides, 1975). Examining Figure 3.6 indicates, for instance, that all activities—aside from activity B—depend on the outcome of activity A.

3.3. ESTIMATING ACTIVITY DURATIONS

Every work activity in most scheduling procedures has a corresponding time frame. When creating a schedule, these times are heavily utilized. Assume, for illustrative purposes, that the project durations depicted in Table 3.3 were estimated. Since the activities directly follow one another and take a total of 1.0 + 0.5 + 0.5 + 1.0 = 3 days, the entire set of activities would then take at least 3 days. This means that the total duration of the four activities may change if another activity is conducted in parallel, as the overall project timeline will depend on the critical path. Parallel activities could potentially shorten the project's duration, depending on their interdependencies.Table 3.3. Durations and Predecessors for a Four Activity Project Illustration

Activity	Predecessor	Duration (Days)
Excavate trench	–	1.0
Place formwork	Excavate trench	0.5
Place reinforcing	Place formwork	0.5
Pour concrete	Place reinforcing	1.0

Source: https://www.cmu.edu/cee/projects/PMbook/09_Construction_Planning.html.

The definitions of the predecessor relationships among tasks and the estimates of the durations of the various project activities are the foundations of all formal scheduling procedures. One may also take into account the variability of an activity's duration. Formally, scheduling may make use of both the expected or most likely duration of an activity and its probability distribution. A probability distribution shows the likelihood that an activity will last for a specific amount of time. We are never quite sure how long a task will take until we actually start working on it.

Keeping historical records of specific activities and using the average durations from this experience to create new duration estimates is a simple method for estimating the duration of an activity. Unit productivity rates are usually used for this purpose because the scope of activities in different projects is unlikely to be the same. For example, the duration of an activity D_{ij} such as concrete formwork assembly might be estimated as:

$$D_{ij} = \frac{A_{ij}}{P_{ij}N_{ij}} \quad (1)$$

where; N_{ij} is the number of crews assigned to the task; A_{ij} is the required formwork area to assemble (in square yards); and P_{ij} is the average productivity of a standard crew in this task (measured in square yards per hour). Unit production time, T_{ij}, is a productivity measure used in certain organizations. It is defined as the amount of time (measured in hours per square yard) needed for a standard crew to complete a unit of work. T_{ij} is therefore a reciprocal of P_{ij}.

An equation like Eqn. (1) is suitable for almost all building tasks. Usually, a thorough analysis of the finished facility design yields the necessary amount of work, A_{ij}. This process of quantity take-off to determine the necessary volumes, areas, and materials is one that contractors frequently use when preparing bids. Specialized quantity surveyors in certain nations give the owner and all possible contractors the details on the quantities that are needed. The planner decides how many crews are working, or N_{ij}. The quantity or number of resources allocated to specific tasks may frequently be changed in light of the final project plan and timeline. In order to apply Eqn. (1), an estimate of the anticipated work productivity, P_{ij}, must be given. Similar to cost considerations, commercial services can offer average productivity numbers for a variety of these kinds of routine tasks. A company's historical records may also contain information useful for estimating productivities.

For several reasons, the duration calculated using Eqn. (1) is only an estimate of the real activity duration. Firstly, it is typically the case that certain activities are more or less difficult to complete due to unique aspects of the project. For instance, it could be difficult to access the forms in a specific location; consequently, form assembly productivity might be less than average for a given project. For this reason, modifications to the computed durations from Eqn. (1) are frequently made based on engineering judgment.

Furthermore, productivity rates can fluctuate due to both systematic factors, such as learning curve effects, and random variations. Systematic improvements in productivity often occur as crews become more familiar with tasks, while random fluctuations may arise from unpredictable conditions, such as weather or material availabilits. The impact of learning on output is one instance of systematic variation. A crew's productivity will usually increase as they get used to a task and each other's work habits. The type of productivity increase that could result from experience is depicted in Figure 3.7; this curve is known as a learning curve. As a result, the length of an activity or project determines the productivity P_{ij}. One typical construction example is that, despite a longer travel time to the active construction area, higher levels of productivity may allow floors in a building to be assembled more quickly. Once more, differences in the average productivity of the learning curve may be represented by subjective adjustments or historical records (Thomas, 1986).

Additionally, random factors will affect productivity rates and introduce uncertainty into the duration estimation process. For instance, a scheduler usually has no idea how skilled the crew and manager assigned to a given project will be when they create the initial schedule. An inexperienced engineer may produce far less than a skilled designer. If specific knowledge is lacking, the estimator can only utilize average productivity values.

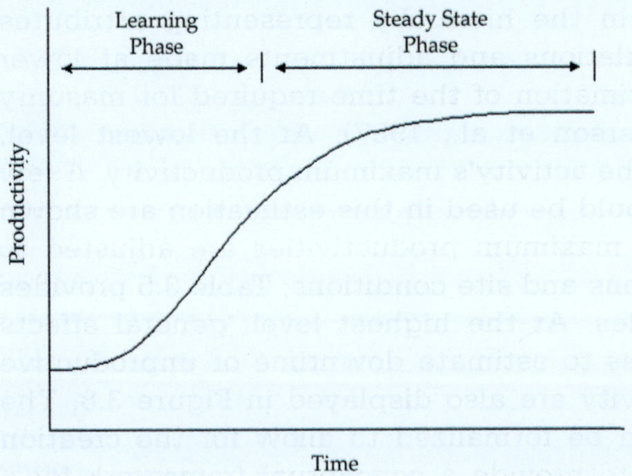

Figure 3.7. Illustration of productivity changes due to learning.

Source: https://www.cmu.edu/cee/projects/PMbook/images/fig9_7.gif.

When estimating durations, weather effects are often very significant and should be given special consideration. The weather affects activity durations in both predictable and unpredictable ways. The likelihood of a rainstorm on a given day is undoubtedly random and affects how productive a lot of things are. On the other hand, the chance of a rainstorm is probably going to change in a predictable way from month to month or site to site. The effects of weather on durations can be taken into account by incorporating adjustment factors for bad weather and meteorological records. As a basic illustration, an activity could take 10 days in ideal conditions, but it couldn't be completed in the rain. Moreover, let's say that 10% of the days in a given month are predicted to see rain. In this instance, 11 days of activity are anticipated, plus one day of anticipated rain.

Finally, the computation shown in Eqn. (1) is flawed due to the use of average productivity factors alone. A variable's multiplicative reciprocal's expected value and its reciprocal's expected value are not exactly equal. For instance, if an activity's productivity is six in favorable weather (i.e., P=6) or two in inclement weather (i.e., P=2) and both favorable and unfavorable weather conditions are equally likely, then P = (6)(0.5) + (2)(0.5) = 4 is the expected productivity, and 1/4 is the reciprocal of that value. But E[1/P] = (0.5)/6 + (0.5)/2 = 1/3 is the expected reciprocal of productivity. Since this example only shows two possible productivity values, it represents an extreme case, but it is always true that using average productivity factors in Eqn. (1) will lead to optimal estimates of activity durations. In this case, the reciprocal of expected productivity is 25% less than the expected value of the reciprocal. This non-linearity issue can be fixed by using tiny adjustment factors or actual averages for the productivity reciprocals.

Eqn. (1), which presents a basic computation of duration, additionally postulates an inverse linear relationship between the total duration of work and the number of crews assigned to an activity. This assumption may not always hold true, even though it makes sense in scenarios where crews can operate independently and don't need extra coordination. For example, multiple architects and engineers may be assigned different design tasks, but the more workers there are, the longer it takes to ensure adequate coordination and communication. As an additional illustration, it might become more challenging to guarantee a seamless supply of materials to every crew on a site as the number of crews rises. Eqn. (1) indicates that the relationship between the number of crews and the duration of the activity is unlikely to be inversely proportional in these latter cases. Consequently, modifications need to be made to the estimated productivity obtained from Eqn. (1). Alternatively, similar to the preparation of nonlinear preliminary or conceptual cost estimate models, more complex functional relationships between duration and resources used may be estimated.

Using a hierarchical estimation framework is one way to formalize the estimation of activity durations. With this method, the estimation problem is broken down into smaller components, with the higher levels in the hierarchy representing attributes that are dependent on the specifics of calculations and adjustments made at lower levels. For instance, different levels in the estimation of the time required for masonry construction are shown in Figure 3.8 (Henderson et al., 1987). At the lowest level, general work conditions are used to estimate the activity's maximum productivity. A few potential maximum productivity values that could be used in this estimation are shown in Table 3.4. At the next higher level, these maximum productivities are adjusted to take into consideration unique crew compositions and site conditions; Table 3.5 provides an example of some potential adjustment rules. At the highest level, general effects like the weather are taken into account. Nodes to estimate downtime or unproductive time related to the masonry construction activity are also displayed in Figure 3.8. The estimation procedure shown in Figure 3.8 can be formalized to allow for the creation of computer tools to assist in the process or to provide a conceptual framework for a human estimator.

Table 3.4. Maximum Productivity Estimates for Masonry Work

Masonry Unit Size	Condition(s)	Maximum Productivity Achievable
8 inch block	None	400 units/day/ mason
6 inch	Wall is "long"	430 units/day/ mason
6 inch	Wall is not "long"	370 units/day/ mason
12 inch	Labor is non-union	300 units/day/ mason
4 inch	Wall is "long" Weather is "warm and dry" or high-strength mortar is used	480 units/day/ mason
4 inch	Wall is not "long" Weather is "warm and dry" or high-strength mortar is used	430 units/day/ mason

4 inch	Wall is "long" Weather is not "warm and dry" or high-strength mortar is not used	370 units/day/ mason
4 inch	Wall is not "long" Weather is not "warm and dry" or high-strength mortar is not used	320 units/day/ mason
8 inch	There is support from existing wall	1,000 units/day/ mason
8 inch	There is no support from existing wall	750 units/day/ mason
12 inch	There is support from existing wall	700 units/day/ mason
12 inch	There is no support from existing wall	550

Source: https://www.cmu.edu/cee/projects/PMbook/09_Construction_Planning.html.

Table 3.5. Possible Adjustments to Maximum Productivities for Masonry Construction/caption

Impact	Condition(s)	Adjustment Magnitude (% of Maximum)
Crew type	Crew type is non-union Job is "large"	15%
Crew type	Crew type is union Job is "small"	10%
Supporting labor	There are less than two laborers per crew	20%
Supporting labor	There are more than two masons/ laborers	10%
Elevation	Steel frame building with masonry exterior wall has "insufficient" support labor	10%
Elevation	Solid masonry building with work on exterior uses nonunion labor	12%
Visibility	block is not covered	7%
Temperature	Temperature is below 45°F	15%
Temperature	Temperature is above 45°F	10%
Brick texture	bricks are baked high Weather is cold or moist	10%

Source: https://www.cmu.edu/cee/projects/PMbook/09_Construction_Planning.html.

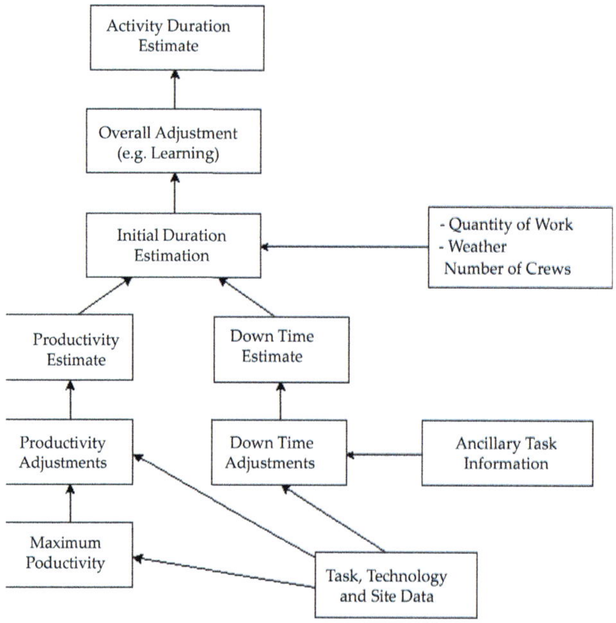

Figure 3.8. A hierarchical estimation framework for masonry construction.

Source: https://www.cmu.edu/cee/projects/PMbook/images/fig9_8.gif.

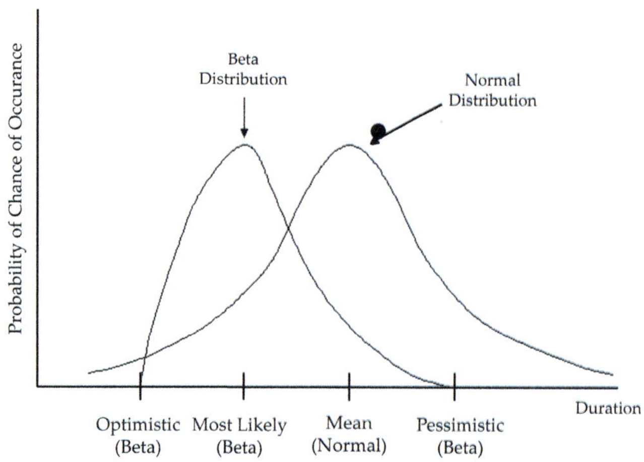

Figure 3.9. Beta and normally distributed activity durations.

Source: https://www.cmu.edu/cee/projects/PMbook/images/fig9_9.gif.

Some scheduling procedures explicitly take into account the uncertainty in activity duration estimates by using the probabilistic distribution of activity durations, in addition to the problem of estimating the expected duration of an activity. In other words, it is assumed that the length of a given activity is a random variable that is distributed in a specific way. As shown in Figure 3.9, one may assume, for instance, that the duration of an activity is distributed as a normal or beta distributed random variable. Based on a probabilistic distribution, this figure displays the likelihood or chance of having a specific activity duration. Since the beta distribution can have both an absolute minimum and an absolute maximum of possible duration times, it is frequently used to characterize the durations of activities. The normal distribution is frequently used as an approximation because it is a convenient and accurate representation of the beta distribution in the distribution's center.

A few parameters are needed to determine the probability of any given duration if the activity duration distribution is characterized by a standard random variable. However, since most of the probabilistic distributions used to represent activity durations require more than one parameter to be characterized, the estimation problem is significantly increased. Whereas the normal distribution only needs two parameters, the beta distribution requires three or four, depending on how general it is.

As an example, the normal distribution is characterized by two parameters, μ and σ, representing the average duration and the standard deviation of the duration, respectively. Alternatively, the variance of the distribution σ^2 could be used to describe or characterize the variability of duration times; the variance is the value of the standard deviation multiplied by itself. From historical data, these two parameters can be estimated as:

$$\mu \approx \bar{x} = \sum_{k=1}^{n} \frac{x_k}{n} \tag{2}$$

$$\sigma^2 \approx \sum_{k=1}^{n} \frac{\left(x_k - \bar{x}\right)^2}{n-1} \tag{3}$$

Where we assume that 'n' different observations x_k of the random variable x are available. This estimation process might be applied to activity durations directly (so that x_k would be a record of an activity duration D_{ij} on a past project) or to the estimation of the distribution of productivities (so that x_k would be a record of the productivity in an activity P_i on a past project) which, in turn, is used to estimate durations using Eqn. (4). If more accuracy is desired, the estimation equations for mean and standard deviation, Eqns. (2) and (3), would be used to estimate the mean and standard deviation of the reciprocal of productivity to avoid non-linear effects. Using estimates of productivities, the standard deviation of activity duration would be calculated as:

$$\sigma_{ij} \approx \frac{A_{ij}\sigma_{1/P}}{N_{ij}} \tag{4}$$

where; σ is the estimated standard deviation of the reciprocal of productivity that is calculated from Eqn. (3) by substituting 1/P for x.

3.4. ESTIMATING RESOURCE REQUIREMENTS FOR WORK ACTIVITIES

For every task, resource requirements are typically estimated in addition to time durations and precedence relationships. Given the comprehensive nature of project work activities, the total resources needed for the project are the sum of the resources needed for each activity. The needs for specific resources throughout the project can be determined by estimating the resources needed for each activity. This makes it possible to identify potential bottlenecks and make adjustments to the schedule, resource allocation, or technology to prevent issues.

Resources that are not always available can impose constraints on many formal scheduling procedures. For example, it might be impossible to carry out certain tasks at a given time if a certain piece of crew or equipment is unavailable. Space is another kind of resource. Usually, a planner will only book one activity for the same time and place. Even though tasks needing the same area might not require technical precedence, they might not be able to be completed simultaneously.

Determining the scope and quantity of resources that could be defined is the first challenge in estimating resource requirements. The quantity of labor (measured in man-hours or dollars), the quantity of materials needed for an activity, and the activity's total cost could be the extent of resource categories at their most general level. The resource estimates could be helpful at this level for cash flow planning and project monitoring. For instance, to identify any issues that may arise throughout a project, the estimated amount of resources needed and the actual expenditures incurred on the activity can be compared. But this broad definition of resource use would not identify bottlenecks related to specific kinds of workers or equipment.

More thorough definitions of required resources would include the quantity and kinds of materials needed for a given activity, as well as the number and kind of workers and equipment needed. It is possible to record and modify the standard resource requirements for specific activities to fit the unique needs of individual projects. Because of this, the kinds of resources needed for specific tasks might already be known. A uniform coding scheme for activities is necessary if one is to rely on past or conventional definitions of these kinds of activities.

Most of the issues with forming duration estimates that were discussed in the previous section also arise when adjusting for the resources needed for a specific task. Specifically, when estimating activity durations in Eqn. (1), work productivity, P_{ij}, will determine how much labor is needed. Mathematically, a typical estimating equation would be:

$$R_{ij}^{k} = D_{ij} N_{ij} U_{ij}^{k} \quad (5)$$

where; R^{k}_{ij} are the resources of type k required by activity ij; D^{ij} is the duration of activity ij; N_{ij} is the number of standard crews allocated to activity ij; and U^{k}_{ij} is the amount of resource type k used per standard crew. For example, if an activity required eight hours with two crews assigned and each crew required three workers, the effort would be R = 8*2*3 = 48 labor-hours.

Determining the kind of technology and equipment to use and how many crews to assign to each task are crucial choices in estimating resource requirements from a planning standpoint. Of course, allocating more crews could enable a specific task to be finished more quickly. Nevertheless, more crews could lead to traffic and coordination issues, which would lower work output.

- Example 5: Resource Requirements for Block Foundations: A typical crew for installing concrete block foundation walls would be made up of two bricklayer helpers and three bricklayers. The completion of the task in proportion to the number of crews could be accelerated if there was enough room on the job site for multiple crews to work on the same task at the same time. Multiple crews may interfere with one another in more restricted sites. In order to ensure efficient and effective work, it may be necessary to assign a bricklayer helper to each bricklayer for specific considerations like intricate scaffolding or large blocks (like 12-inch blocks). Generally speaking, the equipment or technology used and the particular construction task at hand determine the standard crew composition. Then, in response to unique features of a specific site, these standard crews are modified.
- Example 6: Pouring Concrete Slabs: It's crucial to schedule large concrete pours on horizontal slabs so that a full block's worth of work can be finished in a single day. Depending on the technology being used, different resources are needed for the concrete pour. For instance, a typical crew for pumping concrete to the slab might consist of one equipment operator, one finisher, five laborers, and a foreman. The concrete pump itself and vibrators are examples of related equipment. A standard crew for delivering concrete with a chute straight from the delivery truck might include a finisher, four laborers, and a foreman. To ensure that the required amount of concrete could be laid in a single day, the number of crews would be selected. Apart from the resources required for the

placement itself, it would also be important to guarantee that there are enough delivery trucks and that the concrete is available.

3.4.1. Coding Systems

Defining the plan within the limitations of a common coding system for activity identification is one goal of many construction planning initiatives. A pre-defined code unique to each defined activity within a project would serve as its unique identifier. The need for enhanced information flow and better organizational integration is the primary driving force behind the adoption of a common nomenclature or identification system. Coding systems are specifically used to provide a numbering system in place of spoken item descriptions. These codes shorten or simplify the data that needs to be entered. Consistency in definitions and categories between projects and among the various parties involved in a project is also facilitated by a common coding system within the organization. Common coding schemes also make it easier to retrieve past data on the expenses, output, and duration of specific tasks. Lastly, using standard coding systems greatly improves the efficiency of electronic data storage and retrieval operations.

The MASTERFORMAT system, created by the Construction Specifications Institute (CSI) of the United States and Construction Specifications of Canada, is the most widely used standard coding system for built facilities in North America. Following the creation of independent systems, this combined system was first made available as the Uniform Construction Index (UCI) in 1972. A large number of businesses, information sources, trade associations, and professional societies then adopted the UCI for use. The 1978 update of the UCI codes brought in the term MASTERFORMAT. MASTERFORMAT offers a uniform identification code for almost every component involved in building construction.

MASTERFORMAT uses a multi-level hierarchical coding system in addition to keyword-rich text descriptions for every item. The first two digits of the numerical coding system correspond to one of the 16 divisions for work; the 17th division codes the conditions of a constructor's contract. A subdivision within each division is indicated by the addition of a third digit in the most recent version of the MASTERFORMAT. An additional level of subdivision is indicated by a three-digit extension for each division. To identify more precise work items or materials, these subdivisions are frequently further divided with three more digits. For instance, Division 16 (Electrical) and Sub-Division 950 (Testing) are the definitions for the code 16-950-960, "Electrical Equipment Testing." "Electrical Equipment Testing" is a standard term used to describe the activity. Table 3.6 lists the 17 main divisions of the UCI/CSI MASTERFORMAT system. Table 3.7, for instance, displays the second-level divisions for site work.

Although MASTERFORMAT is a very helpful tool for information organization and sharing, as a project coding system, it has certain obvious limitations. First, more precise information may be needed for project cost control, such as the work's location or the accountable organization. Then, in addition to the digits in the basic MASTERFORMAT codes, code extensions are added. For example, a typical extended code might have the following elements:

Table 3.6. Major Divisions in the Uniform Construction Index

0 Conditions of the contract	9 Finishes
1 General requirements	10 Specialties
2 Site work	11 Equipment
3 Concrete	12 Furnishings
4 Masonry	13 Special construction
5 Metals	14 Conveying system
6 Wood and plastics	15 Mechanical
7 Thermal and moisture prevention	16 Electrical
8 Doors and windows	

Source: https://www.cmu.edu/cee/projects/PMbook/09_Construction_Planning.html.

0534.02220.21. A.00.cf34

The project for this activity is indicated by the first four digits; the activity on project number 0534 is referenced by this code. The subsequent five digits denote the MASTERFORMAT secondary division; 02220 "Excavating, Backfilling and Compacting" is the activity in question, according to Table 3.7. The next two digits, in this case 21 for the excavation of column footings, relate to particular activities specified in this MASTERFORMAT code. The block or general area on the site where the activity will take place is indicated by the next character, which in this case is block A. A code could be used in place of the digits 00 to designate the organization in charge of the task. Lastly, the characters cf34 stand for the specific design element number—in this example, column footing number 34—for which this excavation is intended. Therefore, the purpose of this activity is to excavate the site in order to install column footing number 34 in block A. Be aware that column footing 34 would involve a lot of extra tasks, such as formwork and concreting. It may also be necessary to add more fields to the coding systems in order to designate the crew member in charge of this task or to pinpoint the precise location of the activity on the property (which could be defined, for instance, as the coordinates of x, y, and z with respect to a base point). A second issue is that it is challenging to incorporate different construction activities for different types of facilities or activities related to planning or design because the MASTERFORMAT system was initially created for building construction activities. In certain subfields, like power plants or roadways, other organizations have provided different coding systems. However, MASTERFORMAT offers a helpful place to start when arranging data across various construction domains.

There is a constant conflict when creating organizational codes for project activities between implementing procedures that work well for a single project or project manager and procedures that are suitable for the entire organization. The benefits of standard systems for communication and record-keeping typically serve as strong justifications for their adoption. Ad hoc or haphazard coding systems, however, can cause issues when the system is updated and expanded over time, even in small projects.

Table 3.7. Secondary Divisions in MasterFormat for Site Work

Code	Division
02–010	Subsurface investigation
02–012	Standard penetration tests
02–016	Seismic investigation
02–050	Demolition
02–060	Building demolition
02–070	Selective demolition
02–075	Concrete removal
02–080	Asbestos removal
02–100	Site preparation
02–110	Site clearing
02–115	Selective clearing
02–120	Structure moving
02–140	Dewatering
02–150	Shoring and underpinning
02–160	Excavation supporting system
02–170	Cofferdams
02–200	Earthwork
02–210	Grading
02–220	Excavating, backfilling and compaction
02–230	Base course
02–240	Soil stabilization
02–250	Vibro-floatation
02–270	Slope protection
02–280	Soil treatment
02–290	Earth dams
02–300	Tunneling
02–305	Tunnel ventilation
02–310	Tunnel excavating
02–320	Tunnel lining
02–330	Tunnel grouting
02–340	Tunnel support systems
02–350	Piles and caissons
02–355	Pile driving
02–360	Driven piles
02–370	Bored/augered piles
02–380	Caissons
02–450	Railroad work

02–480	Marine work
02–500	Paving and surfacing
02–510	Walk, road and parking paving
02–515	Unit pavers
02–525	Curbs
02–530	Athletic paving and surfacing
02–540	Synthetic surfacing
02–545	Surfacing
02–550	Highway paving
02–560	Airfield paving
02–575	Pavement repair
02–580	Pavement marking
02–600	Piped utility materials
02–660	Water distribution
02–680	Fuel distribution
02–700	Sewage and drainage
02–760	Restoration of underground pipelines
02–770	Ponds and reservoirs
02–800	Power and communications
02–880	Site improvements
02–900	Landscaping

Source: https://www.cmu.edu/cee/projects/PMbook/09_Construction_Planning.html.

3.5. CONSTRUCTION CONTROL

The use of scientific management techniques in managing planning throughout the civil engineering construction process is the main definition of civil engineering management. We will implement comprehensive control and supervision measures for civil engineering construction under modern concepts. According to Coulson (1983), every link in the chain needs to be in place, construction management needs to be put into practice, and project quality needs to be improved. To make civil engineering construction normative and scientific, the relevant personnel should create a general plan early in the construction process. During the building phase, the management staff should equitably distribute engineering resources, maximize the construction framework, ensure the project's smooth progress, minimize expenses, and achieve progress. The quality should meet the requirements of the contract; however, once construction is completed, the management team will need to intensify quality inspections, promptly identify any hidden risks in the project, and enhance overall project quality.

3.5.1. Construction Progress Control Issues

Construction progress, construction quality, and construction are the three main objectives of the project's construction. The three are independent of one another in addition to being bound together by opposition. The main goal of project construction management is to minimize the amount of time spent on each process, which is why the construction schedule generally refers to how each process is arranged and timed throughout the entire construction process (Kim, 2016). The current state of domestic civil engineering construction management work lacks a sound and comprehensive management system, the construction unit's professional abilities are either good or bad, and non-standard engineering subcontracting is a major issue that frequently has a negative impact on the project's construction progress. Additionally, it is rather challenging to strictly control the construction progress in accordance with the scheduled time due to the high susceptibility of civil engineering to external atmosphere, geology, or irregularities during the actual construction process.

3.5.2. Lack of a Perfect Quality Supervision System

Verifying the project's quality of completion is especially crucial during the building phase. Complying with applicable national regulations is the primary objective of civil engineering construction. However, there are glaring weaknesses in the civil engineering quality inspection system within the current construction management process, which has a negative impact on the caliber of civil engineering construction (Rizk, 2018). China's "bean curd dregs" project has become more common in recent years due to economic interests, and the ensuing contradiction crisis has negatively impacted social stability.

Many building materials that did not meet the construction requirements entered the construction site and were used as a result of the absence of a comprehensive quality inspection system for civil engineering. This created serious safety risks for the quality of civil engineering completion and did not meet national safety standards. This not only has severely detrimental effects on the building design unit, but it also costs the construction unit a significant amount of money and prevents the construction industry from growing healthily.

3.5.3. Low Professional Level of Construction Management Personnel

China currently faces a severe shortage of professional and technical workers in the field of civil engineering construction management. Information gathering reveals that construction management staff is frequently made up of foreign workers during the actual construction of civil engineering projects (Kale & Karaman, 2012). Some members of the management team lack formal, systematic learning and training, have relatively low educational backgrounds, and are unable to acquire it. The application of new equipment in inspection, research, and design is woefully lacking in experience. Simultaneously, various issues arise in the actual construction process due to a large deviation in some construction personnel's understanding of the design scheme during the process. One example would be the non-standard electricity used during construction sites. The width of the window and door is insufficient. The building's electrical and water lines are not marked in any particular location. The files have not been created according to the applicable specifications. The data on construction safety is not authentic, which brings about a safety mishap.

3.5.4. Weak Awareness of Safety in Civil Engineering Construction Management

China's construction industry has grown quickly along with the country's ongoing social and economic development, resulting in the emergence of a significant number of small commercial housing buildings. However, the knowledge of safety accidents and civilized construction among civil engineering construction personnel is insufficient, making it impossible to effectively guarantee worker safety during the construction process (Rolfsen, 2011). In certain places, the fire safety signs posted on building sites are too old to be used regularly. Clear safety exit signs have not been installed at the same time. The planning for the construction site does not adhere to the construction specifications. For instance, safety precautions for scaffolding are not always followed in real construction projects, workers did not always wear safety helmets during work in accordance with applicable laws, and overloads frequently happened when getting on and off the elevator. Regarding the current state of construction, there are still a lot of unrecognized safety risks in civil engineering that can arise during the course of construction and have varying degrees of impact on the construction team and the construction unit. This means that improving the knowledge of safety management among managers and construction workers is a pressing issue for civil engineering construction units.

3.6. CONTROL MEASURES FOR CIVIL CONSTRUCTION UNDER MODERN CONCEPTS

3.6.1. Improve the Controllability of Construction Progress

First and foremost, the construction order of each time and process must be rationally controlled, the process and time should be reasonably arranged, and the construction should be completed as early as possible on the basis of ensuring quality in order to ensure that the construction progress can be carried out in accordance with the preset schedule as much as possible. Secondly, a solid construction planning plan needs to be developed by the project management department. The construction planning plan is continuously enhanced and innovated based on the same degree of geographic location and engineering volume in the past, with the underlying tenets of summarizing past construction experience and guaranteeing scientificity. The quality requirements of each construction process and the specific construction links are finally identified and examined, along with any potential logistical issues, material issues, quality issues, or uncontrollable factors, after the various processes in the various periods have been determined, in order to guarantee that managers of construction can efficiently oversee and manage each link in the civil engineering construction process.

3.6.2. Improve Construction Quality Supervision Mechanism

The quality inspection system in civil engineering construction management needs to be further enhanced to better suit the current circumstances. To prevent collusion between construction management staff and material suppliers, first raise the overall caliber of construction management personnel. The drawback of buying subpar building supplies for one's own gain, and enhancing the quality control system can greatly lessen these issues. Second, to ensure that every construction process is carried out strictly in accordance with relevant regulations and that no quality issues arise, it is necessary to improve the work of checking materials in advance, plan and arrange the construction site rationally, strengthen technical supervision and progress during the construction process, and check whether the construction quality meets requirements after completion. In order to ensure that civil engineering has laws and rules to follow in the implementation process, the inspection system's content and requirements for quality inspection will be recorded. Additionally, each requirement will be inspected and supervised through laws and regulations, which will significantly improve the quality of civil engineering construction.

3.6.3. Improve the Professional Level of Construction Management Personnel

To ensure the effectiveness of management, construction management personnel need to possess not only a professional management level and rich management experience, but also a certain understanding of the pertinent technical processes of civil engineering construction. Construction managers should also possess a strong sense of solidarity and accountability. If a safety accident or quality issue arises during a project's construction, accept responsibility, act quickly to mitigate the effects, and actively collaborate with different departments and types of cooperation to improve the project's overall construction efficiency. Nevertheless, the majority of project managers in the real construction industry lack the professionalism necessary to produce quality management outcomes. As a result, construction companies should actively seek out exceptional management talent in the community by offering high salaries; they should also enhance the education of their construction management staff, not only to provide professional training but also to enhance the staff members' sense of accountability and collaboration; and finally, they should raise the bar for civil engineering construction management.

3.6.4. Strengthening Safety Awareness in Civil Construction Management

Civil engineering construction is a high-risk occupation, so it's important to make sure the right employees have the necessary certifications. Before being allowed to enter the construction site, all employees must complete the triple safety training provided by the management department, construction unit, and construction team. Construction accidents can only be prevented to the greatest extent possible by raising workers' awareness of safety procedures. This will lower national losses, protect workers, and guarantee that work is completed on time. This means that safety management at the construction site needs to be improved, and the tenet of "safety first, prevention first" must always be followed during the actual construction process.

CASE STUDY

CONSTRUCTION PLANNING AND CONTROL IN THE MEGA CITY MALL PROJECT:

The Mega City Mall project is a large-scale construction venture aimed at developing a modern shopping complex in the heart of a metropolitan area. The project is a prime example of the application of construction planning and control principles, showcasing how these processes ensure project success through effective resource management, risk mitigation, and quality assurance.

Project Overview:

- **Project Name:** Mega City Mall.
- **Location:** Urban Metropolitan Area.
- **Project Duration:** 24 months.
- **Budget:** $250 million.
- **Scope:** Construction of a 5,00,000 square foot shopping mall including retail spaces, entertainment zones, food courts, and parking facilities.
- **Stakeholders:** Developers, local government, contractors, suppliers, and the local community.

Construction Planning:

1. Defining Project Scope and Objectives:
   - The project aimed to build a state-of-the-art shopping mall with eco-friendly features and high-quality construction standards.
   - Objectives included adhering to budget constraints, meeting the deadline, and ensuring safety and quality.
2. Detailed Project Scheduling:
   - The project was broken down into major phases: site preparation, foundation work, structural construction, interior finishes, and landscaping.
   - A Gantt chart was developed to outline the timeline of each phase, detailing the start and end dates, as well as dependencies between tasks.
3. Resource Allocation:
   - **Labor:** Sufficient workforce was planned for each phase, including specialists for different construction tasks.
   - **Materials:** A procurement plan was developed to ensure timely availability of materials, with considerations for just-in-time delivery to reduce storage costs.

- **Equipment:** High-efficiency construction machinery was scheduled for critical tasks to ensure timely project progress.

4. Budget Planning and Cost Estimation:
   - A detailed cost estimation was performed, including labor, materials, equipment, and contingency funds.
   - Financial milestones were set, and cash flow projections were developed to ensure that funds were available at crucial stages.
5. Risk Management Planning:
   - Potential risks such as delays due to weather, supply chain disruptions, and labor shortages were identified.
   - Mitigation strategies included setting up contingency plans, developing alternative supply chains, and having standby labor teams.
6. Regulatory Compliance and Permitting:
   - Early engagement with regulatory bodies ensured that all necessary permits were obtained.
   - Environmental impact assessments and public consultations were conducted to address community concerns and regulatory requirements.

Construction Control:

1. Progress Monitoring and Reporting:
   - Weekly progress meetings were held with key stakeholders to review project milestones.
   - A project management information system (PMIS) was used to track progress against the project schedule and budget.
2. Quality Control and Inspections:
   - Regular site inspections ensured that construction adhered to design specifications and quality standards.
   - Independent quality audits were conducted at critical milestones to verify compliance with quality benchmarks.
3. Cost Control:
   - Continuous monitoring of expenditures against the budget was performed to detect any deviations.
   - Cost control measures included value engineering to identify cost-saving opportunities without compromising quality.
4. Change Management:
   - A structured process for handling change orders was implemented to manage scope changes effectively.

- Each change request was evaluated for impact on budget, schedule, and quality before approval.

5. Safety Management:
   - Safety protocols and training sessions were conducted regularly to ensure a safe working environment.
   - Incident response plans were in place to address any accidents or emergencies swiftly.
6. Risk Mitigation and Management:
   - Risk reviews were conducted monthly to assess new risks and the effectiveness of mitigation strategies.
   - Contingency plans were activated for risks that materialized, such as delays in material delivery due to supplier issues.

Outcomes:

1. Project Success:
   - The Mega City Mall was completed within the stipulated 24 months and within the allocated budget of $250 million.
   - The project met all quality standards and regulatory requirements, resulting in a high-quality shopping complex.
2. Stakeholder Satisfaction:
   - The project received positive feedback from stakeholders for adhering to the timeline, budget, and quality expectations.
   - The local community appreciated the environmental considerations and the enhancement of local infrastructure.
3. Lessons Learned:
   - Effective planning and continuous control were critical in navigating challenges such as unexpected weather delays and supply chain disruptions.
   - The importance of flexible and adaptive project management practices was highlighted, emphasizing the need for robust contingency planning.

Conclusion:

The Mega City Mall project exemplifies the importance of thorough construction planning and rigorous control processes in ensuring the successful completion of large-scale construction projects. By meticulously planning every aspect of the project and maintaining strict control over progress and quality, the project team was able to deliver a complex, high-quality structure within the specified constraints. This case study underscores the value of integrating comprehensive planning with adaptive control measures to manage risks and achieve project goals effectively.

CLASS ACTIVITY

Students will learn to identify project requirements, develop a project schedule, allocate resources, and perform risk analysis. This activity aims to provide practical experience in construction planning and to highlight the importance of effective project management in civil engineering. Present a hypothetical civil construction project to the class, such as the construction of a community park, a bridge, or a small residential building.

SUMMARY

- Construction planning is a fundamental and challenging activity in the management and execution of construction projects.
- A planner begins with a result (i.e., a facility design) and must synthesize the steps required to yield this result.
- As in the development of appropriate alternatives for facility design, choices of appropriate technology and methods for construction are often ill-structured yet critical ingredients in the success of the project.
- The definition of appropriate work tasks can be a laborious and tedious process, yet it represents the necessary information for application of formal scheduling procedures.
- The proper level of detail will depend upon the size, importance, and difficulty of the project as well as the specific scheduling and accounting procedures which are adopted.
- In most scheduling procedures, each work activity has an associated time duration. These durations are used extensively in preparing a schedule.
- All formal scheduling procedures rely on estimates of the durations of the various project activities as well as the definitions of the predecessor relationships among tasks.
- The initial problem in estimating resource requirements is to decide the extent and number of resources that might be defined. At a very aggregate level, resource categories might be limited to the amount of labor (measured in man-hours or in dollars), the amount of materials required for an activity, and the total cost of the activity.

REVIEW QUESTIONS

1. What are the basic concepts in the development of construction plans?
2. What are the precedence relationships among activities?
3. How to estimating resource requirements for work activities?
4. What do you understand by construction control?
5. Illustrate the control measures for civil construction.

REFERENCES

1. Aras, R., & Surkis, J. (1964). PERT and CPM techniques in project management. ASCE Journal of the Construction Division, 90(CO1), 1–24.
2. Baker, K. R. (1974). Introduction to Sequencing and Scheduling. John Wiley & Sons.
3. Baracco-Miller, E. (1987). Planning for Construction (Unpublished master's thesis). Department of Civil Engineering, Carnegie Mellon University.

4. Christofides, N. (1975). Graph Theory: An Algorithmic Approach. Academic Press.
5. Construction Specifications Institute. (1983). Information on the Master format coding system. The construction specifications institute, 601 Madison St., Alexandria VA 22314.
6. Construction Specifications Institute. (1983). Master Format: Master List of Section Titles and Numbers, 1983 Edition. The Construction Specifications Institute.
7. Construction Specifications Institute. (1983). Master Format–Master List of Section Titles and Numbers. Releasing Industry Group.
8. Coulson, R. (1983). Construction management and design-build/fast track construction: Dispute management under modern construction systems. Law and Contemporary Problems, 46(1), 127–135.
9. Hendrickson, C., Martinelli, D., & Rehak, D. (1987). Hierarchical rule-based activity duration estimation. ASCE Journal of Construction Engineering and Management, 113(2), 288–301.
10. Jackson, M. J. (1986). Computers in Construction Planning and Control. Allen & Unwin.
11. Kale, S., & Karaman, E. A. (2012). A diagnostic model for assessing the knowledge management practices of construction firms. KSCE Journal of Civil Engineering, 16(4), 526–537.
12. Kim, D., & Park, H. S. (2016). Innovative construction management method: Assessment of lean construction implementation. KSCE Journal of Civil Engineering, 10(6), 381–388.
13. Paulson, B. C., Douglas, S. A., Kalk, A., Touran, A., & Victor, G. A. (1983). Simulation and analysis of construction operations. ASCE Journal of Technical Topics in Civil Engineering, 109(2), 89.
14. Paulson, B. C., Jr. (1985). Automation and robotics for construction. ASCE Journal of Construction Engineering and Management, 111(3), 190–207.
15. Rizk, S. M. A. (2018). Stochastic simulation of construction bidding and project management. Computer-Aided Civil and Infrastructure Engineering, 8(5), 343–353.
16. Rolfsen, M. (2011). Co-construction of management concepts: Interpretative viability as opportunity for workplace democracy. Action Research, 9(4), 329–343.
17. Sacerdoti, E. D. (1977). A structure for Plans and Behavior. Elsevier North-Holland.
18. Skibniewski, M. J., & Hendrickson, C. T. (1985). Evaluation method for robotics implementation: Application to concrete form cleaning. In Proceedings of the Second International Conference on Robotics in Construction, Carnegie-Mellon University.
19. Thomas, H. R., Matthews, C. T., & Ward, J. G. (1986). Learning curve models of construction productivity. ASCE Journal of Construction Engineering and Management, 112(2), 245–258.
20. Zozaya-Gorostiza, C. (1988). An Expert System for Construction Project Planning (Unpublished doctoral dissertation). Department of Civil Engineering, Carnegie Mellon University.

CHAPTER

4

Site Management

LEARNING OBJECTIVES

After studying this chapter, you will be able to:

- Learn about construction site management;
- Explore the tips for effective construction site management;
- Describe the site setup and construction processes.

KEY TERMS FROM THIS CHAPTER

Technical knowledge
Communication abilities
Budget
Commercial construction
Construction specifications
Project planning
Construction safety
Leadership skills
Project management
Construction schedule
Construction site inspection
Project requirements
Construction site monitoring
Contract administration

INTRODUCTION

The process of planning, directing, and coordinating all of the moving components—including all parties, resources, and activities—involved in carrying out a construction project at a particular physical site is known as construction site management. It is essential to the effective completion of building projects.

The management of construction sites is a complex task that requires a variety of skills, including technical expertise, communication prowess, leadership qualities, and a deep comprehension of project management concepts. Construction site management actually starts during the preconstruction stage, even though it happens on the actual construction site. Effective preconstruction sets the stage for efficient construction site management by offering a clear project plan, reducing risks, and guaranteeing that all parties involved are aware of the objectives of the project.

4.1. CONSTRUCTION SITE MANAGEMENT

Construction site management is the process of assessing the particulars of a construction site, getting it ready, supervising the workers' day-to-day tasks, and making sure there are enough manpower, supplies, and tools available when they're needed.

The appropriate control of construction activities for the preservation of living and non-living elements of the environmental, ecological, and social landscape is known as construction site management.

Compared to a regular site manager role, construction site management entails and demands a deep understanding of the construction industry, which is rife with particular challenges. Sophisticated construction site managers should employ techniques and tools to help them handle their many responsibilities so that all contractors, architects, and suppliers complete their work on time and within budget (Figure 4.1).

Figure 4.1. Construction-site-management.

Source: https://www.techyildiz.com/construction-site-management-in-civil-engineering/.

4.1.1. Properties of a Construction Site

- Construction companies work outside, which separates them from industrial companies.
- Construction site can be considered as a temporary factory and the construction schedule is specified in the contract; in other words, the construction site exists during the construction.
- Because the construction sector needs a large amount of manpower, the majority of construction workers are unqualified.

- Since construction equipment is subjected to outside factors such as moisture, dust and temperature changes, replacement and maintenance of construction equipment is critical.
- The relationship between worker productivity and production is not the same as in other industries. Construction work is not automated. Workers should be properly supervised in order to improve productivity.
- Workers often violate health and safety policies. This can lead to major issues on construction sites such as loss of life, injuries, or penalties from authorities. Throughout the project, materials like steel and bricks, labor, and equipment are constantly moved in and out of the construction area.
- Construction heavily relies on the skills of the workers, creating a direct connection between workers' mindset and their performance. Therefore, human resources management is a crucial responsibility for a construction site manager.

4.1.2. Types of Construction Sites

We discussed residential and commercial construction sites. Additionally, there are infrastructure and industrial ones. Of course, a construction site can also be defined in other ways, like by owner, sector, occupancy, etc. However, we'll identify four primary categories of construction sites:

- Residential Construction Sites: This type of construction houses people, whether they live in single-family homes or multi-family buildings like row houses or apartment complexes.
- Commercial Construction Sites: The site for commercial construction is where buildings related to businesses, such as offices and warehouses, are built.
- Industrial Construction Sites: Industrial construction sites are places of business, just like commercial ones, but they're factories and large-scale establishments that require more space than just offices. They produce goods, some of which involve large machinery.
- Infrastructure Construction Sites: Infrastructure, which includes anything from highways and bridges to airports, is the final of our four primary categories for construction sites.

4.1.3. Construction Site Management Plan

A written document called a construction site management plan provides detailed information on the supplies, equipment, processes, and knowledge transfer needed to finish difficult construction projects. Project managers can evaluate a project's feasibility and develop resource schedules that support the goals with the help of the plan. It also enables efficient resource management and planning for projects (Figure 4.2).

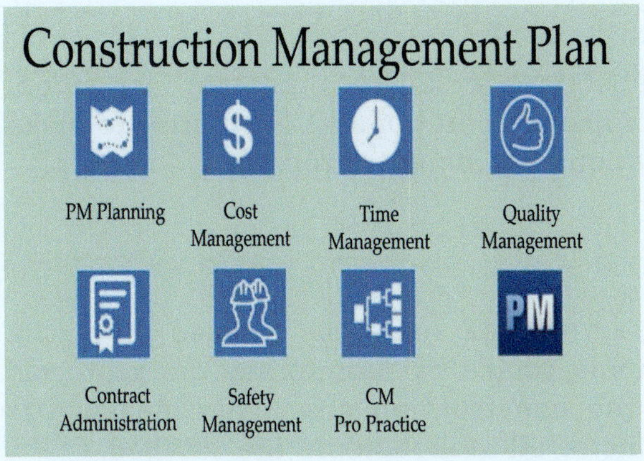

Figure 4.2. Construction site management plan.

Source: https://www.techyildiz.com/construction-site-management-in-civil-engineering/.

The approved construction site management plan is a contract between the council and the builder that covers issues related to site and development management while construction is underway. Every contractor operating on the site is required to abide by the conditions of the approved construction site management plan. Furthermore, in order to guarantee that excavation and demolition on construction sites do not adversely affect nearby traffic, public health, safety, or the environment, a construction site management plan is necessary.

4.1.4. Construction Site Inspection

An inspection of the construction site guarantees that the work is being done in accordance with the project's plans and construction specifications, as well as the client's specifications and code requirements. The job site is routinely inspected during the course of the project; this is not a one-time occurrence. The construction site inspection serves additional purposes, such as guaranteeing the caliber of the job and the safety of its execution with regard to both the working environment and the final product. This involves open communication with stakeholders who have a stake in the project expectations continuing as scheduled and is a part of the monitoring and reporting phase of construction.

4.1.4.1. Construction Site Inspection Checklist

Inspections of construction sites cover a lot of ground, all of which is essential to the project's success. The following highlights the key areas of a typical construction site inspection and explains their significance for the project's success.

4.1.4.1.1. Daily Progress

One of the daily duties at the job site is to conduct a progress inspection to make sure everything is in line with the project specifications. Depending on the size of the task, one or more project members may be able to complete this. These people don't have to be professionals in the same field to perform the inspections; they can come from various trades. Experts can be consulted when necessary for inspections of external matters like environmental policy or waste management.

4.1.4.1.2. Construction Site Safety Hazards

An inspection for safety hazards at a construction site looks for hazards that could endanger the workers or compromise the integrity of the project being built. These dangers are not only noted; suggestions are also given on how to address them, offer protection, or lessen the problem.

4.1.4.1.3. Regulatory Compliance

A job site's performance, quality, and safety inspections include a regulatory inspection. A regulatory or safety inspector examines the project's operations and activities to make sure that all applicable laws, rules, and regulations are being observed.

4.1.4.1.4. Quality Control

This inspection aims to ensure that the project and all materials utilized meet the standards established during the project planning phase by monitoring the protocols and procedures associated with controlling the construction site. This measure is implemented to guarantee that the expectations of all stakeholders are fulfilled.

4.1.4.1.5. Environmental Factors

To measure asbestos and other environmental impacts, an environmental factors inspection measures the amount of soil and existing structures on the construction site by surveying and sampling them. This includes the use of qualified environmental professionals to conduct phase I and phase II site assessments as well as groundwater sampling.

4.1.4.1.6. Draw Inspections

The draw inspection is the easiest of all the inspections specified for the construction site. It involves the builder and the lender in both residential and commercial construction projects. The draw inspection guarantees that the project is moving forward according to schedule for the lender.

4.1.5. Construction Site Monitoring

The practice of keeping tabs on activities occurring on the construction site and addressing any problems that may arise in order to maintain project progress is known as construction site monitoring. For instance, managing project risks is a crucial aspect of construction site monitoring. Construction site managers are required to identify any potential risks, such as machine failure, shortages of materials, insufficient funding, accidents, and even criminal activities like theft that could cause the construction site to come to a complete stop.

Monitoring the construction site also entails keeping tabs on other project-related matters, like reporting on the daily activities of the construction crew to ensure they're finishing the tasks specified in the construction schedule on schedule. Other examples include creating a maintenance schedule to keep equipment on the construction site in top condition and keeping an eye on resource utilization to ensure your team isn't going over the project budget.

4.1.6. Construction Site Safety Management

A distinct feature of a construction site is the constant presence of various hazards that could jeopardize the safety of workers. When organizing a construction project, this must always be taken into consideration.

Planning for construction site safety begins with identifying any potential safety hazards associated with the project and then putting policies, procedures, and guidelines in place to lessen their possible effects. After that, you ought to conduct recurring inspections of construction sites to ensure that safety precautions are being followed.

Safety management at construction sites is a continuous endeavor. During the construction execution phase, it is crucial to set aside funds for the purchase of protective gear or the implementation of any site safety measures.

4.1.7. Key Factors of Construction Site Management

4.1.7.1. Project Planning

Because it outlines everyone's goals and the means of getting there, project planning is essential to the project life cycle. Planning includes developing the project schedule, defining deliverables and requirements, documenting project plans, and conducting some site surveying. It comprises creating a set of plans to guide the working group through the phases of project implementation and completion. The construction site manager will benefit from the plans created during this phase in managing time, quality, money, revisions, risk, and other challenges. They will also help the manager of the construction site oversee employees and outside vendors to ensure that the project is finished on time, on budget, and on schedule (Figure 4.3).

Figure 4.3. Project planning.

Source: https://www.techyildiz.com/construction-site-management-in-civil-engineering/.

Project managers often find that the project planning phase is the most challenging since they have to make educated guesses about the number of workers, supplies, and machinery needed to complete the project. Along with hiring any third parties, the project manager might also need to plan their communications and procurement efforts.

4.1.7.2. Time Management

The project schedule, which outlines what is expected to happen, when it should happen, how long it must happen for, and who is responsible for making it happen, serves as a guide for the construction site manager. Any modifications to the schedule have a cascading effect that delays the completion of subsequent tasks. Even though many commercial construction projects can take weeks, months, or even years, site managers still need to monitor, assess, and adjust the schedule on a regular basis. This could give them a to-do list that they have to finish every day, every week, and every month (Figure 4.4).

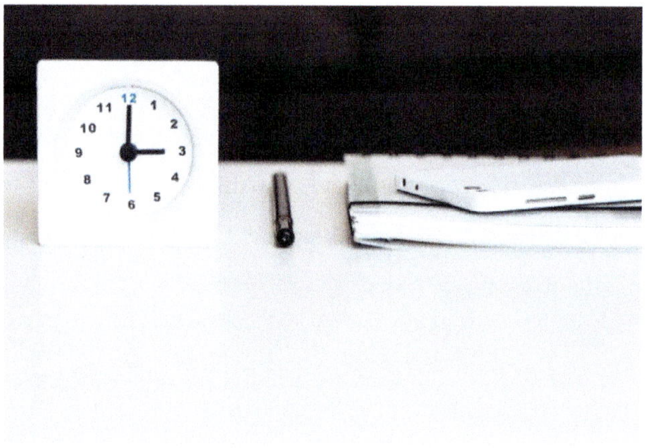

Figure 4.4. Time management.

Source: https://www.techyildiz.com/construction-site-management-in-civil-engineering/.

When time management isn't done well, projects can get delayed and costlier than they should be. This can happen when planned work isn't done well or when unforeseen circumstances aren't handled effectively. Construction companies require a large number of professionals and qualified staff to finish a project that a client requests. Since they are frequently paid hourly, salaries rank among the project's most expensive expenses. In order to keep the project's budget under control, the customer needs to be assured that these resources are being used effectively on the construction site. And one way to make sure that everyone is utilizing their time as efficiently as possible is through time management. In conclusion, one of the most crucial aspects of managing a construction site is time management. With so many people working on the project, it could be difficult to maintain everyone's productivity. Aligning operations with the timeline of the builders and the client's goals, however, might be considerably simpler with the use of modern time management techniques and technologies, such as project management software.

4.1.7.3. Cost Management

Since cost management techniques regulate, establish, and oversee budgets, they are essential components of construction site management. By outlining a specific plan and review process, these programs can help manage cost expectations from the outset of a project. Strategic cost planning and precise cost estimation are essential components of a well-defined cost management strategy (Figure 4.5).

Figure 4.5. Cost management.

Source: https://www.techyildiz.com/construction-site-management-in-civil-engineering/.

The main objectives of cost management strategies are to attain cost predictability and certainty. This can be done by analyzing cash flow and breaking down costs by building or owner in monthly reports that offer crucial details on how closely the budget is followed. Owners may receive comprehensive information on the budget and timeline for the entire project through an ongoing review and estimation process.

4.1.7.4. Quality Management

Construction projects require a wide range of tasks from specialized parties and trades, but they all share the need to deliver high-quality work on schedule and within budget. Construction site management requires quality management, which is essential and often makes the difference between a business's success and failure (Figure 4.6).

Figure 4.6. Quality management.

Source: https://www.techyildiz.com/construction-site-management-in-civil-engineering/.

Construction quality is a measure of how well a project is carried out in relation to the stated specifications found in the work scope. This document serves as a set of project guidelines based on the owner's expectations and provides instructions on how to complete the work in a way that complies with these specifications. The scope of work is a crucial tool in ensuring that a project stays on schedule and serves as a roadmap for achieving the best possible construction quality results. Certain tasks, like the placement of the concrete and the steel anchoring, should be routinely inspected throughout the project. When evaluating a construction job's quality, there are typically many variables to consider. These include whether you fulfilled the owners' requirements and finished the work on schedule and within the predetermined budget. Furthermore, assessing the quality of construction requires knowing if you have complied with the specifications outlined in the project contract and have avoided any conflicts along the way. Last but not least, a critical component of building quality is making sure your finished product fulfills its intended purpose after you've finished a project and given it to its owner.

4.1.7.5. Site Safety Management

Despite being a crucial component of construction site management, construction safety is frequently disregarded on construction sites. Historically, compared to other industries, construction workers have experienced a higher rate of both fatal and nonfatal injuries. To reduce the risk of avoidable accidents resulting in fatalities, injuries, and property damage, the proper measures should be taken to enhance site safety.

To achieve safe and healthy working environments, it is imperative to ensure that health and safety risks are planned, organized, supervised, and monitored. Health and safety are everyone's responsibility when it comes to site work. Planning and organization are necessary to ensure that the intended task won't harm others and to verify that the working conditions are safe and healthy before beginning work. This holds true whether a business is overseeing a small project or a subcontractor is working on a major project that is overseen by a different party (Figure 4.7).

Figure 4.7. Site safety management.

Source: https://www.techyildiz.com/construction-site-management-in-civil-engineering/.

To put it another way, construction sites are dynamic places where employees engage in a range of risky activities, like working from rooftops or scaffolding, falling objects, handling heavy machinery, and using temporary electrical circuits while using electrical equipment and machinery in damp conditions. Therefore, adherence to health and safety policies is mandatory in all construction sites, with construction site managers typically handling their implementation.

4.1.7.6. Contract Administration

Ensuring proper execution of the terms of a construction contract between a contractor and a client is the responsibility or goal of contract administration. The contract administrator may represent a third party or either party, depending on the terms of the agreement.

One of the most important jobs in the construction sector is contract administration, which entails a range of duties that are performed both prior to and following the execution of work orders and contracts. Contract specifications, local, state, and federal laws and regulations, department policy, and terms and conditions must all be followed when performing any work (Figure 4.8).

Figure 4.8. Contract administration.

Source: https://www.techyildiz.com/construction-site-management-in-civil-engineering/.

For contract administration to be effective, it needs to be given top priority from the beginning of the project. A party can guarantee or at least make an effort to address contractual issues and lower risks by comprehending and being aware of their risks under a contract. Contrarily, contract administrators are typically brought on later in the project than is appropriate, by which time a party may have already broken a number of the terms of the contract.

4.1.8. Construction Management Software

Construction management software helps architects, contractors, and house builders oversee ongoing construction projects. Huge corporations and small businesses with employees or independent contractors both use them. By lowering cost overruns and increasing profitability, the program can help general contractors, remodelers, heating specialists, and other professionals.

Construction management software improves efficiency by handling tasks like automating estimates, archiving records, and running reports. These programs, which were especially designed with construction companies in mind, can be a helpful way to save costs and maintain project schedules. Businesses can increase their profitability by monitoring these important parameters. JobProgress, Procore, CoConstruct, Bluebeam, Archdesk, and eCMS are a few of the widely used construction management programs in the sector.

4.1.9. Construction Site Management Documents

Planning, inspecting, and monitoring your construction site can be aided by the creation of numerous documents and reports that are part of the construction site management process. These are a few of the most crucial documents for managing construction sites.

4.1.9.1. Construction Site Plan

An architect, urban planner, or engineer's drawing that shows the current and intended conditions for the construction site is called a site plan, sometimes known as a plot plan. Buildings, sidewalks, parking, drainage, water and sewer lines, lighting, and landscaping are a few examples of these conditions.

A site plan is a drawing that shows how any buildings, parking lots, driveways, and landscaping are to be arranged on the property. The contractor bases their work on a number of construction drawings. However, in order to make sure the drawings comply with the applicable codes, they are also submitted to the local building authority. For historical purposes, copies of the drawings are preserved. This is a crucial document that goes by several titles, such as plot plan, site development plan, site layout plan, and site plan drawing. Plot and site plans differ slightly from one another, though. A site plan also shows the surrounding area, whereas a plot plan only depicts the layout of the construction site itself.

4.1.9.2. Site Inspection Report

After a site inspection, a report known as a construction site inspection report is generated, which enables site inspectors to record any and all concerns and observations they may have regarding a construction site. This report should address a number of topics related to a construction site, including site safety protocols, adherence to building codes and other laws, the state of the machinery and equipment, and the caliber of the materials.

4.1.9.3. Daily Construction Report

The construction project management team can monitor the construction site with the help of a daily construction report. With the help of a construction daily report, you can keep track of all the project tasks completed in a given day, monitor the crew's work hours, record resource usage, and record other crucial information.

To assist you in keeping an eye on the activities occurring at your construction site, we have prepared a free template for a daily construction report (Figure 4.9).

PM ProjectManager

Construction Daily Report

Project	Address	Job Type	Job #	Prepared By	Date
Garage	123 Main St.	Construction build	123	Luc Godard	12/3/2021

Weather Conditions
Sunny, clear, mid-70s

Work	
Task	Status
Paint exterior	Doing
Remove vegetation near foundation	Done

Crew		
Type	Name	Hours
Painter	Joe Smith	8
Painter	Sam Jones	8
Landscaper	Frank Lesser	4

Equipment		
Type	Status	Hours
Paint blower	In Use	8
Brushes, rollers	In Use	8
Pick, shovel, wheelbarrow	In Use	4

Figure 4.9. Construction daily report.

Source: https://www.projectmanager.com/blog/construction-site-planning-inspection.

4.1.10. Construction Site Roles

To build the proposed structure, a large number of construction workers are needed at the construction site. The roles that are involved in construction project management on the construction site are divided into four primary disciplines.

4.1.10.1. Construction Site Manager

Typically, the head of the construction site office is the construction site manager. In order to help guarantee that construction projects are finished on schedule and within budget, this role is in charge of allocating resources, managing costs, and overseeing other areas of the construction site operations. They collaborate with builders, surveyors, and architects to maintain project schedules and guarantee resources are available when needed.

4.1.10.2. Construction Site Supervisor

Site supervisors are less powerful and concentrate on the day-to-day operations of a construction site, whereas site managers are in charge of the site's overall operations. A site supervisor's duties could include organizing daily construction reports, overseeing site paperwork, distributing the crew's workload evenly, or monitoring the productivity and work hours of construction workers.

4.1.10.3. Construction Site Engineer

The engineer responsible for planning, laying out, and leveling the construction site also provides technical advice. This is completed prior to any construction. They are also capable of building drainage systems, roads, and other related projects. Since they both oversee employees, manage the project, and provide onsite support, their duties are comparable to those of a construction site manager.

4.1.10.4. Construction Site Inspector

The work on the construction site must be examined by the inspector to ensure that it passes all inspections, including those for safety and code compliance. Along with checking, they also keep track of the outcomes and oversee the archive, which contains the materials reviewed and the inspection time.

4.1.10.5. Construction Workers

And now we have the construction workers, also known as the construction crew or laborers. They carry out all onsite duties, such as clearing debris, erecting scaffolding, loading and unloading construction supplies, and operating large machinery.

4.2. TIPS FOR EFFECTIVE CONSTRUCTION SITE MANAGEMENT

Best practices for construction site management are put into action to guarantee that projects are finished on schedule, within budget, and in compliance with quality and safety requirements. This section will discuss important best practices that contractors should apply to manage construction sites in order to get the greatest outcomes.

4.2.1. Keep Safety a Top Priority

Safety is the top priority in construction site management because the sector has a higher risk of accidents and fatalities than other professions. It is essential to prevent accidents, guarantee the safety of employees and guests, and uphold an orderly and productive work environment. Effective safety management is crucial to reducing the risks associated with construction sites, which naturally involve a variety of hazards, such as operating heavy machinery and working at heights.

Some key safety considerations include:

- Commitment from Leadership: Leaders in the organization and owners need to prioritize safety in their corporate culture. Through the integration of safety regulations and accident prevention into everyday work routines and communication, all team members take an active role in promoting a safer working environment. It's critical to understand that any safety program's effectiveness depends on ongoing communication with employees and the consistent delivery of messaging.
- Comprehensive Site Safety Plan: Creating a site-specific safety plan facilitates communication and outlines the contractor's safety protocols for various tasks and emergencies. All employees, subcontractors, guests, and pertinent stakeholders should have easy access to these plans. The plan must be updated and reviewed often in order to accommodate any changes in the site's conditions as the project moves through its various phases.
- Ongoing Education and Training: Provide ongoing safety training and education to employees so they are aware of the newest safety rules and procedures.

In addition, mandate specialized training in accordance with the specific tasks, scope of work, or site-specific difficulties to guarantee the safe completion of activities on that specific project.

- Consistent Inspections: Jobsite safety inspections are crucial for methodically identifying possible risks, assessing the efficacy of safety precautions, and guaranteeing compliance with safety guidelines. Safety inspections prevent workplace accidents from taking lives and causing financial and reputational harm by identifying and evaluating potential and current hazards and the risks that go along with them. Prioritizing jobsite safety through the use of external experts, such as safety managers, in conjunction with essential project team members, such as the superintendent and foreman, constitutes a comprehensive approach to site inspection. By combining internal operations with the knowledge of experts committed to upholding the highest safety standards, this method guarantees a 360° evaluation of the workplace. This cooperative effort ensures that a safety culture permeates the entire construction site by combining the knowledge of specialized safety experts with the insights of those directly involved in daily activities, thereby fostering a holistic perspective. The understanding that safety is a shared responsibility that transcends job titles, roles, or specific scope is the foundation of a truly effective safety culture, even though safety managers and superintendents frequently bear the primary responsibility for on-site safety. The "Stop Work Authority" principle, which is commonly emphasized on projects during site orientation, serves to reinforce this ethos.

The Stop Work Authority gives every person present the power, irrespective of their rank, to act without delay in the event that they believe there is an immediate risk to life or property. The Stop Work Authority's power comes not only from its capacity to avert possible mishaps but also from the way it embodies a society that places a premium on safety. This notion fosters candid communication, a readiness to confront safety issues, and an increased level of group awareness.

4.2.2. Understand Site Constraints and Opportunities

Effective management of a construction site requires careful consideration of site-specific factors. Each construction project has its own set of objectives and constraints, and the site's characteristics can have a big impact on how well the project turns out. To recognize and take advantage of both opportunities and challenges, a comprehensive site analysis is necessary during the first stages of project planning.

Project efficiency is significantly impacted by site accessibility, which includes the mobility of construction workers, tools, and supplies to and from the site. Difficulties gaining access to the site can impede project development, add to schedule delays, and raise expenses.

The contractor takes into consideration the site's egress, local traffic patterns, the safety of nearby public and pedestrian pathways, the timing of material deliveries, emergency access, and compliance with relevant regulations when creating the construction execution plan. The contractor assesses the topography, soil, and terrain of the site before beginning construction. Excavation procedures, foundation design, and construction techniques are all impacted by the physical features of the site, such as slopes, rocky ground, and uneven terrain.

4.2.3. Foster a Healthy Jobsite Culture

The values, attitudes, behaviors, and communication styles of all the people working on a construction project come together to form the jobsite culture. The culture of a jobsite is shaped by project leadership and then spreads to all individuals present, including employees and subcontractors. Sustaining a cheerful atmosphere on the job site encourages candid dialogue and efficient cooperation between various crafts, subcontractors, and project participants. Working as a cohesive team on the project fosters a culture of problem-solving and ongoing improvement, which leads to the creation of original solutions and cutting-edge strategies for dealing with obstacles that come up during construction.

A crucial component of efficient construction site management is placing a high priority on creating and maintaining a positive jobsite culture that is consistent with the project's objectives and core values. This creates an atmosphere where all employees can flourish and successfully complete the project.

4.2.4. Map Out the Sequence of Tasks

A key component in defining the project workflow is sequencing. In order to efficiently complete the tasks and activities involved in the project, a strategic sequence must be followed. This is crucial since some tasks might need to be finished before moving on to others or have prerequisites. A thorough project plan that describes the particulars of each task, resource allocation, timelines, and milestones can be created once a clear sequence of tasks has been established.

Work together with members of the internal project team and specialty contractors to make sure that project schedules are realistic and take into consideration any possible delays. Avoid putting team members under undue pressure to meet deadlines; this can cause burnout and lower-quality work.

Effective communication of the sequencing order to specialized contractor partners is essential once it has been established. Confirming everyone's availability and avoiding misunderstandings later in the project are achieved by making sure that everyone is in agreement on the sequence and timeline of tasks and activities.

Review and update the project plan often as it moves forward to account for changing conditions or circumstances. This makes it possible for the group to adapt to changes, take on obstacles, and guarantee that the plan is still applicable. This process usually entails reallocating resources—like labor, supplies, and machinery—to match the crucial path ahead.

4.2.5. Strategically Select Subcontractor Partners

The success of the project is greatly impacted by the critical component of construction site management—the strategic selection of subcontractor partners. Subcontractors are essential to the completion of specialized scopes of work, and their performance has an effect on the final product, schedule, and quality of the project. A list of pre-qualified specialty contractors is often compiled by general contractors for selection. The general contractor evaluates specialty contractors in the prequalification phase based on factors like their track record, safety rating, experience, references, and financial soundness.

When choosing a subcontractor, a general contractor should take the project's objectives and constraints into account. For example, if the project owner places a high value on efficiency and sustainability, the general contractor should give special consideration to choosing subcontractors whose areas of expertise coincide with these aspects of their job description. It is possible to achieve better project outcomes, more client satisfaction, and more seamless project execution by devoting time and energy to selecting dependable, competent, and compatible subcontractor partners.

4.2.6. Document Everything

Maintaining accountability, transparency, and efficient project execution in construction site management requires documentation. Construction projects entail a multitude of stakeholders, intricate procedures, and possible hazards. A historical record of project activities is provided, communication is facilitated, and important information is captured with the aid of proper documentation.

Documenting tasks, responsibilities, and timelines makes it simpler to hold teams and individuals responsible for delivering the contributions that were agreed upon. This keeps people from pointing fingers and makes it easier to solve problems when they do.

Permits and regulatory requirements must frequently be followed in construction projects. Sufficient documentation aids in proving adherence to relevant legal and regulatory requirements. Furthermore, in the event of disputes or legal challenges, properly documented information can be used as evidence to back up legal claims and defenses. This includes things like defining the scope of work and its inclusions and exclusions, defining the mutually agreed-upon scope of work, describing the process for managing change orders, handling payment and invoicing, and setting quality control standards.

4.2.7. Incorporate Rigorous Quality Management Procedures

In order to guarantee that construction projects fulfill the necessary requirements and produce high-quality results, it is crucial to implement efficient quality control and quality assurance procedures. Standardization is the first step in the quality management process. Clearly defined and comprehensive quality standards for all project components, such as materials, labor, and finished products, bring stakeholders together on project results. By routinely carrying out quality tests and inspections at different points throughout the project, possible problems can be found and addressed early on, reducing the impact on budget and schedule as the work is completed.

A thorough approach to quality control lowers rework, raises client satisfaction, enhances the finished product, and improves a contractor's reputation. A contractor's standing in the market is reinforced by a history of precisely executed projects with few problems, which is also a result of good quality management.

4.2.8. Prioritize Open and Clear Communication

A key component of a construction project's successful management is effective communication. Prioritizing polite, honest, and transparent communication can improve teamwork, reduce miscommunication, and increase the success of building projects as a whole. All project stakeholders will be aware of how and when to communicate updates, problems, and progress if dependable communication tools and methods are established. Examples of these include frequent team meetings, email updates, and specialized software or platforms.

Collaboration and trust are further enhanced by transparent leadership that allows team members to feel at ease sharing updates on their work, worries, and recommendations. This helps project teams deal with risks or difficulties before they become a problem rather than just reacting to them after they do.

4.2.9. Embrace Technology

The construction industry has recently incorporated technology to improve efficiency, accuracy, communication, teamwork, and project outcomes overall. Planning, scheduling, cooperation, budgeting, and reporting are made easier with the use of construction management software and tools. All stakeholders can easily access and organize project documents, plans, contracts, and communications thanks to the centralized platform these platforms offer.

Project coordination and visualization are improved through the use of Building Information Modeling (BIM). With BIM, a digital model of a project's physical elements, geometry, materials, and spatial relationships can be created that includes comprehensive information. Throughout the course of the project, it acts as a shared knowledge resource for all parties involved.

Clash detection is another process made possible by BIM that finds inconsistencies between various building components or trades prior to the start of construction. Before putting materials in a structure's plenum space, the mechanical, electrical, and plumbing (MEP) trades ask each other about clash detection, which was formerly known as a coordinated set of drawings. By taking this proactive approach, the need for costly modifications during the construction phase is reduced. For example, think of a location in a building where several systems come together, such as where an electrical conduit and a mechanical duct meet. These systems could conflict if they aren't coordinated, which could result in project delays, cost overruns, and even safety hazards when building is underway. It is clear that the mechanical duct crosses over with an electrical conduit and a structural beam by using BIM's clash detection feature. As a result, specialty contractors work together to modify the path taken by the duct or to reposition the beam and conduit before work ever starts on the project.

Furthermore, members of on-site teams can now communicate and share information in real time thanks to mobile devices and apps. They make it easier to share photos, track progress, send issues straight from the field, and receive immediate updates. These platforms are now cloud-based as well, providing safe storage and remote access to project data, encouraging cooperation and communication among all project participants.

4.2.10. Incorporate Lessons Learned From Past Projects

One of the key components of efficient construction site management is learning from the past. It entails looking back on previous projects, recognizing the triumphs and setbacks, and using those lessons learned to streamline and enhance subsequent projects.

Contractors can identify areas where procedures, tactics, or communication could have been more effective by looking at previous projects. Overall project performance is improved by putting these insights into practice and making improvements. It is possible for contractors to learn how resources were distributed on earlier projects, which helps with better planning and optimization with regard to labor, materials, equipment, and financial concerns.

4.2.11. Create a Project Closeout Checklist

The last stage of construction site management, known as project closeout, makes sure the project is finished and successfully turned over to the client.

This process encompasses various elements including:

- Gather Necessary Documentation: Compile and organize all project documentation, including drawings, operation and maintenance (O&M) manuals, warranties, permits, and inspection reports for handover to the client.
- Address the Punch List: Identify and resolve any outstanding issues, defects, or deficiencies in the punch list before final handover. Work with subcontractor partners to ensure all items are resolved satisfactorily, adhering to predetermined quality benchmarks.
- Perform the Final Inspections: Conduct thorough final inspections to verify that all work aligns with quality standards and regulatory requirements.
- Walk the Client Through the Final Space: Offer the client a tour of the final space to ensure they're satisfied and address any questions or concerns they may have.

A successful construction site is the cornerstone of a well-executed project. Through meticulous planning, resource coordination, communication, and adoption of cutting-edge techniques and tools, contractors can effectively navigate the intricacies of construction projects. Construction site management will surely continue to change in the future as the industry develops, embracing new technologies and methods while appreciating safety, continuous improvement, and teamwork.

4.3. SITE SETUP AND CONSTRUCTION PROCESSES

In order to ensure site security and safety, the land area to be occupied by the construction works should be identified, surveyed, and the necessary fencing or hoardings erected.

Construction processes are the series of actions, stages, or phases required to construct a structure or other infrastructure within a specific time frame.

In order to meet the requirements of the construction project, a construction site's general organization must take into account factors such as the degree of subcontracting, labor force, machinery, equipment, auxiliary and material resource requirements, and intermediate and final completion deadlines.

4.3.1. Site Constraints

Every construction site faces the daily challenge of resolving an important planning activity: construction site layout. Enough space is required for temporary facilities on every construction project, including job offices, material handling equipment, parking lots, maintenance areas, warehouses, workshops, material laydown areas, and batch plants, in order to carry out the construction work. The implementation of temporary facilities and the proper scheduling of tasks depend on the available space, its boundaries, and the specifications of the construction site. According to Illingworth (2000), there are three main types of sites that affect construction planning: linear sites (such as highways, railways, and pipelines), singular or restricted sites (such as building rehabilitations or industrial improvements), and open field or extensive sites (housing and industrial developments and many civil engineering projects). Mawdesley et al. (2002) offer a useful analysis of the elements taken into account when designing and laying out a site.

An effective layout, therefore, needs to take into account the following plot-related factors (Mincks & Johnston, 2010): geographic location, geological conditions, geometry, scale, site topography, ways and paths, proximity to populated areas, existing town planning regulations, land qualification, urban equipment, land acquisition, services needed, and so forth. It also calls for a geotechnical study. Furthermore, the meteorological parameters (temperature, wind, rain, etc.) need to be precisely known. A jobsite's exposure to severe rain and its associated disruptions can be reduced with careful planning. The maritime climate, tides, bathymetric studies, and other factors are also necessary for maritime works.

In addition to needing land for occupation, the proper development of construction projects also requires sufficient space for auxiliary facilities, material supply, and unavoidable temporary works like cofferdams or bypasses. Another essential prerequisite is having access to dumps or quarries. Because of the gravitational force, slopes can facilitate the flow of materials. Furthermore, it is necessary to take into account the installation of a wall or fence, including for access control. Rainwater can impede the construction project's regular progress, necessitating the installation of rainwater evacuation and circulation systems. Roads and work paths need to have sufficient drainage.

As was previously mentioned, there are three types of construction projects (Illingworth, 2000): extensive, linear, and singular. Buildings on small lots of land are examples of singular construction projects, along with the related issues of material storage, makeshift infrastructure, and other things. Both extensive and linear construction projects—such as building airports, harbors, or urban developments—have additional challenges, such as moving supplies and equipment around the construction site, restoring utilities, and maintaining access control, fences, and security. A site setup for a single construction project (building) is shown in Figure 4.10. An example of a site setup for a linear or extensive construction project is shown in Figure 4.11.

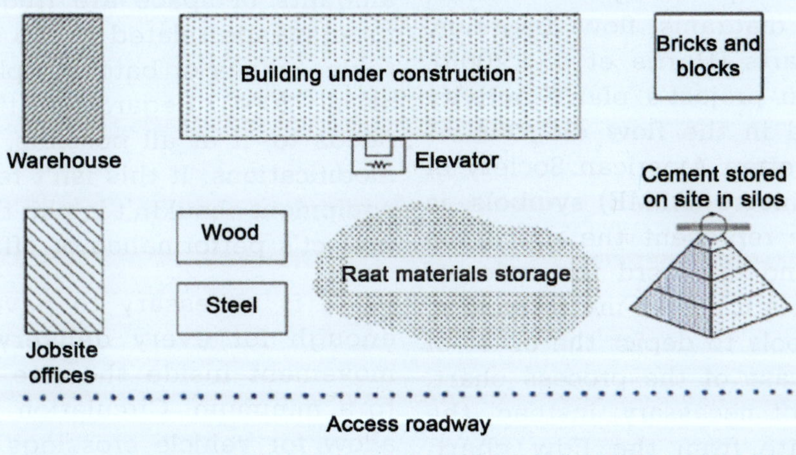

Figure 4.10. Example of the layout of jobsite offices and auxiliary facilities: Singular project.

Source: Eugenio, P., Víctor, Y., José, C. T., Helder, P. M., & Joaquín, C., (n.d.). Construction Management. Wiley-Blackwell. ISBN: 978-1-118-53957-6.

A thorough analysis of the space and temporary facilities needed to carry out a construction project is necessary (Illingworth, 2000). An improper layout can lead to inefficiencies because of safety concerns or a loss of productivity because it requires workers and equipment to make needless trips. Every project is unique and calls for a slightly different strategy. In certain cases, the problem's location and space are fixed variables, meaning there is no other option. For instance, a building project might require a plot of land adjacent to the construction site. Consequently, it is imperative to allocate the requisite time to identify the optimal solution whenever feasible. Keep in mind that supplies and labor must be transported between locations. A well-located jobsite can reduce movement, which affects how much it costs to complete the work.

Space requirements and assembly sequences are often studied using sketches, marked-up blueprints, and cut-out templates (Henderson, 1976). Although there are computer-based tools available, they are not often utilized in this field. These days, construction management can be assisted by event simulation based on building information modeling (BIM). Liao et al. (2011) have recently reviewed the studies that use these tools for project and construction management.

There are numerous easy methods for assessing and planning work sites, including string diagrams, flow diagrams, and process charts (Harris et al., 2006). The construction project's plan's activity flow is depicted in the flow diagram. A set of symbols, often American Society of Mechanical Engineers (ASME) symbols, is used to visually represent the processes being studied and to record the order of events. Sometimes it makes more sense to simply use symbols to depict the order of events. In the case of the process chart, a site plan is not necessary; instead, the symbols originate from the flow chart. When materials are located and temporary roads, access points, and other layouts develop randomly during construction, which frequently causes delays and waste, string diagrams are a suitable tool to use. The work area is mapped out in the string diagram, and the actual movements are drawn on top of it. Reducing the amount that a worker, machine, material, and so forth move is the goal.

4.3.2. Equipment Constraints

According to the construction schedule, the design of the facilities at the construction site necessitates a prior study of the available space as well as the limitations of the machinery, equipment, and facilities. The schedule and site layout are directly impacted by the equipment needed for the construction project. In order to optimize productivity and efficiency, the location of delivery points, machinery, aggregate, concrete or asphaltic manufacturing plants, steel fixing or carpentry workshops, silos and warehouses, prefabrication plants, compressed air production, generator sets, pumps, or other necessary equipment parts and auxiliary resources at the construction site need to be planned for. Keep in mind that rigging for crawler cranes and certain large mobiles needs a sizable space. Large amounts of space are frequently needed for craneage related to the construction of tower cranes or batching plants (Elbeltagi et al., 2004). Regardless, their placement needs to, if at all possible, impede future modifications. If this isn't feasible, moving equipment shouldn't impair the construction project's performance or efficiency.

It is necessary to have an area big enough for every delivery vehicle, and movement inside the site must be kept to a minimum. Circulation pathways that allow for vehicle crossings or designated bypass lanes, with turning radii and slopes tailored to the specific type of machinery, are necessary for the adequate transportation of machinery, equipment, and materials through the interior. The price of building and maintaining haul roads on the property needs to be taken into account. These need to be in good shape, permit airflow in all weather conditions, and have artificial lighting available if needed. It is significant to remember that normal trucks cannot operate on haul roads due to the presence of snow, mud, dust, and frozen ground. It's also important to make sure there are no structures obstructing traffic, such as low bridges, overhead power lines, or other structures. Likewise, contingent on the nature and attributes of the works, alternative transport systems (such as fixed cranes, compressed air, conveyor belts, and cranes) need to be investigated. Planners must create a plan that minimizes the total distance driven by vehicles in order to lower the cost of such movements. Actually, most distribution and transportation systems' financial outcomes are impacted by the severe issue of vehicle scheduling and routing (Yepes & Medina, 2006). The shortest route cut and fill problem (SRCFP), for instance, looks for a path that minimizes the total distance driven between cut and fill locations by a single earthmoving vehicle (Henderson et al., 2003; Lim et al., 2005).

The choice of entry and exit points to the construction site will be based on the current communications network, traffic, link opportunities with facilities, access track, and road. Well-thought-out temporary traffic management plans are necessary because construction activities shouldn't substantially disrupt the flow of traffic. These internal communication channels must also facilitate the profitable movement of people, goods, machinery, and equipment. Most sites require truck weights, so using the site's own balances—which are by the entrance—makes it simple to manage supplies based on weight. It might be essential to provide a suitable area for trucks to wait in order to prevent traffic congestion, particularly on streets. When the construction site's dimensions permit it, it is also convenient to install a gas station or fuel storage facility.

4.3.3. Material Storage and Handling

Storage, identification, retrieval, transportation, and construction techniques are all part of material management. To maximize task productivity and project profitability, effective material handling at the construction site is crucial. Although it rarely happens, materials should ideally arrive at the worksite just before they are needed. The ease of flow is contingent upon the manner in which materials are transported and allocated. Thomas et al. (2005) points out that inadequacies in the movement and supply of building materials will lead to significant declines in worker productivity and performance, as well as monetary losses.

It is common for materials brought onto a construction site to be left undisturbed. Certain materials must be kept safe from theft because they are valuable, and others must be kept secure so that people are not harmed. In any event, they need to be kept in stockpiles or under cover to provide a safe reserve and prevent work at the site from being interrupted. The warehouses associated with cement, aggregates, fuels, and lubricants, along with those housing tools and auxiliary resources, are the most frequented ones at construction sites. It's critical to calculate the required storage capacity. To prevent the negative effects of a shortage of materials or an excessive inventory of materials on site, planning and carrying out the procurement and storage of materials on construction sites is an essential task. A slower supply rhythm may be the cause of the shortage of materials, i.e., supplies run out before fresh supplies arrive, or because of other issues like bad weather, mis scheduled trips, or transportation delays.

First, the cumulative supply and consumption curves dictate the amount of storage that is needed. The amount of volume stored and the amount of anticipation needed before construction can begin increase with decreasing supply. On the other hand, work shouldn't begin until all materials are acquired if supplies are not guaranteed. The supply volume will therefore be determined by the likelihood and significance of the disruption, accounting for the impact on the economy and delivery schedule, in order to prevent shortages.

However, material storage should be minimized as it impacts not only the cost of necessary facilities but also the costs associated with transportation, general expenses, loading and unloading processes, surveillance, and potential material deterioration. In this situation, just-in-time material deliveries are ideal, but they necessitate closer supplier coordination (Polat & Arditi, 2005). Heuristic optimization is a viable solution to the non-trivial problem resulting from the critical and mutually interdependent relationships among material procurement planning, dynamic material storage, and site-layout planning (Said & El-Rayes, 2011).

4.3.3.1. Good Practices in Relation to the Handling and Stocks of Materials

- Minimize the number of movements of materials at the construction site;
- Store the materials as close as possible to the place where the tasks have to be carried out;
- Place the materials in their storage area;
- Use specialized personnel to transport the materials;
- Coordinate shop fabrication drawings with the supplier;
- Sort the material upon its delivery to the construction site;
- Provide appropriate space for sorting and storage waste material;
- Do not store materials next to the building as these can obstruct access routes;
- Keep all storage areas tidy;
- Estimate the storage volume required;
- Ensure that there is adequate access to warehouses and stocks;
- Compact the storage surface to guarantee the support capacity;
- Keep the material on pallets or packaged to avoid damp;
- Protect sensitive material from weather effects;
- Designate special storage areas for flammable substances such as foam plastics, flammable liquids or gases.

4.3.4. Temporary Facilities and Auxiliary Works

Temporary facilities are needed to meet operating requirements during the construction project because the facilities needed to carry out the project, like heating, drinking water, or electricity, are not available until the works are almost complete (Doran, 2004). Until the infrastructure is turned over to the owner, the contractor needs to request temporary supply connections for each service at the construction site. If the business only offers a portion of the service, it's crucial to supply the necessary tools and supplies for the remaining portion. During the construction phase, a weatherproof, grounded electric power service and distribution system with adequate size, capacity, and power characteristics must be offered as temporary power supply.

Making connections to the various external supply networks is the first task. The proprietors of the networks that supply water, electricity, telephone service, sewerage, and other utilities typically set these points. Before laying slabs or foundations, the connections and channels to the site—whether above ground or below—must be planned. Water service and distribution, temporary electric power and light, temporary heat, ventilation, telephone service, sanitary facilities, including drinking water, and storm and sanitary sewerage are examples of temporary utilities.

The construction site installs its own supply system when there isn't a sufficient water supply nearby. In this instance, it is crucial to sanitize the temporary water piping before using it. The topography of the terrain will be utilized, with the water supply network being located at the highest points of the site. If residual water is to be connected to the current networks, it needs to be sufficiently treated and drained.

The facilities needed to carry out the primary construction works are also considered auxiliary works or provisional facilities (Illingworth, 1987). Thus, temporary facilities or auxiliary works – such as work paths, temporary bridges, access slopes to particular areas of the site, bypasses, cofferdams, auxiliary foundations for cranes or scaffolding, quarries with installations for the manufacture of aggregates, and sometimes even housing for workers and their families during the construction of major works (e.g., a dam) – are all necessary to plan before beginning work and to install and then disassemble after the work is finished. Building and maintaining temporary roads and paving to suitably support the specified load and endure exposure to traffic during the construction phase is another crucial duty. In order to facilitate material delivery and storage, equipment use, administration, and supervision, it is also essential to extend temporary paving inside and outside the construction site.

While they are not strictly required to finish the main works, temporary facilities or auxiliary works must be completed at other times to prevent work disruptions. These actions are necessary to maintain the current services that are affecting third parties as a result of the construction works (Doran, 2004). This means that while planning the construction project, temporary amenities like irrigation channels, private plot access, noise-proof screens, temporary traffic signage, dewatering facilities, drains, and the like must be considered. The possible demolition or dismantling of these temporary facilities or services must be taken into account when planning the construction schedule and estimating the associated expenses.

4.3.5. Construction Jobsite Offices

Before the construction project begins, the contractor sets up jobsite offices. As there are projects and contractors, there are as many different kinds of jobsite offices. On the other hand, typical varieties might consist of pre-existing structures, mobile offices, trailers, or offices constructed on the jobsite. A number of considerations should be made when choosing the ideal jobsite office, including availability, cost, and jobsite space (Mincks & Johnston, 2010).

The job site office houses, among other things, the contractor's locations, resources for communication technology, an archive, employee information, and supplies for first aid (Mincks & Johnston, 2010). To put it briefly, the superintendent, other field management staff, and the manager of the construction site are housed in this office. Typically, one or more prefabricated offices are installed temporarily, along with a meeting room of suitable size and offices for the manager of the construction site, field engineers, supervisors, foremen, and departments responsible for technical support and administration. An authorized copy of the contract, the project documentation, and a construction diary will be kept at the facilities. Furthermore, these facilities might house the project manager's offices, who acts as the owner's representative. All of the contemporary ICT resources, such as computers, printers, fax machines, internet, and phones, must be present in these offices. Certain areas are designated for the purpose of occupational health and safety, such as dining rooms, canteens, restrooms, locker rooms, medical or first-aid facilities, first-aid kits, storage, and so forth. The approximate number of employees determines the size of these spaces, which must provide a high level of comfort in terms of livability, safety, health, and other factors. Parking areas, perimeter fences, drainage, the removal of leftover water and rainwater, water connections, electrical energy, communications, and other factors must all be considered in addition to the needs of the staff and the environment (Mincks & Johnston, 2010).

The offices should be far enough away from the site to prevent issues with noise, machinery circulation, and other problems. But they also shouldn't require staff members to travel excessively or unnecessarily. Furthermore, the ability of these installations to oversee the work areas and access the construction site makes it convenient to manage the construction site from that location. Additionally, the contractor has the option to rent a building close to the construction site, such as private residences, office buildings, and warehouses. The costs of adapting these structures for use as jobsite offices should be considered in these cases. Another option would be to use prefabricated modular offices or trailers, which offer great flexibility in both assembly and disassembly and the potential to relocate them in the future (Doran, 2004). Finally, office buildings are constructed when construction projects have a long duration. These buildings may eventually be demolished or repurposed.

Illustrations 4.10 and 4.11 provide examples of various office layouts and additional facilities found on job sites. While Figure 4.11 is more suitable for a linear or extensive project, Figure 4.10 is better suited for a single construction project.

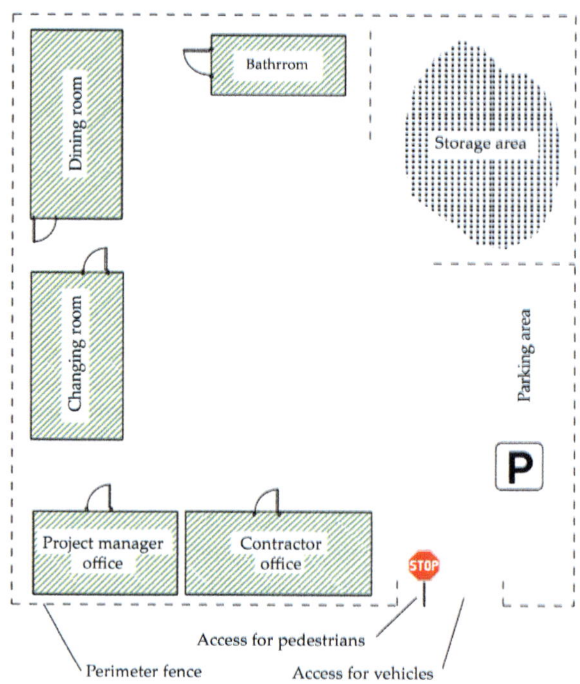

Figure 4.11. Example of the layout of jobsite offices and auxiliary facilities: Linear or extensive project.

Source: Eugenio, P., Víctor, Y., José, C. T., Helder, P. M., & Joaquín, C., (n.d.). Construction Management. Wiley-Blackwell. ISBN: 978-1-118-53957-6.

4.3.6. Security on Construction Sites

It is necessary to secure the construction site for people, things, and machinery. The contractor bears the responsibility of barring public access due to various hazards, including but not limited to large machinery, excavations, structures, concealed obstacles, and so forth. The general public is more at risk than the workers at the construction site, who are better informed, skilled, and have the necessary experience to lower the risks. The following are just a few examples of security and protection features (Mincks & Johnston, 2010): temporary fire protection; barricades, warning signs, and lights; enclosure fences for the site or sidewalk bridge; safety and health; and environmental protection.

Vandalism and the theft of supplies, tools, and machinery from construction sites are additional factors that need to be taken into account. There have been numerous construction site thefts involving large pieces of equipment like trucks and bulldozers. This issue has the potential to lower project profitability and have an impact on productivity. Accurately estimating the losses' magnitude is challenging, though. According to a 2005 survey by Berg and Hinze of 97 businesses, vandalism losses were greater for small and medium-sized businesses than for large businesses (Figure 4.12). According to this study's findings, for every $1 million of self-performed work, the annual direct costs of thefts and vandalism incidents come to $1,682. Roughly 82% of these expenses are related to theft.

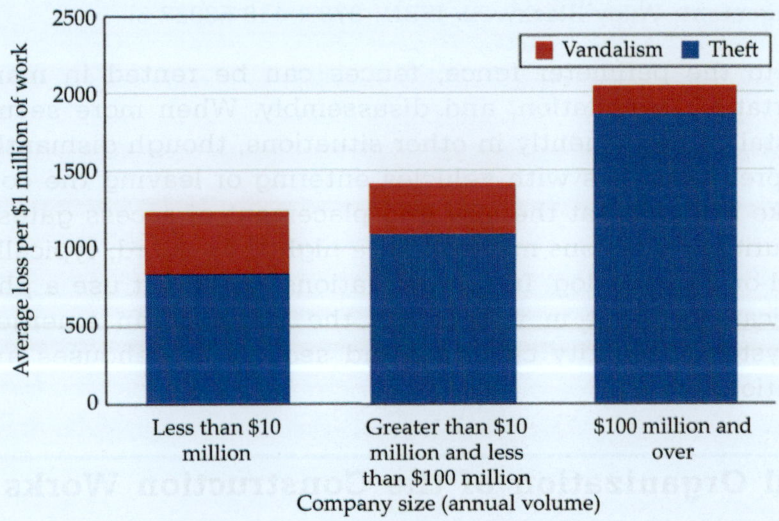

Figure 4.12. Cost of theft and vandalism per $1 million of construction work.

Source: primary data obtained from Berg & Hinze (2005); Eugenio, P., Víctor, Y., José, C. T., Helder, P. M., & Joaquín, C., (n.d.). Construction Management. Wiley-Blackwell. ISBN: 978-1-118-53957-6.

As such, the contractor has to keep people who aren't supposed to be there from entering the construction site. Moreover, the following steps must be taken to stop materials from being stolen: perimeter fencing, security monitoring, controlled access gates, night lighting, coordination with local authorities, equipment kept in secure locations, and so forth (Berg & Hinze, 2005). A few standard security procedures used by contractors on building sites are listed in Table 4.1, which was taken from Berg & Hinze (2005).

Table 4.1. Most Common Security Practices Employed on Construction Sites

Ensure jobsite security	Prevent theft of tools	Prevent theft of machinery and equipment
Lockboxes	Maintaining a secure storage area	Parking of equipment and machinery in well-lit areas
Security fencing	Marking of tools	Parking of equipment in a specific formation at the end of the day
Posting of warning signs	Maintaining tool inventory	Including additional identification on equipment and machinery
Use of exterior lighting	Minimizing tools left on site	Using a distinctive color to mark machinery and equipment
Strategic parking of large equipment	Making workers responsible for tools	Modifying the ignition or fuel lines

Source: Eugenio Pellicer, Víctor Yepes, José C. Teixeira, Helder P. Moura, and Joaquín Catalá. Construction Management. Wiley-Blackwell. ISBN: 978-1-118-53957-6.

In reference to the perimeter fence, fences can be rented in many places; this includes transportation, installation, and disassembly. When more security is needed, fences can be installed permanently in other situations, though dismantling them costs more money. To prevent issues with vehicles entering or leaving the construction site, it is crucial to take into account the size and placement of access gates (Doran, 2004). Certain extra security precautions may call for a nighttime guard, typically accompanied by a safety patrol or a guard dog. In both situations, you must use a phone or another form of communication to get in touch with the police or an emergency service, if needed. Alarm systems, security cameras, and security warehouses are examples of additional precautions.

4.3.7. Internal Organization of the Construction Works

The construction company can effectively and efficiently meet the economic, time, quality, and safety objectives by adhering to a set of rules and codes of conduct and organizing the project in a hierarchical manner (Mustapha & Naoum, 1998). The activities are divided into departments or sections with a clear distribution of roles and responsibilities so that everyone is aware of their specific role and how it relates to other tasks in order to achieve these goals in a coordinated manner.

An organizational chart delineates the functional levels of the construction project's internal organization (Mustapha & Naoum, 1998). This organizational chart shows how posts or positions relate to one another, as determined by a set of guidelines, policies, or internal rules necessary to meet the organization's goals. Every construction company organizes a construction project differently, tailoring their operations to the unique features of each construction site.

A well-planned project offers benefits in terms of cost, schedule, safety, and quality. However, no two projects are the same because they are all developed on separate locations with a large number of temporary workers. Additionally, due to unanticipated events, flaws, or other types of circumstances, the projects undergo various changes throughout the construction phase. Thus, the project's success depends critically on how the construction site is set up.

If a regulated organization is founded on strict and predetermined norms, the dynamics of the construction project render it unfeasible. Although a structure with these traits, which are typical of public administrations, has the advantage of solving similar problems in a similar way, production is typically bureaucratic and slow, meaning that it lacks the flexibility to adjust to the unique circumstances that arise at each construction site.

A linear approach is typically used to organize construction projects (Halpin & Woodhead, 1998). The organization's communication among its members follows the established hierarchical line, and orders, tasks, and responsibilities are communicated in a clear and precise manner. This simpler structure is defined by the linear authority principle. A linear organization has the benefit of having a simple implementation and a definite, stable unit. This kind of structure works particularly well for small- to medium-sized construction projects, which have regular, standardized tasks with clear execution guidelines and are not overly specialized.

Other construction projects require consultants or specialized departments due to their size, complexity, or extended timelines for completion. This is an example of a functional organization with a knowledge-based management style in which no manager or supervisor has complete control over every employee. Decentralized decision-making and direct, middle-man-free communication are facilitated by this kind of organization (Robbins & Coulter, 2010). On the other hand, purely functional construction sites may result in a loss of authority over control, multiple departments that are subordinate to one another, and possibly unclear goals.

Construction projects with a determined complexity propose a hierarchical-advisory organization to avoid these issues and increase their organizational advantages (Reve & Levitt, 1984). These organizations uphold the principle of sole authority, with advisory or support bodies assisting line managers in decision-making. Specialists offer the consulting and advisory services, while the hierarchy (line) ensures control and discipline.

The construction site manager oversees medium-sized construction projects and is typically subordinate to three departments: technical support, operational or productive (i.e., construction-related), and administration. A typical organizational chart is displayed in Figure 4.13.

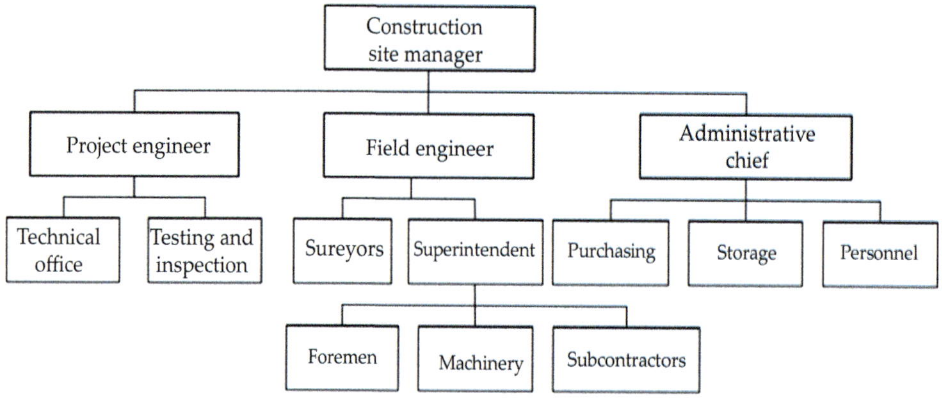

Figure 4.13. Organizational structure of the construction site.

Source: Eugenio, P., Víctor, Y., José, C. T., Helder, P. M., & Joaquín, C., (n.d.). Construction Management. Wiley-Blackwell. ISBN: 978-1-118-53957-6.

The manager of the construction site takes on the duties and goals allocated to the project. They typically answer to the director of construction operations at the construction company or the portfolio manager. The following duties fall under the purview of the construction site manager (Edum-Fotwe & McCaffer, 2000):

- Represents the company in front of the client, the subcontractors, and any other third parties.
- Leads the resources at the site, including the personnel.
- Clarifies any issues regarding the definition of the project, together with the project manager representative of the owner.
- Drafts the bill of quantities for work units, materials, subcontractors, and so on.
- Decides about the layout of plants, stocks, warehouses, and workshops.
- Analyzes the construction processes.
- Schedules the construction tasks.
- Coordinates and monitors the execution of the construction project.
- Cooperates with the project manager in the drafting of valuations and final payment.

The manager of a construction site is also in charge of administrative management duties, such as material intake and storage, material consumption, inventory control for the job site, hiring and supervising staff, monitoring subcontractor activity, maintaining machinery, and managing fuel and electricity usage, among other things. In terms of project execution, the construction site manager oversees the pre-construction activities, scheduling personnel and machinery use as well as material implementation, prior to the commencement of construction works (Mincks & Johnston, 2010). They are also in charge of work reports, occupational health and safety, and site and subcontractor control.

However, in order to achieve these goals, the construction site manager needs to have sufficient organization (Halpin & Woodhead, 1998). The group that develops the construction project is completed by the superintendent, foremen, and field engineers. The superintendent is one of the most crucial individuals; they should have extensive knowledge of organizing, managing, and closely monitoring construction-related activities and procedures. The manager team and the workforce are connected hierarchically by the superintendent. Major projects involving multiple locations or specialized units may have a general superintendent overseeing a group of superintendents.

The intermediary between the superintendent and the site operators is the foreman (Mincks & Johnston, 2010). Their extensive experience and responsibility are typically the reasons they are hired. They oversee employee performance, adherence to schedules, worker order, and, as needed, employee training. They complete the daily labor, machinery, and material reports. They also suggest to the management any suitable adjustments, modifications, and controls.

The technical support services are not directly on the production line but rather report to the construction site manager. This department is in charge of:

- Technical office: detailed design, studies, calculations, valuations, and cost control.
- Quality and environmental management: laboratories and technical control.

The administrative services also report directly to the construction site manager, out of the executive line. They are responsible for:

- Material and tool purchasing orders, storage, distribution, and control.
- Management of facilities and equipment: workshops, maintenance, machinery plant and so on.
- Administrative and legal matters related to the personnel.
- Registry of accounting operations.
- Collections and payments.
- Other support tasks: correspondence, typing, filing, and so on.

4.3.8. General Approach to Construction Processes

The construction process is a collection of procedures, systems, and organizational techniques that enable a project to be completed with the labor force, equipment, and material resources at hand (Illingworth, 2000). The process entails planning, carrying out, running, and maintaining the project, as well as occasionally closing and dismantling the site. It's critical to recognize the differences between building and civil engineering construction processes and procedures.

Executing the tasks involved in a building project requires a set of methods known as construction procedures. Numerous building techniques are also applied in civil engineering projects. But building also calls for particular methods for some constructions, like installations, partition walls, and roofs. As a result, tasks essential to construction operations, like formwork installation and removal, must be included in traditional construction processes linked with foundation laying activities (screen walls, pile driving, etc.).

The latter obviously impacts loads on each slab as well as deadlines, necessitating periodic reapplying of the formwork.

This field also includes industrial buildings. These structures include industrial production facilities, raw material reserves, engine rooms, chimneys, and warehouses as well as buildings used for extraction or transportation. Furthermore, significant building initiatives (sports arenas, convention centers, etc.), as well as extremely tall buildings, call for particular building techniques; new lightweight materials, industrialization, and prefabrication of work execution all play a significant part (Van Bueren & De Jong, 2007).

Building systems should be constructed in a way that complies with all legal requirements and enhances the building's sustainability (Van Bueren & De Jong, 2007). This can be accomplished by utilizing conventional methods and processes, which are typically not very effective due to their heavy reliance on manual labor and labor-intensive construction methods. As an alternative, more industrialized and logical methods can be employed, where prefabrication plays a bigger role.

The type of construction project determines the organization and arrangement of construction elements and systems in civil engineering as well as the set of techniques needed to build the various units (Illingworth, 2000). As a result, the factors that determine a harbor differ from those of a structure (a bridge) or an underground project (a tunnel). However, it is worthwhile to take into consideration a general overview of the most typical construction practices in this field.

Any kind of project needs to analyze the land using drilling and boring, but it's also common practice to use soil improvement techniques (preloads, grouting, stabilizations, etc.) to boost support capacity, prevent deformations, or regulate groundwater levels. Dredging or earthworks are the next stage, and these are the most crucial procedures in certain projects (roads, beach restoration, etc.). There are times when blasting is necessary to remove rocks from the site or from the quarry that provides the building materials. While specific equipment, such as shields or tunneling machines, as well as alternative methods, like the Austrian method, can also be utilized, blasting procedures are sometimes utilized in the construction of tunnels and subterranean works (Rabcewicz, 1964). Compaction of the soil occurs following the earthwork activities. This procedure is especially crucial for civil works projects like building rock-fill and earth-fill dams. It is also essential for linear construction projects like roads, where concrete or asphalt road execution protocols are needed.

The construction process may entail the laying of both shallow and deep foundations, as well as caissons, sheet walls, piles, and piles, and may require additional frameworks, like formwork and falsework, to hold and pour concrete (Harris et al., 2006). Prefabrication is important in other processes as well, like building bridges with precast sections.

Lastly, it is important to think about disassembling and dismantling any temporary or even finished construction projects. In addition to special projects like landscaping an abandoned quarry or building a concrete caisson for a nuclear power plant's reactor, these activities might involve the use of explosives for demolition (Seike & Akita, 2008).

4.3.9. Temporary Works

In the construction industry, "temporary works" refers to any "parts of the work that allow or enable construction of, protect, support, or provide access to the permanent works, and which might or might not remain in place at the completion of the works." This term is commonly used in the industry. As supplemental construction works are often required for the completion of the main construction project, they must often be designed, carried out, and maintained during the project's execution. For instance, in order to maintain traffic flow on a road while a bridge is being constructed, traffic detours must be constructed. In some cases, a temporary foundation slab in a port is needed to support a crane while a breakwater is being built. Implementing a sheet pile with metallic profiles, which helps with soil containment during earthwork movements, is another example. Activities required to stabilize or safeguard an existing building or structure that are not meant to be permanent are also included in the category of temporary works.

Any construction process is likely to include a significant amount of temporary work (Harris et al., 2006). For instance, the preload approach lessens the soil's deformability while simultaneously increasing its bearing capacity. This is accomplished by allowing material to build up over several months and then removing it. The construction of temporary structures, such as the formwork and shuttering required to pour concrete for an on-site bridge, is another example of temporary works. Furthermore, one of the most important aspects of risk prevention and mitigation in the construction industry is the proper design and implementation of temporary works. However, if temporary works are not properly designed, planned, built, supervised, or managed, they can expose projects to risks, which can lead to delays and higher costs.

Maintaining the services or supplies that are impacted by the main works is sometimes the goal of the interim works (Illingworth, 1987). Some of the works or facilities that are implemented for the duration of the project include sections of channels or irrigation ditches, access roads, temporary drains, or noise-proof screens. Examples of temporary works include:

- Earthworks: Trenches, excavations, temporary slopes and stockpiles.
- Structures: Formwork, false work, scaffolding, temporary bridges, propping, facade retention, needling, shoring, edge protection, site hoarding and signage, site fencing, cofferdams.
- Equipment/Plant Foundations: Tower crane bases, supports, anchors and ties for construction hoists and mast climbing work platforms, ground works to provide suitable locations for plant erection.

Despite being temporary, this kind of construction work needs to be properly planned for during the design, execution, and dismantling stages. It is customary to rely on the knowledge acquired in comparable situations, but this can result in construction-related mishaps or unacceptably high risks. To ensure the safety of people and property while the main construction activities are underway, these construction works necessitate both site supervision and a specific design project signed by a certified technician, unless they are specifically included in the main design project (Illingworth, 1987).

Determining in advance how the temporary works will be removed, dismantled, or assembled, as well as how the construction-affected areas and services will be restored, is also essential. It is important to keep in mind that utilizing reusable or modular materials that are easy to disassemble can minimize implementation expenses and lessen the environmental effects of these kinds of temporary building projects and infrastructure. Temporary construction works must obtain an official license and, if needed, have their activities recalculated to ensure they have a longer useful life than initially planned if they are not dismantled or disassembled after the construction project is finished (Seike & Akita, 2008). This is because they should not be dismantled or disassembled after the construction project is completed. Actually, compared to the requirements for final construction works, many standards call for lower partial safety coefficients. A review of the materials' functionality (unacceptable deflection, crack appearance, etc.) and durability factors (metallic oxidation, inadequate coating on concrete frameworks, etc.) is also necessary.

Items to be addressed in temporary works:

- Appointment of a temporary work coordinator for health and safety matters.
- On larger sites, appointment of temporary site supervisors.
- Assessing/ensuring competence of design and site-based staff responsible for temporary works.
- Preparation of adequate design briefs.
- Designing to include calculations, sketches, drawing, specifications and design risk assessments.
- Independent checking of the design.
- Issuing of design checking certificates.
- Procurement of temporary works materials and equipment (in accordance with the designer's specifications).
- Site control of erection, use, maintenance and dismantling of temporary works.
- Checking of erected temporary works (and control of their use) to ensure compliance with the design.
- Issuing of the 'permit to load' and 'permit to dismantle' when required.

CASE STUDY

ISLAND TUNNELLING PROJECT OF HONG KONG-ZHUHAI-MACAO BRIDGE:

- **Background:** The Hong Kong-Zhuhai-Macao Bridge (HZMB) is the longest sea-crossing bridge in the world with a length of 50 km. The designed service life of the bridge is 120 years. A scheme using a bridge and tunnel combination has been adopted for the offshore bridge and tunnel part to avoid interfering with navigation. The tunnel section is approximately 6.7 km long across the Lingding West Channel and Tonggu Channel, and the bridge section is about 22.9 km. For the transitions of the bridge to tunnel sections and to accommodate tunnel ventilation shafts, artificial islands are being built at each end of the tunnel. The bridge and tunnel sections provide a dual three-lane carriageway with a design speed of 100 km/h. The length of the immersed tunnel is 5,664 m with a tunnel width of 2 × 14.25 m and a vertical clearance of 5.1 m. The immersed tunnel is assembled using 33 tunnel elements with a typical tunnel element length of 180 m and a weight of 75,000 t. The maximum water depth of tunnel element immersion is 44.5 m.

All the huge tunnel elements are prefabricated by a flow line, and involved standardized production in a factory located on an island near the construction site. Tugboats will transport the prefabricated tunnel elements to the construction site. After immersion, the tunnel elements will be connected undersea by positioning deep water unmanned automatic element immersion and positioning adjustment system.

- **System Framework of Smart Construction Site in HZMB Project:** Based on the idea of smart construction site, an instantiated framework of smart construction site is adopted to support the manufacturing, transportation, and assembly of the tunnel element.

In the offsite manufacturing stage, the detailed design information for the tunnel elements is obtained from the BIM for prefabrication production. Several automatic production systems, including full section automatic hydraulic segment prefabrication form system, belt conveyor, and distribution system for full section concrete cast of tunnel element, and multiple-point hydraulic support, and synchronous jacking system, work collaboratively for the prefabrication production of tunnel elements. All production information is updated to the data center.

In the transportation and assembly stage, related information is acquired through sensors, GPS, sonar systems, and weather systems, and is transmitted to the data center. To ensure the quality and safety of transportation and assembly, multiple monitoring and control systems (e.g., meteorological and hydrological forecasting system, real-time attitude of tunnel element monitoring system, deep water unmanned automatic element immersion and position adjustment system, and command system) can work together to support the tunnel element transportation and assembly processes. The data center consists of the data center of the construction site command center in Zhuhai and the meteorological and hydrological database of the national marine environmental forecasting center in Beijing.

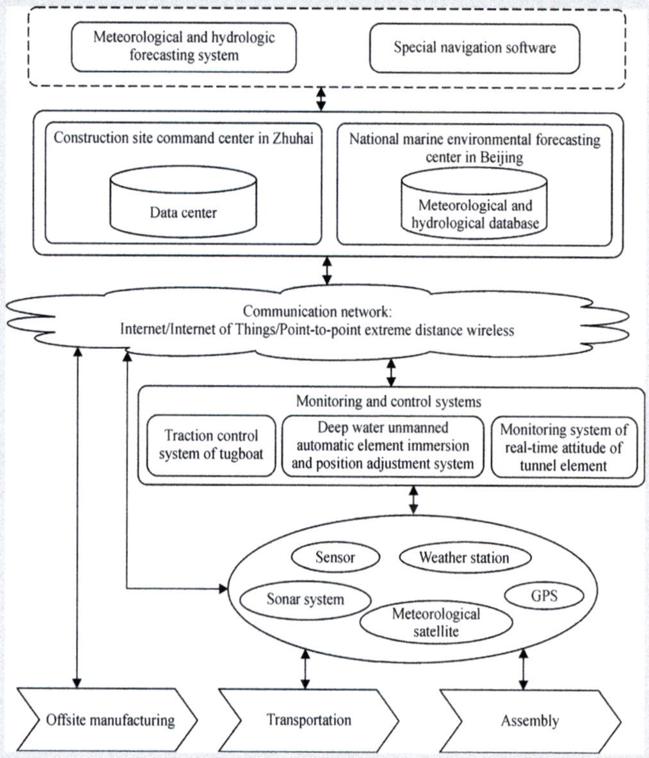

Source: https://journal.hep.com.cn/fem/EN/10.15302/J-FEM-2018075.

The following subsection will focus on the details of the construction-related information acquisition and transmission and collaborative operations in the transportation and assembly process and on illustrating the preliminary practice of smart construction site.

- **Construction-Related Information Acquisition and Transmission:** Meteorological and hydrological information at construction sites and attitude data of tunnel elements are very critical for the successful transportation and assembly of tunnel elements. Various sensor technologies and communication networks have been adopted to acquire this data and information.
- **Acquisition and Transmission of Meteorological and Hydrological Information:** The transportation and assembly of tunnel elements are significantly affected by meteorological and hydrological conditions. Therefore, multiple meteorological and hydrological information monitoring equipment and ocean weather stations have been placed at the construction site to continuously observe ocean currents, waves, and weather information 24 hours a day. Meteorological satellites are

also used to monitor the meteorological information of the construction area.

All meteorological and hydrological information acquired at the construction site is transmitted to the data center located on land. Long-distance wireless networks combined with optical fiber have been implemented for communication between the construction site at sea and the command center on land. In particular, point-to-point extreme distance wireless broadband transmission is used to communicate between the construction site and the wireless transmitting/receiving end on land, which is connected to the command center via optical fiber.

- **Real-Time Attitude Data of Tunnel Element Acquisition and Transmission:** During the suitable meteorological window, the prefabricated tunnel element will be transported to the assembly site by tugboats after outfitting. A real-time attitude tunnel element monitoring system has been developed. In the monitoring system, 3D positioning information of the immersed tunnel element is acquired by GPS, RTK (Real-time kinematic), and sonar systems. Micro-mechanical gyro angle sensor, high-speed angle sensor, and a six-dimensional measurement system, such as movement direction, speed, accelerated speed, swing frequency, oscillation amplitude, monitor the movement state data of the immersed tunnel element in real-time. The real-time attitude data of the tunnel element is transmitted to the data center through the above-mentioned communication network.

In this case, construction environment information (e.g., meteorological and hydrological information) and construction information (e.g., attitude data of tunnel element in transportation and assembly process) are successfully acquired and transmitted to the data center using several kinds of sensor technology and the above-mentioned communication network. According to the idea of a smart construction site, these construction-related data and information acquisition and transmission are the foundation of a smart construction site that makes the achievement of collaborative operations in the transportation and assembly work of tunnel elements possible.

- **Collaborative Work in Transportation and Assembly Process:** Three kinds of participants are involved in the transportation and assembly process of the tunnel element:

1. Decision makers, consisting of three members who determine when to start transportation, the closure of navigation, and assembly.
2. Construction site commanders consisting of nine captains who send shipping and assembly instruction to operators.
3. Operators who manipulate various machines, equipment, and control systems.

Based on the acquired construction-related information, several systems including meteorological and hydrological forecasting system, comprehensive situation analysis system, and command system are utilized to support collaborative operations among the different participants in the transportation and assembly process of tunnel element.

According to the acquired meteorological and hydrological information, the suitable transportation and assembly time window will be analyzed. In particular, the acquired meteorological and hydrological information is transmitted via the internet to the National Marine Environmental Forecasting Center in Beijing. After computing and analyzing by the supercomputer at the National Marine Environmental Forecasting Center, the meteorological and hydrological forecasting service is provided to the command system. The decision-makers determine the suitable transportation and assembly time window.

In the transportation process, the huge tunnel element will be transported by tugboats from the prefabrication factory to the construction site. The commanders synthesize the acquired meteorological and hydrological information and real-time attitude data of the tunnel element and sensor data of the tension force of the tugboat's mooring ropes, and provide shipping instructions to the operators. The operators manipulate the traction control system of the tugboats to adjust the attitude of the tunnel element, which avoids deviation from the voyage route or grounding.

In the assembly process, the commanders will provide assembly instructions to operators supported by the acquired real-time attitude data and positioning data of immersed tunnel elements. The operators handle the deep-water unmanned automatic element immersion and position adjustment system to control the immersion and position process of tunnel elements.

- **Intelligent Construction Management:** Solving the extensive collaboration and coordination problem is the main purpose of a smart construction site. In the case based on the acquired construction-related information sharing, several decision support systems are developed including a meteorological and hydrological forecasting system and special navigation software, to help decision-makers determine suitable time windows for transportation and assembly operations. The meteorological and hydrological forecasting system can proactively acquire meteorological and hydrological information of the construction site, forecast the trend of meteorology and hydrology using the calculation function of a supercomputer. Considering the actual situation of the construction site, a tugboat towing transportation scheme is adopted to transport tunnel elements on the sea, which includes the combination of longitudinal and horizontal towing. The special navigation software has been used to coordinate assignments between different tugboats to ensure the collaborative transportation of tunnel elements.
- **Discussion:** By the end of May 3, 2017, the entire length of the immersed tunnel had been successfully assembled on the seabed. The smart construction site mode is inspired by this case study, which works as evidence to support the practicability of the proposed mode. In this case, ICT has been widely used in the construction processes of immersed tube assembly to realize perception and interconnection of construction information, collaborative works of equipment, systems, and participants. Smart construction site management in the Hong Kong-Zhuhai-Macao Bridge project could be partially realized. The smart construction

site mode is inspired by the successful practice of the Hong Kong-Zhuhai-Macao Bridge. Although many improvements of the smart construction site mode should be made in the future, it has some implications for other projects. First, in this case, all immersed tunnel elements are prefabricated on an island near the construction site. These immersed tunnel elements are installed at the bottom of the sea. The command center of immersed tube assembly is located on land. Construction sites often have the characteristics of geographical distribution. In addition, construction processes of immersed tube assembly are often affected by meteorological and hydrological conditions. How to construct the information infrastructure of the construction site is critical, as it can overcome negative construction site environmental conditions. Second, the information support platform of immersed tube assembly incorporates real-time attitude data and meteorological or hydrological information. Relying on this data or information, commanders will provide assembly instructions to operators, and operators can follow instructions to manipulate various devices for controlling the immersion and position process of tunnel elements. Without a good information support platform, it is difficult for all participants to work together. Third, to realize collaboration work among participants, all participants must actively perceive and share construction information in the immersed tube assembly process. Subsequently, smart management would be realized through the decision management system. In other words, construction site information must be more transparent for all participants, which is the foundation of smart management. In the future, a smart construction site would transform the model of project construction, reinforcing collaborative work among participants for the entire construction process, and providing information support for the operation and maintenance of the facility after the project is put into operation.

CLASS ACTIVITY

Student outline the health and safety requirements relating to the provision of welfare, fire prevention and first aid facilities on construction sites.

SUMMARY

- Construction site management is the process of orchestrating, supervising, and harmonizing the moving parts, including all activities, resources, and parties, involved in executing a construction project at a specific physical location. It serves as the backbone of the successful execution of construction projects.
- Residential construction site is a construction that serves as a home for people, whether it's single or multi-family homes, such as apartment buildings or row houses.
- Commercial construction site is a place to erect structures related to business including offices, warehouses, etc.
- Industrial construction sites are places of business, but they're factories and large-scale facilities that need more than office space. They manufacture products and include heavy machinery.
- A construction site management plan is a written document that contains comprehensive information on the procedures, supplies, equipment, and knowledge transfer required to complete challenging construction projects. The plan gives project managers the tools they need to assess a project's viability and create resource schedules that complement the objectives. It also makes effective resource planning and administration for projects possible.
- Construction site monitoring is the process of tracking the activities that take place on the construction site and resolving issues that might occur to keep projects on track.
- Construction processes are the set of steps, phases, or stages necessary to erect a building or infrastructure in a certain time period.
- Construction site layout is an important planning activity that is solved daily on each construction site. Each construction project needs enough room for temporary facilities such as job offices, material handling equipment, access roads, parking lots, maintenance areas, warehouses, workshops, lay down area for materials, and batch plants to execute the construction works.
- The construction project is organized in a hierarchical structure, subject to a series of rules and codes of behavior that enable the construction company to reach the economic, time, quality, and safety objectives with efficacy and efficiency.

REVIEW QUESTIONS

1. What is construction site management?
2. List the four main types of construction sites.
3. What is construction site management plan?
4. What is site setup in construction?
5. What are construction processes?

REFERENCES

1. Berg, R., & Hinze, J. (2005). Theft and vandalism on construction sites. Journal of Construction Engineering and Management, 131(7), 826–833.
2. Doran, D. (2004). Site Engineer's Manual. Whittles Pub.
3. Edum-Fotwe, F. T., & McCaffer, R. (2000). Developing project management competency: Perspectives from the construction industry. International Journal of Project Management, 18(2), 111–124.
4. Elbeltagi, E., Hegazy, T., & Eldosouky, A. (2004). Dynamic layout of construction temporary facilities considering safety. Journal of Construction Engineering and Management, 130(4), 534–541.
5. Halpin, D. W., & Woodhead, R. W. (1998). Construction Management (2nd ed.). Wiley.
6. Harris, F., McCaffer, R., & Edum-Fotwe, F. (2006). Modern Construction Management (6th ed.). Blackwell.
7. Henderson, D., Vaughan, D. E., Jacobson, S. H., Wakefield, R. R., & Sewell, E. C. (2003). Solving the shortest route cut and fill problem using simulated annealing. European Journal of Operational Research, 145(1), 72–84.
8. Henderson, E. M. (1976). The Use of Scale Models in Construction Management (Technical Report No. 213). Department of Civil Engineering, Stanford University.
9. Illingworth, J. R. (1987). Temporary Works–Their Role in Construction. Thomas Telford.
10. Illingworth, J. R. (2000). Construction Methods and Planning (2nd ed.). E & FN Spon.
11. Liao, T. W., Egbelu, P. J., Sarker, B. R., & Leu, S. S. (2011). Metaheuristics for project and construction management: A state-of-the-art review. Automation in Construction, 20(5), 491–505.
12. Lim, A., Rodrigues, B., & Zhang, J. (2005). Tabu search embedded simulated annealing for the shortest route cut and fill problem. Journal of the Operational Research Society, 56(7), 816–824.
13. Mawdesley, M. J., Al-Jibouri, S. H., & Yang, H. (2002). Genetic algorithms for construction site layout in project planning. Journal of Construction Engineering and Management, 128(5), 418–426.
14. Mincks, W. R., & Johnston, H. (2010). Construction Jobsite Management (3rd ed.). Thomson.
15. Mustapha, F. H., & Naoum, S. (1998). Factors influencing the effectiveness of construction site managers. International Journal of Project Management, 16(1), 1–8.
16. Polat, G., & Arditi, D. (2005). The JIT materials management system in developing countries. Construction Management and Economics, 23(7), 697–712.
17. Rabcewicz, L. (1964). The New Austrian Tunneling method: Part one. Water Power, 15(11), 453–457.
18. Rad, P. F. (1982). A graphic approach to construction job-site planning. Cost Engineering, 24(4), 211–217.

19. Reve, T., & Levitt, R. E. (1984). Organization and governance in construction. International Journal of Project Management, 2(1), 17–25.
20. Robbins, S. P., & Coulter, M. (2010). Management (11th ed.). Pearson.
21. Said, H., & El-Rayes, K. (2011). Optimizing material procurement and storage on construction sites. Journal of Construction Engineering and Management, 137(6), 421–431.
22. Seike, T., & Akita, N. (2008). Dismantlement of buildings and recycling of construction waste. In K. Hanaki (Ed.), Urban Environmental Management and Technology (pp. 107–128). Springer.
23. Thomas, H. R., Riley, D. R., & Messner, J. I. (2005). Fundamental principles of site materials management. Journal of Construction Engineering and Management, 131(7), 808–815.
24. Tommelein, I., Levitt, R., & Hayes-Roth, B. (1992a). Sight plan model for site layout. Journal of Construction Engineering and Management, 118(4), 749–766.
25. Tommelein, I., Levitt, R., & Hayes-Roth, B. (1992b). Site-layout modeling: How can artificial intelligence help? Journal of Construction Engineering and Management, 118(3), 594–611.
26. Van Bueren, E., & De Jong, J. (2007). Establishing sustainability: Policy successes and failures. Building Research & Information, 35(5), 543–556.
27. Yepes, V., & Medina, J. R. (2006). Economic heuristic optimization for heterogeneous fleet VRPHESTW. Journal of Transportation Engineering, 132(4), 303–311.

CHAPTER

5 Time Management

LEARNING OBJECTIVES

After studying this chapter, you will be able to:

- Discuss on time management in construction;
- Understand the aspects of time management;
- Define the time schedule adjustments;
- Describe the need for time reduction.

KEY TERMS FROM THIS CHAPTER

- Time management
- Critical path method (CPM)
- Time tracking
- Construction management software
- Critical path method (CPM)
- Accelerated bridge construction (ABC)
- Effective time management
- Construction project
- Poor time management
- Progress analysis
- Project time reduction
- Project extension

INTRODUCTION

The process of planning and carrying out a strategy regarding the amount of time needed for work activities on a project is known as time management. Achieving profitability and successfully and efficiently meeting program and budget targets depend on effective time management. Due to insufficient time management, projects may be at risk of unneeded expenses and delays. This can occur from not accounting for all of a project's complexity or from improperly managing planned work or unforeseen circumstances. If a contractor has not been chosen, the client may designate a program consultant for large projects in order to create a comprehensive project schedule that includes an outline schedule for construction. The contractor will be in charge of scheduling the construction work after they are hired, but the program consultant may still work with the client to create an overall program. All work activities should be thoroughly understood and planned during the planning stage in order to maximize resource allocation and lower the possibility of "unknowns." After that, an estimation of the duration of each task can be made. This is essential for establishing goals and dates, allocating resources, figuring out contract prices, and figuring out cash flow needs.

5.1. TIME MANAGEMENT IN CONSTRUCTION

Managing and organizing construction tasks and activities in a way that maximizes efficiency and optimizes performance is known as time management. It entails establishing priorities, making and adhering to schedules, allocating responsibilities and resources, tracking advancement, and resolving problems. Construction companies can make sure that projects are finished on schedule and within budget by managing time and resources well.

Effective time management is crucial in the construction industry as it guarantees that all deadlines are fulfilled and the final product fulfills the requirements of the job. It also helps to reduce any risks related to construction and guarantees the safety of both the general public and workers. Effective time management in construction plays a crucial role in enhancing customer satisfaction by ensuring timely completion of projects at superior quality. Various tools and techniques can be utilized for efficient time management in construction, such as the Critical Path Method (CPM) which visually outlines the necessary steps in the construction process to determine the most efficient approach. Additionally, tools like Gantt charts can provide a clear representation of task durations, while project management software can assist in resource management and communication of project updates. The success of a project can be significantly impacted by efficient time management in the construction industry. It can assist in making sure that tasks are finished on schedule, to the greatest possible quality, and with the fewest possible risks involved. Construction companies can increase customer satisfaction and profitability by completing projects more quickly when they manage time and resources effectively (McGraw & Leonoudakis, 2009).

5.1.1. Benefits of Effective Time Management in Construction

Effective time management is a crucial skill for building projects to be successful. It can support the completion of tasks quickly, the meeting of deadlines, and the timely and cost-effective delivery of projects. But efficient time management in the construction industry has a lot more advantages. Saving money is one of the key advantages of efficient time management in the construction industry. Contractors can minimize or completely avoid expensive overruns and avoid wasting time and resources by keeping a careful eye on and managing deadlines. By maximizing their productivity, construction teams can shorten project durations and finish tasks faster thanks to time management.

The stress and strain of approaching a deadline can also be lessened with the aid of time management strategies. Project participants can maintain organization and work methodically toward deadlines by segmenting larger tasks into smaller ones and setting reasonable deadlines. This helps keep workers from feeling overburdened and enables team members to concentrate on the task at hand. Team members that manage their time well also communicate clearly with one another. Project participants are more likely to talk about how to approach tasks when timelines are closely monitored. This can boost team morale, improve collaboration, and foster creativity. Teams are able to function more productively and successfully as a result.

Lastly, proficient time management can aid groups in minimizing unforeseen issues and improving risk management. Contractors who adhere to strict timetables are less likely to experience expensive delays, which can avert possible issues and enhance the success of the project. In addition to saving money and time, efficient time management in the construction industry can lower stress levels, promote improved communication, and improve risk management. Construction teams can achieve optimal efficiency and timely and cost-effective project delivery by employing efficient time management techniques and strategies.

5.1.1.1. Reducing Project Costs

In the construction industry, time management is critical because it can significantly lower project costs. Costs associated with labor and materials can be decreased by using effective time management techniques to ensure quality, meet deadlines, and keep projects on schedule. Tasks can be effectively planned and scheduled by project managers to guarantee that resources are used effectively and economically. Project managers can contribute to the overall cost reduction of a project by making sure that the necessary resources are available to finish each task on schedule. Project managers can also avoid financial setbacks by keeping an eye on each task's progress. This allows them to identify potential pitfalls or delays early on and make the necessary adjustments to keep the project moving forward.

5.1.1.2. Increasing Productivity

In order to increase productivity in the construction industry, time management is crucial. Construction workers and project managers can complete more work in less time without sacrificing quality by developing their time management skills. Tasks can be completed effectively by a team by organizing and defining expectations clearly. Time can be saved and productivity can be raised by taking small actions like shortening breaks, minimizing distractions, and simplifying procedures. Setting quantifiable objectives and monitoring progress can also inspire teams and guarantee that deadlines are fulfilled. For construction projects to be successful and profitable, efficient time management is essential.

5.1.1.3. Improving Quality Control

Projects are completed on time and to the highest quality control standards when time management is practiced. Time management done well keeps the project moving forward and guarantees that deadlines are met. It also aids in making sure that resources are used effectively, cutting down on waste and raising the project's standard overall. Because time management enhances resource management, it also facilitates more accurate budgeting. Tasks can be more effectively planned and arranged to improve team coordination and reduce delays brought on by misunderstandings. Effective time management can be critical to any construction project's success.

5.1.1.4. Maintaining a Safe Working Environment

It is an essential component of efficient time management in the building industry. It is imperative to allocate time for safety measures and routine upkeep of equipment, machinery, and work areas when scheduling. Employees are more likely to take the required safety precautions and abide by regulations when they are aware of the safety protocols they are expected to follow. Furthermore, the risk of illness or injury is decreased by properly managed construction sites that follow safety procedures and offer comfortable working environments. Enough time should be set aside for training sessions and safety check-ins in an efficient time management plan. To guarantee that workers understand their responsibilities and are not overworking themselves or endangering their health or safety, it is also important to communicate clear job expectations and safe working capacity to them.

5.1.2. Time Management Strategies for Construction

Efficient time management in the building industry requires allocating time for safety measures, routine equipment upkeep, and machinery. Employees are more likely to follow safety protocols and regulations when aware of expectations. Properly managed construction sites with safety procedures reduce the risk of illness or injury. Training sessions and safety check-ins should be included in time management plans. Clear communication of job expectations is crucial to prevent overworking and ensure worker safety.

1. Prioritize and Plan: When it comes to projects, planning and prioritizing tasks is important. Constructing a detailed plan enables the team to identify which items should be addressed first and allows them to allocate resources accordingly. This can help ensure that resources are used efficiently and goals are met on time.
2. Set a Realistic Timeline: It is important to set a realistic timeline to keep projects on track. This allows teams to schedule resources more effectively and estimate the time needed for completion. It is also important to allow flexibility in case unanticipated issues arise.
3. Establish a Reporting System: An effective reporting system helps construction teams track progress. This is especially important when multiple projects are running simultaneously. A reporting system can help identify potential problems early, allowing teams to adjust the timeline and budget accordingly.
4. Utilize Technology: Technology has made tracking progress easier for construction management teams. Utilizing tools such as project management software can help teams monitor multiple projects' progress, ensuring that deadlines are met, and budgets are adhered to.
5. Automate Tasks: Automating tasks can help streamline the construction process. By automating mundane tasks such as invoicing and scheduling, construction teams can save time and resources. This allows them to focus on more pressing matters and ensure that projects are completed on time.

By following these time management strategies, construction teams can ensure that projects are completed on time, within the allotted budget. Additionally, by utilizing the right tools and processes, teams can ensure that resources are used efficiently and effectively. Time management is an essential part of success in the construction industry.

5.1.2.1. Creating a Master Plan

Efficient time management is crucial for the success of any construction project. To optimize productivity and effectiveness, it is important to develop a comprehensive plan for each project. This plan should outline clear objectives, a timeline for completing tasks, anticipated resources, and cost estimates. It is also important to anticipate and plan for potential delays or unforeseen circumstances, such as inclement weather or staffing issues. By incorporating these factors into a well-thought-out master plan, construction projects can be completed on time and within the client's specifications, minimizing setbacks and mistakes.

5.1.2.2. Time Tracking

This is an additional efficient method for appropriately managing time in construction. Building professionals should use a reliable time-tracking system to precisely measure the amount of time spent on tasks and activities in order to effectively track and manage time. This covers both tasks that are completed on- and off-site. Building teams can quickly determine where they are most and least productive by doing this and making any required adjustments. Building teams can also see precisely how much time is spent on each task by using this method of time tracking, which helps reduce the possibility of cost overruns. Finally, time monitoring enables groups to more effectively arrange their calendars and guarantee that tasks are finished on schedule and within budget.

5.1.2.3. Prioritizing Tasks

If construction projects are to be finished on schedule and within budget, time management is crucial. Setting work priorities based on their urgency and importance is essential for efficient time management. Make a list of all the tasks you need to do and rank them in order of importance and urgency before beginning any project. Establish deadlines and goals to ensure that every task is finished on time, but also allow for flexibility in case any tasks take longer than expected. To further ensure that the most crucial tasks are completed, consider assigning some of your work to colleagues. Finally, to make sure you are staying on schedule, it's critical to periodically review your list of tasks.

5.1.2.4. Proper Planning

Effective time management is crucial for construction projects to be successful. The foundation of efficient time management in the construction industry is careful planning. It entails outlining the project's layout, determining the labor and material requirements, and estimating when it will be finished. A thorough risk assessment is another essential component of proper planning since it allows for the early detection and resolution of any possible problems. To prevent expensive delays, accurate estimates of the time required for each task should be made. Building strong working relationships with suppliers and subcontractors is also crucial to ensuring that the required supplies and services are provided on schedule and within budget. Projects can be finished on schedule, within budget, and with the least amount of disturbance with the correct planning.

5.1.2.5. Effective Communication

For construction projects to be effective and successful, it is necessary to comprehend their responsibilities and roles. Team members must be able to communicate clearly. They also need to be able to foresee and resolve problems as they come up. Clear communication between all parties involved can help achieve this. There's more space for interpretation and clarity when everyone is in agreement. Furthermore, maximizing productivity and reducing time spent in meetings or reworking plans is possible with a system in place for monitoring progress and updating stakeholders. Effective time management in the construction industry requires effective communication.

5.1.3. Importance of Time Management in Construction

Among other things, adopting effective time-management strategies will help you accomplish the goals discussed in subsections.

5.1.3.1. Maximize the Quality of Construction Projects

Keeping built assets of the highest caliber should always be your first concern. You can establish a solid reputation in the construction sector, draw in new clients, and expand your company by continuously providing for your clients. Effective time management results in higher-quality projects because your team won't be rushing to meet deadlines and will have enough time to finish important tasks.

5.1.3.2. Save Time and Cut Costs

Prioritizing time management can help you save time, reduce project costs, and get rid of waste in your workflows. Throughout the duration of your project, even modest weekly work reductions of one to two hours will add up to substantial savings. There are a number of drawbacks to not utilizing fleet tracking software to your advantage in the construction industry, according to Hill, including workers taking longer breaks than necessary or paying workers excessive wages for time they weren't even on site. Moreover, you can lessen the possibility of scheduling conflicts (such as two crews arriving on the same day), prevent delays on the job site, and improve team collaboration.

5.1.3.3. Accelerate the Journey to Project Success

Making a thorough plan and taking the initiative to manage your time well will also hasten the completion of your project. You can avoid the chance of unforeseen delays and stay within your budget. Because they will have a clear plan that specifies what needs to be done when, your crews will even collaborate better.

5.1.3.4. Build Trust with Clients

Building trust with your clients is essential, particularly when embarking on long-term projects. Building trust with clients through efficient time management in the construction industry will benefit you because it will demonstrate to them that your company is well-organized, dependable, and results-driven.

5.1.3.5. Avoid Construction Delays

Your budget is at risk from even seemingly small delays, which could end up costing you thousands of dollars in needless expenses. A time management plan gives your project a roadmap, which helps you avoid construction delays. You can focus on the task at hand and break down large projects into smaller milestones with the correct plan. You can assign more workers and resources to catch up if you get behind schedule.

5.1.4. Impacts of Poor Time Management in Construction

Poor time management in construction can have severe and far-reaching impacts, affecting not only the project's completion but also its overall success and the satisfaction of stakeholders. One of the most immediate and noticeable consequences is the delay in project completion. When activities are not properly scheduled or monitored, tasks can overrun their allotted time, leading to a cascading effect that delays subsequent tasks. This often results in a failure to meet project deadlines, which can erode client trust and damage the contractor's reputation.

Cost overruns are another significant impact of poor time management. Delays often lead to increased labor costs, as workers need to spend more time on site than originally planned. Additionally, extended project durations can lead to higher costs for equipment rentals and materials, especially if prices fluctuate. The need for rework due to rushed or poorly coordinated activities further exacerbates these costs. In extreme cases, cost overruns can make projects financially unviable, leading to significant losses for the contractor.

The quality of work also suffers under poor time management. When projects fall behind schedule, there is often pressure to expedite tasks, which can compromise the quality of workmanship. Rushed work increases the likelihood of errors and defects, which can lead to safety issues and the need for costly repairs or rework. Poor quality can result in failing to meet regulatory standards and can also lead to legal disputes and penalties.

Safety risks are heightened when time management is poor. In a rush to meet delayed deadlines, workers may overlook safety protocols, leading to a higher incidence of accidents and injuries on site. This not only jeopardizes the well-being of workers but also can result in work stoppages, further delaying the project. Increased accidents and unsafe practices can attract regulatory scrutiny and lead to fines and increased insurance premiums.

Poor time management also adversely affects team morale and productivity. Persistent delays and the pressure to catch up can lead to worker burnout and dissatisfaction. High stress levels and a demotivated workforce can further decrease productivity, creating a vicious cycle of delays and inefficiencies. This can result in higher turnover rates, as skilled workers seek more stable and well-managed projects.

Finally, client relationships and future business prospects are severely impacted by poor time management. Clients expect timely completion of projects, and failure to deliver on time can damage trust and lead to strained relationships. Negative word-of-mouth and poor performance records can make it difficult to secure future contracts, impacting the long-term viability of the construction firm.

Poor time management in construction has dire consequences, including project delays, cost overruns, compromised quality, increased safety risks, reduced worker morale, and damaged client relationships. Effective time management is thus crucial to ensuring project success and maintaining a competitive edge in the construction industry.

5.1.5. The Role of Construction Management Software in Time Management

The construction management software is an essential tool for time management in construction projects because it provides numerous benefits to all parties involved (Liberatore, Pollack-Johnson, & Smith, 2001).

5.1.5.1. Centralized Project Management

Software for construction management offers a centralized system that handles every aspect of the building project. This enables users to collaborate with the team in real-time using shared data, monitor project progress, and access the necessary project information. This method ensures that everyone is working with the most recent information, avoids errors caused by decentralization, and does away with the need for manual issue tracking.

5.1.5.2. Efficient Resource Allocation

Allocating resources in the most efficient way is necessary for effective time management. Construction managers can now efficiently manage labor, tools, and supplies thanks to modern construction management software. Construction teams can more easily address bottlenecks, prevent resource overload, and facilitate project execution when they have a clear picture of available resources and how they are allocated.

5.1.5.3. Streamlined Communication

Effective and unambiguous communication is essential for construction project time management success. Through the provision of tools for sharing project updates, notifications, data, and rapid team communication, construction management software facilitates effective communication. This reduces the need for numerous channels of communication, the possibility of misunderstandings, and the chance of information sharing gaps.

5.1.5.4. Real-Time Tracking and Reporting

Construction managers can track the status of their projects, identify bottlenecks, and make decisions based on project data with the help of construction management software, which offers real-time monitoring and reporting capabilities. The construction team can now benefit from real-time access to project data, which allows them to identify any discrepancies in the project timeline early on and take appropriate action before the project schedule as a whole is impacted.

5.1.5.5. Document and Data Management

Effective time management is largely dependent on the management of the project's data and documents. Plans, specifications, and project documents can all be organized, stored, and retrieved with the help of construction management software. This guarantees that all pertinent data is efficiently retrieved, thereby decreasing the amount of time lost on document searches when needed.

5.1.6. Key Features of Time Management in Construction Management Software

A number of essential features for efficient time management are offered by construction management software. These functions make project procedures simple, automate tedious tasks, and aid in project timeline optimization. Some of the key features and elements of construction management software are discussed in subsections.

5.1.6.1. Project Estimate

Each project starts with an estimate or budget that serves as the foundation for the development. As a project advances, modifications, additions, or deletions to the original project plan increase complexity, necessitating the use of software to manage it where necessary. The estimate takes into consideration every little detail regarding the labor and materials. It helps the builder determine profitability by giving them accurate construction costs in addition to the client.

5.1.6.2. Project Scheduling and Planning

Project scheduling and planning features should be strong points of construction management software. This will facilitate the creation, management, and updating of project schedules that include resources, milestones, and essential tasks linked to a specified work breakdown structure. The Gantt chart, resource leveling, and critical path are examples of additional tools that improve project planning and time management (Mubarak et al., 2015).

5.1.6.3. Task and Workflow Management

Construction project time management is largely dependent on effective task and workflow management. One should be able to assign tasks, track them, and keep an eye on the dependencies that overlap with the construction management software. Workflow management can streamline task assignments, automate approval procedures, and ensure that project activities are carried out on schedule and in the correct order.

5.1.6.4. Resource Management and Allocation

To maximize time management in construction projects, effective resource management is essential. Labor, equipment, and material management functions should be included in the construction management software. This includes managing availability, keeping track of resources, and planning tasks. This software should also support sensible resource allocation, enabling resources to be used effectively and meeting project deadlines.

5.1.6.5. Time Tracking and Timesheets

Time management and project progress monitoring depend greatly on accurate timekeeping. The program should have the capacity to track and record workers' working hours, handle timesheets, and produce timesheet reports based on the various tasks that make up a project. This will help evaluate productivity, identify areas in need of improvement, and ensure that project deadlines are met.

5.1.6.6. Collaboration and Communication

One of the key elements in a construction project's successful time management is effective collaboration. This implies that the construction management software should, of course, allow for the sharing of project updates, enabling team collaboration and communication. Everyone will stay in sync with improved communication tools like document sharing, instant messaging, and commenting.

5.2. ASPECTS OF TIME MANAGEMENT

A construction project has time constraints, and those deadlines must be fulfilled in a constantly changing environment. In order to achieve this, the time management process depicted in Figure 5.1 is repeatedly carried out during the project's duration. The time management system is supported by the current operating plan and schedule, which are compliant with project time constraints that have been established. The most recent version of the job network is used to calculate forward- and backward-passes, which result in a work schedule with calendar dates for the beginning and end of each project activity. The project's daily time management is done using this schedule. When and where the project may be running behind schedule, such a system serves as a useful early-warning tool. If project objectives are to be successfully achieved, the work plan must adapt to changing conditions. An established system of progress feedback from the field must be used to quickly identify time slippages if the overall time schedule of a construction operation is to be met. Time management's monitoring phase entails measuring actual job progress in the field on a regular basis and comparing it to the predetermined goals.

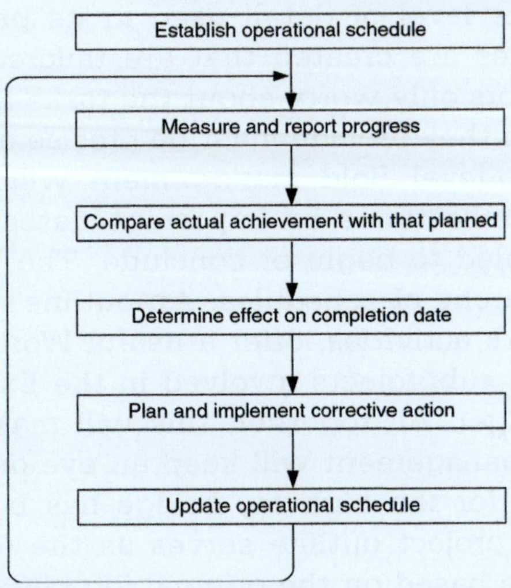

Figure 5.1. Time management cycle.

Source: http://dspace.vnbrims.org:13000/jspui/bitstream/123456789/4250/1/Construction%20Project%20Management%20A%20Practical%20Guide%20to%20Field%20Construction%20Management.pdf.

Therefore, determining the amount of work implemented and reporting this information in a format that allows for comparison with the planned job schedule constitute project monitoring. Network activities provide a practical and helpful foundation for tracking and reporting progress.

The progress of the project is tracked and reported on a regular basis. Notes are kept on the activities that have been finished and the level of completion of the ongoing tasks as of each deadline. Project management reviews this data to determine where and how much the project is currently behind schedule. Because they are crucial to the project's strategic success and have low float values, critical activities are closely watched. After options are identified through analysis of the reported progress data, corrective action is taken to expedite lagging work items. There can never be a perfect project plan or schedule, and as the project moves forward, deviations will unavoidably occur. As a result, as modifications, slippages, and other logical and schedule aberrations happen, the baseline version of the schedule will grow more and more erroneous and unrealistic. In order to ensure that the current work schedule accurately reflects actual job experience to date, the network must be adjusted as necessary and computations must be updated on a regular basis. These updates frequently show changes in critical paths and significant adjustments to the activity floats. The most recent revised schedule serves as the foundation for project time control and accurately depicts the state of the work at hand.

5.2.1. Key-Date Schedules

The person using the information must be taken into account when creating time schedules for project control. A hierarchy of schedules might be required for a big project like the Example Project. Depending on the level of management for which a job schedule is intended, the level of detail used in its preparation can vary greatly. As a result, various schedules are created that are tailored to the recipient's specific requirements. Craft supervisors only worry about the tasks that fall under their purview, so the scheduling information they provide must be precise and comprehensive. Owners, architect-engineers, and top-level field management will typically receive schedule information in the form of milestones or important dates that indicate when major project segments are scheduled to begin or conclude. The project outline can be used as one way to create a hierarchy of schedules. An outline's levels, which are made up of different groups of network activities, offer a useful Work Breakdown Structure. Key dates for each of the several subprojects involved in the Example Project will probably be sent to the project manager. All together, this will make up a master schedule of time objectives that senior management will keep an eye on while work is being done. An initial key-date schedule for the highway bridge has been prepared and is shown in Table 5.1. Level 2 of the project outline serves as the foundation for this key-date schedule. The presentation is based on the original job plan and a period of 70 working days. These activity times are often referred to by different names by contractors and scheduling software. For instance, early starts and ends are frequently denoted as "scheduled" starts and ends. In a similar vein, late starts and finishes are usually indicated as "needed."

Table 5.1. Highway Bridge, Initial Key-Date Schedule

Key-Date Schedule								
Project Highway Bridge					Project No. 200808-05			
		Scheduled		Required		Actual		
No.	Operation	Start	Finish	Start	Finish	Start	Finish	
1	Procurement	Jun-14	Jun-19	Jun-14	Jun-19			
2	Field mobilization & site work	Jun-25	Jul-14	Jun-30	Jul-19			
3	Pile foundations	Jul-6	Jul-20	Jul-9	Jul-30			
4	Concrete abutments	Jul-20	Aug-17	Jul-20	Aug-30			
5	Deck	Aug-13	Aug-30	Aug-13	Sep-7			
6	Finishing operations	Aug-31	Sep-13	Aug-31	Sep-13			

Source: http://dspace.vnbrims.org:13000/jspui/bitstream/123456789/4250/1/Construction%20Project%20Management%20A%20Practical%20Guide%20to%20Field%20Construction%20Management.pdf.

5.2.2. Detailed Schedules

The immediate work supervisors require a great deal more information than what was shown in Table 5.1. An extensive timetable that offers a daily forecast of field operations will be necessary for the highway bridge site superintendent. Activities are often used as the foundation for creating comprehensive project schedules. Regarding the activities in these schedules, different times can be given; however, early start and end dates are typical. This kind of schedule is highly optimistic, and there are typically plenty of situations in which it will not be fulfilled due to scarce resources, unavoidable delays, and scheduling errors. However, the most common method for determining project time objectives is to use a project schedule that is predicated on activity early starts. The comprehensive work plan ought to specify the essential tasks and provide information on the free float values. For those who oversee the work in the field, this information is extremely important. Designating an activity as critical emphasizes to managers at lower levels how important it is. Understanding free float is also beneficial because it allows field supervisors to use that extra time to handle unforeseen circumstances on the job.

It's not always thought of as best practice to provide field supervisors with access to total float values. Unlike free float, total float is typically shared by strings or groups of activities. The float values of other activities are typically impacted when total float is used in conjunction with one activity. Total float data may not be accurate for people working on limited project components. The only person in charge of overall scheduling is able to assess and manage how total float is used and distributed. While total float needs to be carefully distributed by experienced planners, free float is a resource that can be used immediately.

For every craft supervisor and subcontractor on a large project, a detailed schedule is prepared. The estimated time schedule for the tasks that each of these schedules represents is the responsibility of that specific supervisor. These schedules are very short, usually spanning the next two weeks to 30 days. The type of work involved will determine how long it takes in large part. Developing comprehensive work schedules more than a month ahead of time is rarely beneficial. Longer periods of time are unnecessary if a large portion of such a schedule could be superseded by later updates and modifications. Schedules are released as work progresses, revised, and updated. The most popular formats for short-term work schedules are computer-printed bar charts and tabular listings. The network diagram is unquestionably an essential supplement to the work schedules. The various supervisors may receive copies of the network, or it may be retained in the field office.

Due to its small size, the highway bridge only needs one comprehensive work schedule, which is for the job superintendent. Table 5.2 displays the schedule of events for the first month of the highway bridge construction. The job superintendent will update the Table 5.2 schedule with the dates that the activities actually begin and end as the work is being done in the field. Field supervisors are given work schedules such as the one depicted in the figure, which may or may not contain details about the delivery of job supplies.

Table 5.2. Highway Bridge, Initial Detailed Schedule

Detailed Schedule

Project Highway Bridge
Project No. 200808-05

Prepared by G.A.S.
Date June 14

Activity	Activity Number	Scheduled Duration (Working days)	Scheduled Start Date	Scheduled completion Date	Free Float (Working days)	Actual Start Date	Actual Completion Date
Move in	40	3	25-Jun	29-Jun	0		
Prefabricate abutment forms	80	3	30-Jun	2-Jul	10		
Excavate abutment #1	90	3	30-Jun	2-Jul	0		
Mobilize pile-driving rig	100	2	30-Jun	1-Jul	1		
Excavate abutment #2	120	2	6-Jul	7-Jul	1		
Drive piles, abutment #1	110	3	6-Jul	8-Jul	0		
Forms & rebar, footing #1	130	2	9-Jul	12-Jul	0		
Drive piles, abutment #2	140	3	9-Jul	13-Jul	0		
Pour footing #1	150	1	13-Jul	13-Jul	0		
Demobilize pile-driving rig	160	1	13-Jul	13-Jul	3		
Strip footing #1	170	1	14-Jul	14-Jul	0		
Forms & rebar, footing #2	190	2	15-Jul	16-Jul	0		
Pour footing #2	210	1	19-Jul	19-Jul	0		
Strip footing #2	230	1	20-Jul	20-Jul	8		
Forms & rebar, abutment #1	**240**	**4**	**20-Jul**	**23-Jul**	**0**		
Critical activities in bold							

Source: http://dspace.vnbrims.org:13000/jspui/bitstream/123456789/4250/1/Construction%20Project%20Management%20A%20Practical%20Guide%20to%20Field%20Construction%20Management.pdf.

Although practices differ in this area, the supervisor is still very interested in knowing what materials are ordered and when they will arrive, so this information does not make them unimportant. But material delivery scheduling is typically managed independently of field operations. Periodic reports regarding the condition of job materials are given by the project expeditor to the project manager and site superintendent; these reports are then included in the weekly progress reports for the project. As a result, work schedules created for field supervisors frequently overlook material delivery information and only include tasks that are physically involved in the job. Since this is the process that is being followed here, Table 5.2 does not contain any material delivery information.

5.2.3. Progress Measurement

Network activities are incredibly handy packages of work for measuring progress on a regular basis in the field. The advancement of an activity in progress can be expressed in different ways. Three commonly used methods are:

1. Estimated number of working days remaining to complete the activity;
2. Estimated percentage completion of the activity in terms of time;
3. Quantities of work units put into place.

How a contractor chooses to express activity completion depends on the type of work involved and whether the same data are also used to check field costs. However, the number of working days remaining to finish an activity is fundamental to the workings of project time management. Progress data in other forms are readily converted into days to complete by using these relationships:

$$\text{Working days to complete} = d\,(1 - P/100)$$

$$\text{Working days to complete} = d\,(1 - W/T)$$

where; 'd' is the total activity duration in working days; 'P' is the estimated percentage of completion; 'W' is the number of work units put into place; and 'T' is the total number of work units associated with the activity.

The majority of scheduling programs enable users to enter progress in any of these formats, and they automatically convert them. The relationships are predicated on the idea that time and work accomplishment vary linearly. It is expected that 25% of the work quantities to be installed will be finished each day in an activity that takes four days to complete. If an activity has a narrow scope, this assumption is typically valid. Where activities are carried out to a significant degree, a more realistic relationship between time and work accomplishment may need to be used. It is clear that the quality of the input data determines how effective a time management system is. Decisions about management will be based on made-up scenarios if field reports on progress are not accurate. Progress measurements must be completed diligently and with a reasonable degree of accuracy. Rather than focusing on what ought to have happened, management decisions must be made based on what actually occurred. It is imperative that the person in charge of progress reporting understands the significance of accurate and factual conclusions. In order to resolve disagreements about project delays in the future, project progress records are frequently crucial. Progress reports actually serve as the foundation for lawsuits and claims.

When evaluating the conclusion of an activity, it is important to consider its substantial completion rather than its absolute finalization. In many cases, there may be minor deficiencies that are addressed at a later time when the opportunity arises. Progress reports should take into account these temporary shortcomings, and finish dates are typically recorded based on substantial completion rather than technical completion of activities.

5.2.4. Progress Reporting

How much time control is thought to be desirable and practical for the specific task at hand will determine how frequently field progress should be measured and assessed. Within reason, there's a positive correlation between the frequency of feedback and response and the likelihood of meeting project time objectives. This rule needs to be balanced with other factors, though. For instance, a balance needs to be established between the effort and expense incurred and the management advantage realized. Another thing to keep in mind is that field cost management and time management are frequently handled by the same field progress report. As a result, both management applications' cycle times frequently match. Projects with a high pace and many shifts might need daily progress reports. Large-scale projects with a narrow scope of work, like building earth dams, may require reporting every month or even longer. Because management control must be consistent with the unique features and characteristics of each project, generalization is challenging. Progress reports on projects the size, scope, and nature of the highway bridge would likely be completed once a week. This is the idea at play here. It should be noted that while official progress reports might be sent out on a weekly or monthly basis, more frequent and quicker forms of cost and production control might be required to monitor critical operations.

The project manager is responsible for ensuring accurate progress measurement, reporting, and timely management review and analysis of the progress data. It is necessary to establish a standard operating procedure for gathering and sending the weekly progress data. Measuring progress in the field necessitates direct visual observation by an individual knowledgeable about the nature of the work being done. This could be as simple as calculating the actual number of work units or as complex as estimating the amount of work completed based on the project drawings. Sometimes delivery tickets for materials like concrete or load slips for earth moving can provide an appropriate measurement. The measurement and reporting tasks are frequently completed by the project manager in person. If not, a staff member—like the field engineer—will carry out these tasks. In any case, it is preferable to have an impartial evaluation of the work completed. Due to their hectic schedules and aversion to handling this type of paperwork, field supervisors are typically not the best candidates for progress reporting. Furthermore, a field supervisor might withhold supplies for use on a less fortunate day or neglect to report unfavorable progress in the hopes of resolving issues later.

5.2.5. Bar Charts

Management must verify the data with the most recent project schedule after receiving field-based progress measurements. Various approaches can be taken, based on the policies and preferences of management. For each activity, for instance, a tabular listing displaying the actual and scheduled start and end dates can be created. While progress data in this format might be helpful in some situations, it is typically not the most effective way to conduct a thorough assessment of the state of construction operations at the moment. A graphic display of some kind is useful and practical for managing a project's daily time. Bar charts are a popular tool for tracking job progress because of their general features. The bar chart serves as a highly effective tool for documenting progress data and illustrating the present state of individual tasks or other project components. It is not, however, a suitable tool for planning corrective actions when work lags behind schedule or for assessing the project's overall time status.

5.2.6. Progress Analysis

Determining the impact of this most recent information on the project completion date and any intermediate time goals that have been set is the main focus of the job progress analysis. The primary goal of the time management system is, after all, the successful completion of predetermined time objectives. The fact that the critical path's length affects how long it takes to reach a particular network event has received a lot of attention. As a result, practically speaking, the main goal of progress data analysis is to ascertain the precise length and location of the relevant critical path at this time.

The status of the currently identified critical activities is probably the first thing to note when a progress report is received from the field. This check is short and straightforward. Insofar as the current critical path is concerned, there is no issue with the project time goal if these crucial tasks have been completed by their planned completion times. If a crucial task is started later than planned, there may be a delay in finishing the project unless the missed time can be made up in some way. The completion date is postponed proportionately if a crucial task is completed after the deadline.

Analyzing whether a new critical path has formed is the next step in the process. You can accomplish this by putting the noncritical activities through one of two distinct checks. Using a late-start activity sort or a late-finish activity sort is one of these checks. An activity listing arranged in accordance with the latest permitted starting times (LS) is known as a late-start sort. Comparable in terms of the latest permitted finish dates (LF) is a late-finish sort. These kinds can be acquired manually or automatically. The project is behind schedule if an activity has not been completed by the deadline specified in the current action plan. A new critical path is created and the job is delayed by the amount that the completion of the activity trailed the LF time if it did not finish by the LF date.

Total float (TF) values can be used in a similar manner to perform the check. The TF of an activity is decreased by the same amount if it takes longer to complete than its early-finish (EF) time. Another critical path forms if the delay equals the TF. The project's completion is automatically delayed by the amount of the overrun if more than the TF is used. This results in a new critical path that is longer than the initial one.

5.2.7. Corrective Action

A determination of the necessary corrective action, if any, must be made following the analysis of each progress report. Small delays—actual or possible—typically don't call for special corrective measures as long as the project's duration accounts for the customary contingency allowance. The problem with driving the piles for the abutment will be of great concern because the highway bridge schedule only allows for a contingency of six working days. When there are deviations from the estimated duration of the network time contingency, project managers need to seriously consider implementing corrective measures. In general, remedial steps are indicated when situations like these arise:

- Activities begin to fall appreciably behind their late start or late finish dates.
- Substantial delays of resource availability are indicated.
- The time durations of future activities appear to have been materially underestimated.
- Logic changes in work yet to be performed have become necessary.

If remedial action is required, it must be taken quickly to ensure that the intended goal is met. One way to reduce the critical path, or paths, is to either compress individual tasks or rework the network logic locally. As before, this frequently entails experimenting with potential different approaches. Management is only provided with a broad understanding of the job's time status by the examination of the weekly progress reports if network calculations have not been updated recently. As such, unless a corrective measure is simulated by integrating it into a network update, its effectiveness can only be estimated.

Regular job progress meetings can significantly improve the project time management process as a whole. These site conferences take place every week, every two weeks, or every month. Project managers, field supervisors, large subcontractors, material suppliers, owner representatives, and other relevant parties attend. While these meetings can and often do provide a forum for discussing a range of work-related issues, one of the main factors taken into account at these sessions is the current state of construction operations. Meetings in person can be highly effective in generating ideas and gaining cooperation from those who are crucial to completing the work on time.

5.2.8. Network Updating

Divergences from the predetermined plan and schedule are inevitable when construction moves forward. Changes in project logic, activity delays, and durations are the result of unforeseen job circumstances. Similar results can be obtained from corrective measures like time expediting and resource leveling. The actual job status deviates more and more from what is predicted by the planned plan and schedule as these deviations happen and mount. Therefore, if the working operational program is to continue offering practical management guidance, it will occasionally need to be updated to reflect the changes and deviations. This is achieved through a process known as network updating. An update's primary goal is to rearrange the work that still needs to be done, starting with the project's current status. Updating shows the job's current time posture, suggests whether expediting measures are necessary, and offers suggestions for how to maintain the job's timeliness. In order to evaluate the efficacy of suggested time-recovery measures, an update is also very helpful. Updating entails recalculating activity and float times as well as performing any necessary network corrections. Its sole focus is on figuring out how schedule delays and plan modifications will affect the remaining project portion. Included here are both unanticipated program withdrawals and remedial measures taken to address particular issues with resource availability and time progress.

Only from the perspective of time management efficacy, it might be helpful to provide an update with every weekly progress report. It would guarantee timely and informed remedial action when necessary and keep job management continuously informed about the status of the work. These updates, however, can be costly and time-consuming. When it comes to network update timing, there are no hard and fast rules. Recalculation is required based less on the quantity of schedule and plan deviations and more on their severity. Maintaining an antiquated schedule and plan beyond a certain point will no longer yield benefits. Making the project fit into an out-of-date program is a fruitless endeavor that will actually cause more harm than good. For time control to function at its best, it needs to be founded on an accurate and up-to-date job model. The degree to which the plan and schedule continue to align with the way the work is actually done is more significant than the frequency of network recalculations. The network needs to be adjusted to reflect the most recent data regarding the logic and times of all work that still needs to be done before an update can be released. This is a crucial point to make. The completed tasks are now part of history and cannot be changed. Starting from the current project status as of the specified cutoff or data date, new calculations must be made to ascertain the time condition of the remaining activities.

To make a network update, information as of the cutoff date is required concerning the work that has yet to be completed for:

- New activities that must be added to the network;
- Existing activities that are to be deleted;
- Changes in job logic;
- Changes in original material delivery or other resource availability dates;
- Estimated times to complete all activities presently under way but not yet completed;
- Changes in estimated activity durations;
- Changes in the scope of the work.

Subjective feedback from the job expediter and field supervisors can be especially helpful when it comes to the information required for a project update. For example, based on recent work experience, the project manager can provide updated information about future progress. This data could be related to future logic modifications or anticipated activity durations. Information about labor disputes, business disruptions, and other relevant factors relating to supplier plants and transportation facilities may be input by the expediter. Conditions that weren't anticipated when the original plan was created might make it feasible to expedite deliveries by combining material shipments for one job with those of another. The revised schedule must take all of this information into account.

5.3. TIME SCHEDULE ADJUSTMENTS

Time schedule adjustments are an integral part of managing construction projects, ensuring that projects stay on track despite inevitable changes and unforeseen circumstances. These adjustments require a proactive approach, continuous monitoring, and a clear understanding of the project's critical path. The critical path method (CPM) is particularly useful in identifying which tasks directly affect the project completion date, allowing managers to focus on these key activities when making schedule adjustments.

One of the primary reasons for adjusting the time schedule is dealing with delays. Delays can stem from various sources, such as adverse weather conditions, supply chain disruptions, design changes, or unexpected site conditions. When delays occur, it's essential to reassess the project schedule to minimize their impact. This often involves re-sequencing tasks, reallocating resources, or extending work hours to catch up. For example, if a critical delivery is delayed, other tasks that do not rely on this delivery can be prioritized to keep the project moving forward.

Resource leveling is another technique used in time schedule adjustments. This process involves redistributing tasks to avoid overloading specific resources or crews. By balancing the workload, construction managers can prevent bottlenecks and ensure a steady progress rate. Resource leveling can also help in accommodating changes in resource availability, such as workforce fluctuations or equipment breakdowns, by adjusting the schedule to match the available resources.

Incorporating buffer times or contingency plans into the original schedule is a proactive strategy that makes future adjustments easier. These buffers act as a cushion for unplanned delays, allowing the project to absorb minor setbacks without significant schedule disruption. When such buffers are utilized, it's essential to review and adjust them regularly to ensure they remain effective throughout the project lifecycle.

Effective communication and collaboration among project stakeholders are crucial when adjusting schedules. Regular progress meetings, clear documentation, and open lines of communication ensure that everyone is aware of schedule changes and their implications. This collective understanding helps in aligning all parties towards the adjusted timelines and prevents miscommunications that could lead to further delays.

Technological tools, such as construction management software, play a vital role in facilitating schedule adjustments. These tools offer real-time tracking and analysis of project progress, making it easier to identify when and where adjustments are needed.

They provide a centralized platform for updating schedules, notifying stakeholders, and reassigning tasks efficiently.

In some cases, schedule adjustments may require formal contract amendments, especially when changes affect the project scope or timeline significantly. It's important to manage these adjustments legally and contractually to avoid disputes and ensure all parties are on the same page regarding the new schedule and any associated costs.

Time schedule adjustments in construction are essential for managing delays, resource fluctuations, and unforeseen events. By using techniques such as resource leveling, incorporating buffer times, leveraging technology, and maintaining effective communication, construction managers can navigate these adjustments effectively. These practices ensure that projects remain on track, meet deadlines, and stay within budget, ultimately leading to successful project completion and satisfied stakeholders (Vanhoucke et al., 2013).

5.4. NEED FOR TIME REDUCTION

There are numerous real-world scenarios where cutting the duration of specific job components can prove advantageous in terms of meeting significant project deadlines. For example, the owner may set a deadline for job completion that the current project plan cannot meet due to the terms of the construction contract. If the contractor fails to meet this time commitment, they will be in breach of contract and liable for any damages the owner incurs from the project's delayed completion. During a project, the owner may wish to expedite the work beyond what the contract initially specifies and request a quote from the contractor. It is possible that the contractor's needs may not align with the project's scheduled duration. For various reasons, such as avoiding bad weather or completing work before the annual spring runoff, the contractor may aim to finish the job by a specific date to free up personnel and equipment for other projects. Financial arrangements may require completing specific tasks within a given fiscal period. The prime contractor may aim to finish the project earlier than expected to receive an early completion bonus from the owner. Time acceleration due to delays causing a significant loss of time that must be made up by the project's completion is a common reason for expediting work.

Even though it doesn't involve the entire project, trying to reach a set milestone can lead to a similar circumstance. The computed early times of milestone events frequently happen later than intended. Similar circumstances can occur in relation to events that occur on network interfaces. One of the many benefits of being able to create advance construction schedules with a reasonable degree of accuracy is evident from these examples. With this kind of information, the project manager can identify specific time issues well in advance and take the necessary corrective action. This is undoubtedly better than being unaware of these issues until it is too late to take significant action.

At first glance, thinking about cutting a project's duration after adding a contingency allowance to the estimated normal time may seem contradictory. A six-day contingency was added to the 64-day estimate for the highway bridge project in order to arrive at a likely completion time of 70 working days. This project time represents the most accurate preliminary estimate of the total amount of work that will be needed. This means that the highway bridge will probably take 70 working days to complete. Whether or not the project needs to be shortened should be determined by how this time compares to the agreed-upon contract period or other time constraint. Eliminating the contingency from the project will not guarantee an earlier completion date.

5.4.1. General Time-Reduction Procedure

It might be appropriate at this point to point out that the process of cutting project time durations is referred to by a number of different names. The terms "project compression," "least-cost expediting," and "time-cost trade-off" are all used in relation to the methods covered in this chapter. It doesn't really matter what the exact terminology is called long as you understand the method and how to use it. It is unfortunate that the term "expediting" is being used in reference to cutting project time requirements when, in fact, the same term is being used to describe steps taken to guarantee timely resource support for construction operations. Nonetheless, the term "expediting" has two meanings, which are commonly used in the construction sector and are used here. One need only worry about cutting the durations of a specific set of activities in order to reduce the amount of time needed to reach a milestone, interface event, or to complete the project. As demonstrated previously, the longest path from the current project advancement stage to any future network event—terminal or not—determines the time required to reach that event. Therefore, the only way to shorten the time needed to arrive at a particular event is to cut down on the lengthiest path that leads there. This is a really significant and illuminating observation. When a project is running behind schedule and such management information is unavailable, the typical response is to haphazardly speed up all ongoing activities in an attempt to make up for the lost time. Such expediting actions may be far more costly than necessary due to the inability to distinguish between those activities that truly control and those that have little to no time consequence.

The network critical path needs to be shortened if the project completion date needs to be moved forward. In order to shorten the time required to complete a milestone or interface event, there may be no involvement at all from the diagram critical path. It is necessary to shorten the longest time path that leads to the milestone or interface event. The critical path that serves the entire network may not be connected to this path at all. When discussing how to shorten the project's duration, the event's critical path is the one that takes the longest to occur. It is important to note at this point that shortening the longest path results in a corresponding reduction in the floats of other activity paths that lead to the same event. Therefore, it is inevitable that new critical paths and critical activities will eventually form as a result of the original critical path being continuously shortened. If there are several critical paths involved, they must all be shortened at the same time in order to move the event forward as planned. All that is achieved when one critical path is shortened but not another is to give the shortened path extra float that is not required.

When time is reduced manually, it is necessary to assess the impact of each action taken to see if it resulted in the creation of new crucial tasks. Typically, this is accomplished by recalculating the network after each stage of the time-reduction procedure. Recalculations of this kind can be performed manually or by computer. In small networks, manual computations can be quick and easy. Numerous iterative recalculations can grow to be a significant chore for a large network.

5.4.2. Shortening the Longest Time Path

It is now established that the longest time path leading to an event must be shortened if the date of a specified project event is to be advanced. There are essentially just two methods for achieving this. One is to alter the job logic in a way that reduces the length of the longest route. In order to achieve this, the initial job plan must be locally revised. Time can be gained by changing the sequence in which job activities are completed or by lengthening lead and lag times in order to maximize activity concurrency. This time-saving technique saves time by choosing certain job operations in a more advantageous order, rather than shortening the durations of the activities themselves. Reducing the duration of one or more of a critical path's component activities is another potential strategy for shortening its length. To determine whether any critical activities can be shortened, they must first be examined. Depending on the nature of the activity, there are several ways to compress it. It may be necessary to employ extra crews, overtime, or multiple shifts. You could be able to subcontract it. It may be necessary to temporarily bring in more equipment and designate it for that task. Using air freight or special handling, allowing the fabricator to work overtime, or dispatching one of the contractor's trucks to pick up and deliver the material are some ways to get the material delivered earlier. Naturally, there are some activities whose durations cannot reasonably be shortened.

5.4.3. Project Direct Costs

Before the project's time can be further reduced, the nature of its direct and indirect costs must be discussed. Since building costs are a function of time, costs must be associated with time reduction. Although occasionally management discretion may require a different approach, efforts are typically made to maximize project cost reduction while achieving gains on schedule. It is essential for project managers to comprehend how construction operations costs change over time in order to make schedule modifications at the lowest possible expense. The price of labor, tools, supplies, and subcontracts make up an activity's direct cost. Every activity has a typical price and duration. When estimating a job, the "normal cost" is the least amount of money that must be spent directly on the task at hand. It is also the cost that is typically assigned to the work. "Normal duration" is the amount of time that an activity takes, as decided upon in the scheduling stage. A normal time is not exact, but it is still a fairly distinct datum or reference point for completing a task at the lowest possible direct cost. It is clear that the normal project duration is determined by adding the normal durations of the individual activities, and that the direct cost of the entire project is the same as the total of those costs. Thus, the cost related to the typical project duration will be the lowest overall direct cost of the entire job.

If the estimated activity times are accepted as being closely related to the minimum direct cost, then any deviation from the estimated activity times—whether by more or less—must lead to an increase in the direct cost that is proportionate. This rule's practical applicability is much more certain when it comes to the higher costs incurred when an activity is shortened as opposed to prolonged. On the other hand, contractors almost never worry about purposefully prolonging the time of a job. It is a well-established practical fact that reducing the duration of an activity will typically result in an increase in its direct cost. It goes without saying that working overtime or multiple shifts incurs additional labor costs. Increasing the number of work crews or pieces of equipment results in higher production costs, less operational efficiency, and more difficulty supervising the job. Early material delivery entails paying the vendor premiums or incurring higher handling and transportation expenses. All of this points to the conclusion that if the project's time requirement is to be lowered, the activities that are actually shortened will typically result in higher direct costs.

5.4.4. Variation of Activity Direct Cost with Time

Although there are many ways in which the direct costs of an activity can change over time, in this case, it is assumed that these costs always change in an inverse manner. One typical example is a continuous linear, or straight-line, change in direct cost over time. This is the outcome of an expediting measure where the additional cost for each day gained is nearly constant, like working overtime or multiple shifts. In Figure 5.2(a), the typical activity duration of 15 days can be shortened by up to 3 days. This activity's constant $100 daily increase in direct costs is known as the activity's cost slope. The contractor has the option to expedite the work by one, two, or three days at an additional cost of $100, $200, or $300, accordingly.

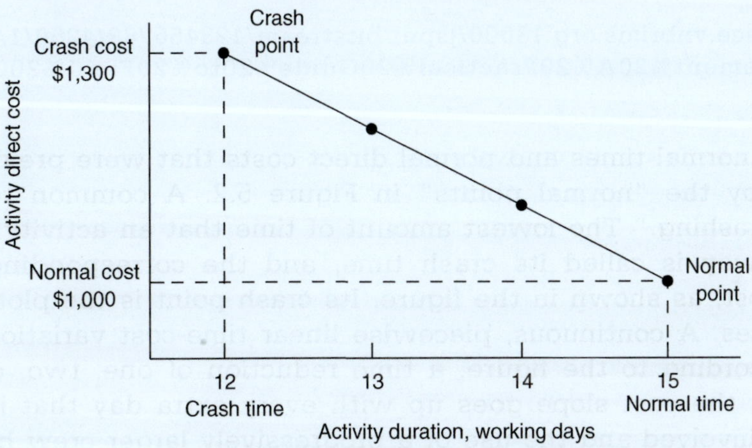

(a) Continuous Linear Variation

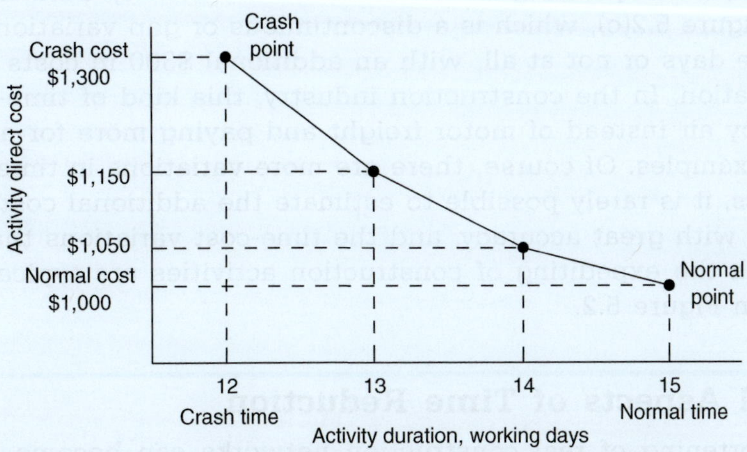

(b) Continuous Piecewise Linear Variation

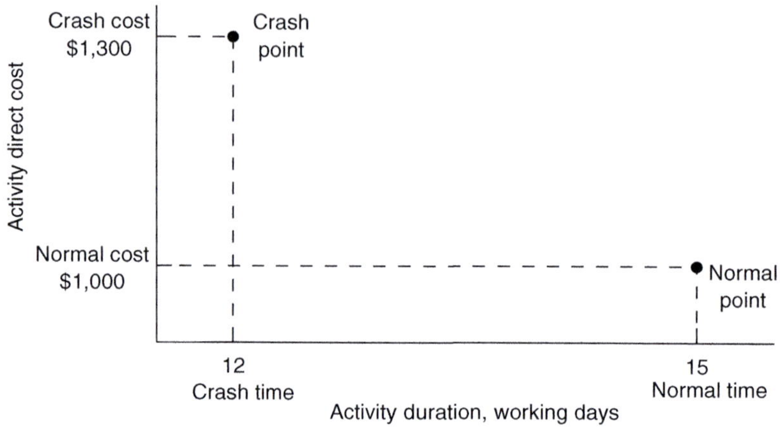

(c) Noncontinuous Variation

Figure 5.2. Activity time-cost variation.

Source: http://dspace.vnbrims.org:13000/jspui/bitstream/123456789/4250/1/Construction%20Project%20Management%20A%20Practical%20Guide%20to%20Field%20Construction%20Management.pdf.

The activity's normal times and normal direct costs that were previously discussed are represented by the "normal points" in Figure 5.2. A common term for activity acceleration is "crashing." The lowest amount of time that an activity can be reduced to in a realistic way is called its crash time, and the corresponding direct cost is called its crash cost, as shown in the figure. Its crash point is the plotted intersection of these two values. A continuous, piecewise linear time-cost variation is depicted in Figure 5.2(b). According to the figure, a time reduction of one, two, or three days is feasible; however, the cost slope goes up with every extra day that is gained. When crew overtime is involved and the use of a progressively larger crew becomes feasible as the work progresses, this is an example of a piecewise linear variation. With no time options in between, the expediting action shortens the activity duration from 15 to 12 working days in Figure 5.2(c), which is a discontinuous or gap variation. The activity is shortened by three days or not at all, with an additional $300 in costs associated with the shortened duration. In the construction industry, this kind of time-cost variance is typical. Shipping by air instead of motor freight and paying more for an early material delivery are two examples. Of course, there are more variations in time-cost that could occur. Nevertheless, it is rarely possible to estimate the additional costs resulting from expediting actions with great accuracy, and the time-cost variations that are employed in conjunction with the expediting of construction activities are typically restricted to the three shown in Figure 5.2.

5.4.5. Practical Aspects of Time Reduction

The least-cost shortening of real construction networks can become a very intricate process. It is possible for multiple critical paths to emerge, which would complicate the shortening process further. In order to find an ideal solution, a very large number of expediting combinations may need to be tested. It is imperative to acknowledge that the standard manual time reduction approach may not always yield truly optimal expediting combinations for project management. But in a process like this, mathematical accuracy in the face of imprecise data is neither the only nor always the most crucial factor to take into account. Naturally, achieving time reduction in practice requires taking into account a variety of pragmatic factors in addition to the issue of getting the most time for the least amount of money.

Even though they might not be the best, manual methods for cutting project duration do give the project manager important insight into whether expediting is feasible and, if so, how to move forward. Generally speaking, advice on how to choose time-reduction strategies wisely is just as useful as a theoretically ideal answer. Construction is just not an exact or totally predictable process, conditions vary daily, and input data are imprecise. As work is done, even a network's critical path may occasionally alter its routing. Instead of aiming for perfection, project managers look for workable, reasonable solutions. When a project is expedited manually, value judgments can be added to the process and the project manager can get a feel for how actions will affect other aspects of the project intuitively. Furthermore, a critical assessment of the time saved by reducing individual tasks as well as by making changes to the job logic can be simply incorporated into a project time-reduction study. The goal of the manual project time reduction effort is to shorten the relevant critical path or paths as much as possible. This is a methodical procedure that makes use of time-saving techniques deemed most appropriate for the particular work environment. These could involve modifying the work schedule, cutting back on specific tasks, or doing both. Typically, the process aims to maximize time increments while minimizing direct cost increases. In cases where other job-related factors take precedence over incremental costs, the sequence in which shortening steps are implemented is determined by project management to be optimal for the project as a whole.

5.4.6. Reduction of the Highway Bridge Duration

Let's say, for the sake of illustration, that the highway bridge contractor finds the estimated duration of 70 working days or 98 calendar days to be inadequate. This job has not yet started, and research will be done to see if it is possible to shorten the project's duration by up to 10% while also examining the associated costs. Of course, the most important question is how to shorten the project's critical path from its current duration. First, common sense dictates that the current operating plan be reexamined, with the potential goal of gaining time without incurring additional direct costs. Even though the job plan was meticulously created by skilled and informed individuals, it is unrealistic to expect the initial planning to be flawless. A reexamination of the operational sequence is probably going to help refine the planning strategy and possibly point out areas where more time efficiency can be achieved. The contractor will be forced to forgo project cost in favor of time if reexamining the original job logic fails to yield the desired time gain at no additional cost. Almost any construction task can be completed faster if someone is prepared to incur the extra cost. There are definitely several ways to reduce the length of the highway bridge, either by altering the logic of the job or by cutting down on the amount of time required to complete specific crucial tasks. Project shortening when additional costs are involved is accomplished by assessing the viable options and typically selecting the least expensive combination of those that will result in the required time adjustment.

The following five sections aim to provide detailed discussions of potential ways to shorten the highway bridge's overall duration at little or no additional direct cost to the contractor. Although there's no guarantee that such a reexamination of the planned plan will produce a time gain, it's a possibility worth exploring. However, a word of caution is in order. There is always a risk of becoming overly optimistic when such restudies are conducted to gain some much-needed time. Those who decide how to shorten the project's duration must make thoughtful, practical decisions.

5.4.6.1. Restudy of Critical Activity Durations

Examining the time estimates of each critical activity is a clear first step to take when reexamining the critical path of a project that needs to be shortened. Since mistakes can happen, it's important to confirm that the initial time estimates were reasonable. It is also possible to review an additional option. Which of the tasks would turn out to be crucial was unknown when the initial time estimates were created. It's possible that the initial time estimates for some tasks took into account the potential shortage of laborers or construction equipment in the future. Because of this, some of the activity duration estimates for what turned out to be important work items were based on crews or equipment spreads that were not as large as they should have been. Now that the critical activities have been identified, it might be possible to temporarily reallocate resources to the critical activities in order to complete them faster while postponing action on some noncritical activities and utilizing their floats.

5.4.6.2. Restudy of Project Plan

A re-examination of the project with the goal of gaining time at no extra direct cost is essentially a critical second look at the current operational plan. Of course, the goal is to shorten the critical path by reworking or improving the logic of a specific portion of the network. Here, new ideas and a creative approach could lead to significant advancements in working techniques. All too frequently, conventional wisdom dictates that there is no other option for resolving issues in the field. Occasionally, they ought to be questioned by curious minds looking for a better answer. Sometimes a good old-fashioned brainstorming session yields some creative ideas for novel approaches. Depending on the situation, the contractor may be able to alter the project's materials or design without the owner's consent. A project manager may choose to shorten the project's duration overall, for instance, in a design-construct contract, by redesigning a portion of the project to benefit from a quicker construction technique. While most opportunities to shorten construction time and cut costs arise during the design phase of a project, this is rarely achieved with a standard design-bid-build contract.

5.4.6.3. Critical Activities in Parallel

One potential modification to the project plan that could shorten the critical path would be to carry out some critical tasks in parallel or concurrently rather than sequentially. The highway bridge's time-scaled diagram is cited as an example of how this might operate. Examining the crucial tasks will reveal whether painting can be completed simultaneously with deck form stripping, as opposed to after. This modification to the project schedule is depicted in Figure 5.3, where it is possible for activity 370, "Paint," to begin concurrently with activity 350, "Strip deck," thereby reducing the critical path by two days. Painting would no longer be critical if this were done; instead, as Figure 5.3 illustrates, the guardrail installation would replace the painting activity as the activity on the critical path.

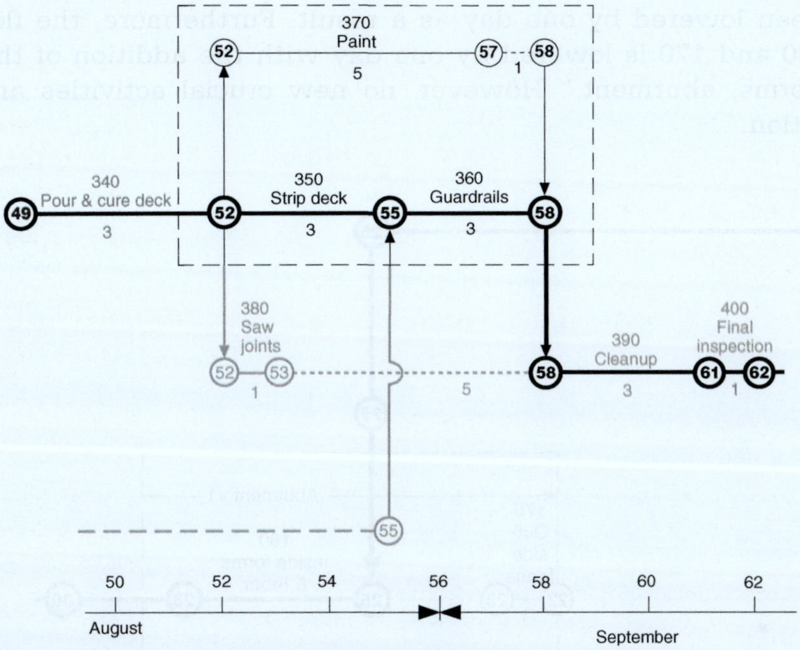

Figure 5.3. Highway bridge, critical activities in parallel.

Source: http://dspace.vnbrims.org:13000/jspui/bitstream/123456789/4250/1/Construction%20Project%20Management%20A%20Practical%20Guide%20to%20Field%20Construction%20Management.pdf.

Before implementing the method to shorten the critical path by completing activities 350 and 370 concurrently, it is essential to address the potential safety risks that may arise from this approach.

5.4.6.4. Subdivision of Critical Activities

It is possible to shorten a critical path by checking to see if every critical task must be finished before moving on to the next. One crucial activity may be able to overlap another if it is divided wisely. Stated differently, it could be feasible to split up a crucial task and carry out a segment of it concurrently with another crucial task. Assuming that it takes a day to erect one side of these prefabricated forms, the completion of activity 60, "Fabricate and deliver abutment & deck rebar," may therefore coincide with the beginning of activity 180 by one day. The original time-scaled diagram would need to be adjusted to account for this change, as shown in Figure 5.4. It is evident that this modification would result in a one-day reduction in the critical path and, consequently, in the time required for any subsequent critical activity, including the terminal activity. Whenever the critical path is shortened, it is necessary to look into the potential that other critical paths could form. The floats displayed at the conclusion of activities 230 and 260 have been lowered by one day as a result. Furthermore, the float that comes after activities 80 and 170 is lowered by one day with the addition of the new activity 175, "Outside forms, abutment." However, no new crucial activities are formed as a result of this action.

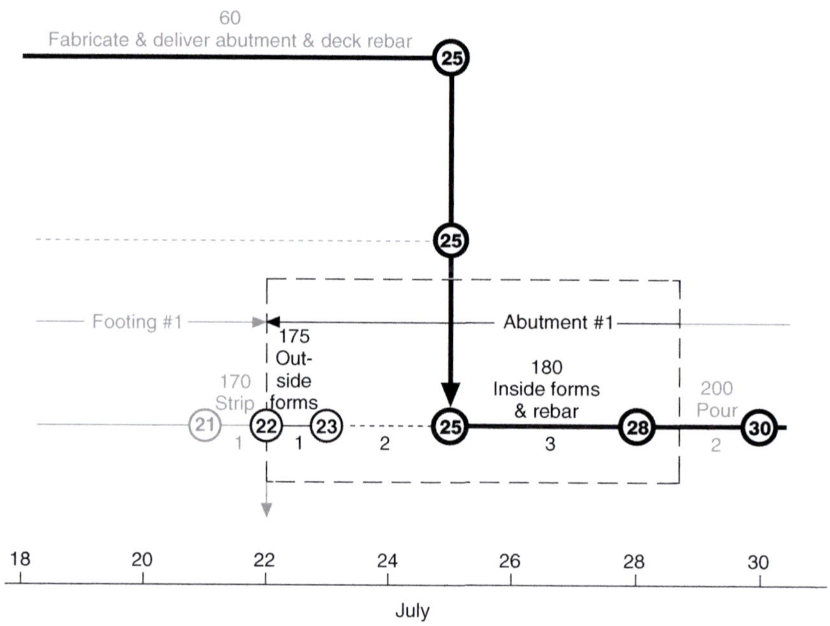

Figure 5.4. Highway bridge, subdivision of critical activity.

Source: http://dspace.vnbrims.org:13000/jspui/bitstream/123456789/4250/1/Construction%20 Project%20Management%20A%20Practical%20Guide%20to%20Field%20Construction%20 Management.pdf.

One more way to reduce a project's critical path could be to subcontract some of the work that the general contractor had planned to complete using its own workforce. The project plan may indicate that some essential tasks should be completed in sequence, not because of the physical order in which they must be completed, but rather because they share a single, scarce resource. The tasks could be completed concurrently rather than sequentially, saving a significant amount of time, if all or some of the work is subcontracted to a specialty contractor with the necessary tools and personnel. It is common for a dependent task to be included in the critical path along with a labor or equipment constraint. Generally speaking, this indicates a restriction on the general contractor resource's availability. The prime contractor may think about assigning that portion of the project to a company whose labor forces or tools would be available sooner than its own. It would take research to see if the work could be subcontracted for the same price as the prime contractor had estimated. If there are additional costs, this becomes an expediting measure that must be weighed against the other potential sources of additional expenses.

5.4.7. Time Reduction of Highway Bridge by Expediting

Reducing the time required to complete the construction of a highway bridge through expediting involves a comprehensive approach that leverages advanced planning, efficient resource management, innovative construction techniques, and effective stakeholder coordination. This approach not only helps meet stringent deadlines but also minimizes the economic and social impacts of prolonged construction periods.

One of the primary strategies for expediting highway bridge construction is the use of prefabricated bridge elements and systems (PBES). Prefabrication involves manufacturing bridge components, such as beams, decks, and abutments, off-site in a controlled environment. These components are then transported to the construction site and assembled rapidly. PBES reduces on-site construction time, enhances quality control, and minimizes the impacts of weather delays. The Accelerated Bridge Construction (ABC) method, which often incorporates prefabrication, has been widely adopted to significantly reduce bridge construction times.

Another critical aspect of expediting construction is optimizing project scheduling through techniques like the Critical Path Method (CPM) and Program Evaluation Review Technique (PERT). These scheduling tools help identify the sequence of critical activities that directly impact the project timeline, allowing project managers to focus on expediting these tasks. Overlapping activities, or fast-tracking, can also be employed, where tasks typically done in sequence are performed concurrently. For instance, while foundation work is being completed, superstructure components can be prefabricated simultaneously.

Efficient resource allocation is essential for expediting construction. This involves ensuring that labor, equipment, and materials are available exactly when needed to avoid delays. Just-in-time delivery systems can be implemented to synchronize material deliveries with the construction schedule, reducing on-site storage requirements and minimizing the risk of material deterioration. Additionally, employing a skilled and experienced workforce capable of working extended hours or in multiple shifts can further accelerate the construction process.

Innovative construction techniques and materials can also contribute to reducing construction time. High-performance materials, such as self-consolidating concrete and high-strength steel, allow for faster construction and curing times, enhancing the overall speed of the project. The use of advanced construction equipment, such as automated rebar tying machines and modular formwork systems, can streamline labor-intensive processes and improve efficiency.

Effective communication and coordination among all stakeholders, including contractors, engineers, government agencies, and the community, are crucial for successful expedited construction. Regular progress meetings, transparent reporting, and proactive problem-solving help address potential issues before they cause delays. Stakeholder engagement ensures that all parties are aligned with the project goals and timelines, facilitating smoother implementation.

In some cases, regulatory and procedural adjustments may be necessary to expedite construction. This can include fast-tracking permit approvals, adopting flexible contracting methods such as design-build or construction manager/general contractor (CM/GC) approaches, and employing incentives for early completion. These adjustments can streamline administrative processes and reduce bureaucratic delays.

Expediting the construction of a highway bridge requires a multifaceted strategy that incorporates prefabrication, optimized scheduling, efficient resource allocation, innovative techniques, and strong stakeholder coordination. By implementing these measures, construction timelines can be significantly reduced, leading to earlier project completion, minimized disruptions, and enhanced public satisfaction.

5.4.8. Limitations on Time-Reduction Steps

A number of limitations apply to how much time reduction can be accomplished in any one step. These five limitations are summarized next.

1. Physical Limit of a Critical Activity: This is the maximum shortening of a given activity considered to be practical. Although most activities can be shortened to some extent, some are considered to be intractable on practical grounds.
2. Logical Limit of a Critical Activity: The reduction in duration of a critical activity reduces the total floats of other activities, which sometimes causes another chain of activities to become critical. This can, at times, prevent expediting an activity to its full potential.
3. Logical Limit of a Network Logic Change: A network logic changes results in a set number of days being gained with no time change possibilities in between. A logic change that reduces the highway bridge duration by six days. The six days is an irreducible time reduction—it is either that or nothing.
4. Shortening is Limited by a Parallel Critical Path: Parallel critical paths, or subpaths, are common. If one branch of parallel critical paths is to be decreased in length, a commensurate decrease must also be made in the other branch. If only one branch is reduced, it simply becomes a floater, and neither the remaining critical path nor the project is reduced in duration.
5. Shortening is Limited by an Irreducible Critical Path: When any given critical path has been shortened to its full capability, no further reduction in project duration is possible. A critical path that cannot be compressed makes fruitless the shortening of noncritical portions of the network.

5.4.9. Milestone and Interface Events

The reduction of the overall project duration has been the main focus of the previous discussions. A current project that has experienced delays in the past will be expedited as part of this reduction.

However, as was already mentioned, there are situations when pushing forward the anticipated dates of milestone and interface events is necessary. Should the event's scheduled time not meet a predetermined window of time, the contractor must take appropriate action. With the possible exception of the project critical path, which may or may not be involved, the process for advancing miles.the process is essentially the same as that for cutting project duration. Any event's early time is determined by the longest path from the project start to that event; therefore, any efforts to advance the event time must focus on reducing this longest path. Initially, the lengthiest route would be reexamined to determine if the intended reduction could be accomplished without incurring additional direct expenses. Should this not be the case, the contractor will have to expedite at additional direct cost.

5.4.10. Project Extension

The goal of this chapter has been to shorten a project in its entirety. On occasion, however, it might be preferable to extend the project or just some of the activities. A project whose initial cost estimates were made with the assumption that expediting measures would be required in order to meet the owner's time requirements is an example of this. Usually, a contractor will account for time constraints in their initial estimate and include the additional cost of overtime, working multiple shifts, and other methods of completing the work more quickly. Regretfully, the contractor typically hasn't determined the essential tasks or produced a precise estimate of the project's duration when the project is up for bid. The contractor can only really account for the additional expenses associated with expediting actions if they speed up most or all of the job operations or if they make some educated guesses about which work items might end up being crucial. Occasionally, it turns out that some or all of the anticipated expediting procedures will not be required after a thorough network analysis. This case is the opposite of the one that was discussed earlier in the chapter. Now that the savings are apparent, the contractor wants to loosen the job plan. It goes without saying that the contractor wants to complete this as profitably as possible. The contractor can typically revoke the expediting actions for at least some of the noncritical activities, even if the project's total duration must stay as scheduled. The length of each activity's total float would be the maximum duration increase that could be granted. Additional relaxation would lead to the emergence of a new subpath or critical path. This is why it's necessary to keep track of how each activity's time change affects the floats of the other activities once more. The most expensive actions should be stopped first if it is not possible to reverse all expediting actions of noncritical activities. Certain essential activities can also be loosened if the project's total duration can be extended beyond what was initially planned. The only outcome of this relaxation is to increase the floats of all noncritical activities by lengthening the critical path. In this situation, the most expensive tasks should be completed first in order to free up time for the most important ones. The non-essential tasks can then be handled normally.

CASE STUDY

REDUCING INEFFICIENCIES IN ROAD CONSTRUCTION:

Corruption plagues many developing countries where the world's poorest live, and combating it continues to be an arduous task. It adds significant costs to the provision of public services, often making them inaccessible to those who need them most. Many observers suggest that the right combination of government-led monitoring and punishment schemes, such as audit-based accountability efforts, can control misappropriation of funds. But often the very individuals tasked with monitoring and enforcing punishments are themselves prone to corruption. Another approach, which has gained prominence in recent years, is to utilize community-level monitoring to reduce "leakages." Local community members have the most to gain from a successful anti-corruption program and are thus believed to have better incentives to appropriately monitor government programs in their locality. However, there is little empirical evidence regarding the effectiveness of such strategies. If community-level monitoring is used to combat corruption in government programs, will we see a reduction in leakages? Furthermore, is this approach more effective than centralized, government-led audits? What experimental designs could test the impact of these interventions?

- *The Kecamatan Development Program (KDP):* KDP is an Indonesian Government program established in 1998, supported by a loan from the World Bank. As of 2004, KDP funded projects in approximately 15,000 villages each year. Each village received an average of Rp. 80 million (US $8,800), which they usually used to surface existing dirt roads. KDP funded projects are large relative to ordinary local government activities. In 2001, the average annual village budget was Rp. 71 million (US $7,800), so implementation of a KDP project more than doubled average local government expenditures. The influx of such a large amount of money through KDP creates opportunities for a high level of leakages.
- *The Problem of Leakages:* KDP project implementation teams are often complicit in facilitating program leakages through several methods. The teams may inflate expenditures by, for instance, paying a certain amount for road materials (e.g., rock, gravel, sand), but reporting having paid 50% more on official project budgets. Implementation teams may also inflate quantities of reported material procurement. For example, teams claim to procure enough rock, sand, and gravel to make a road that is 20 cm thick but instead build a road that is only 15 cm thick. Another strategy employed by the teams is to report inflated costs for labor, and/or report expenses for volunteer labor provided by villagers. In each of these examples, the implementation teams pocket the leakages or share them with complicit partners, such as a supplier that produced an inflated receipt to corroborate a false budget report. Ultimately, such practices deteriorate the intended benefits of KDP projects. Leakages may result in insufficient funding for other village priorities and poor-quality roads that require expensive ongoing maintenance and do not optimally facilitate local economic development.

- *Program Accountability Structure:* Two checks on corruption are built into KDP. First, communities are given an official role in monitoring the flow of KDP money going into the village and its utilization. Specifically, KDP funds are disbursed in three installments. To receive the second and third payments, village implementation teams must produce accountability reports and attend an open village meeting where they present how the previous installment was spent. Only after village members approve accountability reports at the meeting is the next payment released.

Though it is difficult to identify inflated quantities through this approach, it can be particularly useful to catch over-reporting of the cost of materials and labor, given that people in the village are typically aware of market prices and wages. Critically, however, in practice only village leaders who control project funds are likely to attend the accountability meetings. Many of these leaders may be complicit in siphoning off project funds through kickbacks. Second, each project is subject to audits by an independent agency within the central government to scrutinize expenditure reports and monitor the quality of roads constructed, and to punish culprits where appropriate. While legal prosecution is rare, officials found to have stolen money are forced to publicly return the money, which can result in shame and substantial social sanctions. Further, the village would be less likely to win KDP projects in the future. However, many distrust the willingness and ability of the independent auditors to uphold accountability controls. Indeed, they are often complicit in corruption. For example, the auditors may accept bribes from the implementation team or village leader. Notably, even in the absence of auditor corruption, projects had only about a 4% chance of being audited by the government. Overall, in spite of measures to promote accountability, corruption accounted for a large percent of reported project expenses.

- *Proposed Interventions Stimulating Community Participation:* Many policymakers and researchers believe that social sanctions wield a lot of power over the behavior of decision-makers. In the case of KDP, social sanctions could potentially be harnessed by making the village accountability meetings more inclusive and democratic, and hence more effective in catching any misappropriation of funds. However, two barriers may be obstructing community participation: (i) villagers are not aware of accountability meetings, or believe they are not invited; (ii) fear of retaliation stops potential "whistle blowers" from exposing questionable activities. Researchers therefore proposed two interventions:
  - **Invitations:** Invitations to accountability meetings would be distributed throughout the community, encouraging direct participation in the monitoring process and reducing elite capture of the process. There were two methods of distributing invitations: (a) by sending them home with school children; and (b) by asking the heads of hamlets and neighborhood associations to distribute them throughout their respective village areas.
  - **Comment Cards:** An anonymous comment form would be distributed along with the invitations, providing villagers an opportunity to relay information about the project to be shared at the meetings, without fear of retaliation.

- *Increasing the Probability of Audits:* The effectiveness of audits at reducing leakages is likely tied to the probability of being audited. With only 4% of KDP projects being audited, implementation teams and village leaders are quite willing to take the risk of misappropriating funds. To test this assumption, researchers proposed the following: after funds are awarded, but before construction begins, villages would be told that their project would be audited by a government agency with 100% certainty. These audits would carry the possibility of criminal action for perpetrators. To harness the potential power of social sanctions, the results of the audits would be read publicly at an open village meeting.
- *Addressing Key Experimental Issues Through Evaluation Design:* To examine the role of enhanced community monitoring and government audits on leakages, researchers conducted a randomized evaluation in 608 villages (in 156 subdistricts) in East Java and Central Java, Indonesia's most populous provinces. Each village in the study was about to start building a village road with KDP funding. There was a concern that the audit treatment might be likely to spill over from one village to another, since officials in other villages might worry that when the auditors came to the subdistrict, their villages might be audited as well. The participation treatments, however, would be less likely to have similar spillover effects, since the treatment was directly observable in the different villages early on.

CLASS ACTIVITY

1. Find out the completion time and the critical activities for the following project:

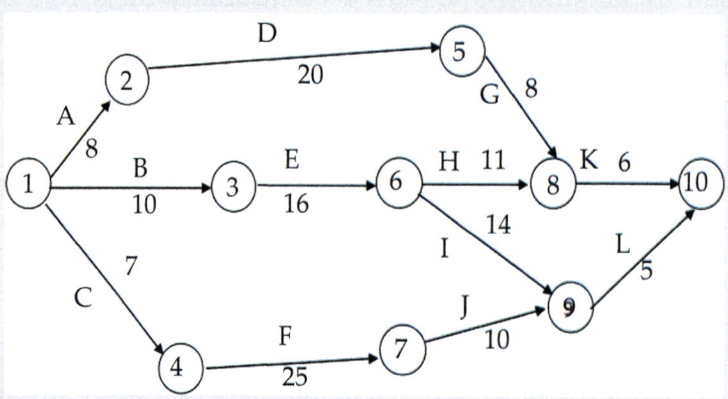

SUMMARY

- Time management is the process of organizing and implementing a strategy related to the time required for work activities on a project. Effective time management is essential to successfully and efficiently meeting budget and program targets, as well as achieving profitability.
- Time management is organizing and managing construction tasks and activities to maximize efficiency and optimize performance. It involves setting priorities, creating and maintaining schedules, assigning tasks and resources, monitoring progress, and resolving issues.
- Effective time management is essential in the construction industry as it can have a major impact on reducing project costs. Keeping projects on track, meeting deadlines, and ensuring quality are all facets of effective time management that can help reduce both labor and material costs.
- Time management is essential in the construction industry if projects are to be completed on time and within budget. The key element to managing your time effectively is prioritizing tasks according to their importance and urgency.
- Construction management software provides a centralized system that takes care of all functions of the construction project. This allows users access to the needed project information, track project progress, and work with the team in real-time and shared data.
- Construction management software provides real-time monitoring and reporting functionality to construction professionals, enabling them to monitor their project progress, spot bottlenecks, and make decisions based on the project data.
- For project time reduction to proceed further, it is necessary to discuss the nature of project direct costs and indirect costs. Costs are necessarily involved with time reduction because construction expense is a function of time.

REVIEW QUESTIONS

1. What are the primary benefits of effective time management in construction projects?
2. How do Gantt charts and the Critical Path Method (CPM) assist in project scheduling?
3. Discuss the impact of effective time management on project quality and safety.
4. How can delays and cost overruns resulting from poor time management affect a construction project?
5. Discuss how software can improve communication and coordination among project stakeholders.
6. Describe the importance of real-time updates and notifications in construction management software.

7. Explain the advantages of using bar charts for tracking project progress.
8. Describe the role of optimized scheduling and resource management in expediting the project.

REFERENCES

1. Abdel-Wahhab, O., & Elazouni, A. (2010, May). Progress monitoring of construction projects using statistical pattern recognition. In Construction Research Congress (Vol. 2, pp. 1204–1213).
2. Ashworth, A. (2001). Contractual Procedures in The Construction Industry. UNITEC.
3. Baldwin, A., & Bordoli, D. (2014). Handbook for Construction Planning and Scheduling. John Wiley & Sons, Ltd.
4. El-Razek, A. M. E., Bassioni, H. A., & Mobarak, A. M. (2008). Causes of delay in building construction projects in Egypt. Journal of Construction Engineering and Management, 134, 831–841.
5. Fewings, P. (2005). Construction Project Management: An Integrated Approach. Taylor and Francis.
6. Liberatore, M. J., Pollack-Johnson, B., & Smith, C. A. (2001). Project management in construction: Software use and research directions. Journal of Construction Engineering and Management, 127(2), 101–107.
7. Mubarak, S. (2015). Construction Project Scheduling and Control (3rd ed.). John Wiley & Sons, Inc.
8. Pierce, D. R. (2014). Project Scheduling and Management for Construction (4th ed.). John Wiley & Sons, Inc.
9. Sawalhi, N., & Enshassi, A. (2012). Application of project time management tools and techniques to the construction industry in the Gaza Strip. Australasian Journal of Construction Economics and Building, 5(1), 1–8.
10. Scott, S., & Assadi, S. (1999). A survey of the site records kept by construction supervisors. Construction Management & Economics, 17(3), 375–382.
11. Simpson, B. G. (1978). Effective time management. Parks & Recreation, 13(9), 61–63.
12. Vanhoucke, M. (2013). Project Management with Dynamic Scheduling: Baseline Scheduling, Risk Analysis, and Project Control (2nd ed.). Springer.
13. Westland, J. (2006). The Project Management Life Cycle: A Complete Step-By-Step Methodology for Initiating Planning Executing and Closing the Project. Kogan Page Limited.

CHAPTER 6 Cost Management

LEARNING OBJECTIVES

After studying this chapter, you will be able to:

- Understand types of cost in construction projects;
- Define the term cost planning;
- Describe cost control, monitoring and accounting;
- Deal with cost reconciliation in construction projects.

KEY TERMS FROM THIS CHAPTER

Cost budgeting	Site acquisition
Construction planning	Environmental taxes
Cost planning	Quotation costing
Insurance costs	Price negotiations
Cost expenditures	Financial accounting system
Financial reports	Accounting system
Cost reconciliation	Financial accounts

INTRODUCTION

The process of organizing and managing a company's budget is known as cost management. An organization can more easily estimate and distribute its budget if it has a sound cost management system in place. A type of management accounting called cost management makes more accurate projections of future expenses, which helps a company lower the likelihood of going over budget. Cost management strategies are employed by many companies for both individual projects and the entire company.

When projected costs are applied to a project, they are determined during project planning and authorized in advance. Every project expense is tracked and documented to make sure it follows the budgetary plan. The actual total costs and the projected costs are compared once the project is completed. This analysis improves cost budgeting and future cost management projections.

Among the many metrics that a general contractor must monitor are construction costs. Of course, there are more, like the timetable, the caliber, the safety, and much more. However, the contractor will bear the additional costs if the construction project exceeds its budget.

Effective cost control in construction is essential to the success of any project. Every stage, from site acquisition to handover, has unique financial implications that call for careful planning. While careful planning is necessary, leave a little amount of money unaccounted for to cover unforeseen costs. This is the reason that construction cost control is crucial. Understanding construction costs and construction cost management in detail can make all the difference in a project's success or failure.

6.1. TYPES OF COST IN CONSTRUCTION PROJECTS

Starting a construction project is a difficult undertaking that involves numerous expenses to be taken into account at every stage. A thorough understanding of these different kinds of construction costs is essential to the project's effectiveness and success. Our goal is to shed light on the different expenses related to construction projects, ranging from the preliminary planning and briefing costs to the ultimate closeout and handover costs.

It is crucial to understand the Project Lifecycle in order to effectively manage costs during construction. In this series, we will explore the various costs associated with each stage of construction (Hang Xiaojian & Ni Zirong, 2005).

Put simply, every construction project goes through a marathon of stages, each with its own set of expenses. Pre-construction usually marks the beginning of the process, and handover and closeout mark its conclusion.

Let's break it down:

1. Pre-Construction: This stage establishes the foundation by completing all required planning and setup prior to the start of construction.
2. Construction: This is the project's core, where ideas become projects. It is frequently the longest and most noticeable stage.
3. Handover and Closeout: The final lap where the project is completed and delivered to the client, with all requisite documentation in place.

The importance of efficient cost management throughout the project lifecycle is emphasized by the fact that each phase has a bearing on the total project cost. A greater comprehension and cautious stewardship allow one to more confidently traverse the cost landscape.

6.1.1. Pre-Construction Costs

Costs are associated with construction projects even before the first brick is laid. These Pre-Construction Costs, which fall into three primary categories—Site Acquisition, Pre-Planning Expenses, and Planning—lay the groundwork for the project.

1. Site Acquisition: This initial stage involves securing the project site. Costs may include:
    - Purchase price;

- Legal fees;
- Stamp duty;
- Site surveys.

2. Pre-Planning Costs: This stage encompasses activities that determine project feasibility. Costs usually include:
    - Feasibility studies;
    - Site investigations;
    - Consultations;
    - Outline proposals;
    - Project viability – establishing a budget and securing funding.
3. Planning: This stage involves obtaining necessary permissions and approvals. Costs can consist of:
    - Application fees;
    - Professional fees (architects, engineers, consultants);
    - Community infrastructure levy.

6.1.1.1. Planning

In construction projects, the planning phase is crucial and entails numerous expenses. This stage calls for expert guidance from consultants, engineers, and architects, as well as consideration of application fees and the price of adhering to local laws:

1. Application Fees: These are obligatory payments made to local authorities when submitting your project plans.
2. Professional Fees: Architectural, engineering, and consulting services are vital in this phase. Their expertise ensures your project is viable and compliant with regulatory standards.
3. Community Infrastructure Levy: A charge levied on developers to help fund infrastructure enhancements such as roads and schools in the local area.

Understanding these costs is a crucial aspect of effective cost management during the planning stage of construction projects.

6.1.1.2. Developed and Technical Design

The phase of Developed and Technical Design holds the highest authority in construction planning. This includes a variety of expenses that have a significant impact on how the project turns out.

1. Architect Fees: This is the remuneration for the architect who breathes life into the building design. This cost is a cornerstone of the project budget.
2. Professional Fees: Beyond the architect, other experts also grace the stage. This includes engineers, Quantity Surveyors, and project managers. Their skills don't come free, adding to the project's financial forecast.
3. Planning Conditions: Adhering to planning conditions in a construction project isn't just a requirement—it also comes with its own set of financial considerations.
4. Building Regulations: Building regulations ensure that construction projects comply with safety and environmental standards, but this compliance comes at a cost.
5. Specialist Surveys or Tests: Sometimes, a project requires a more in-depth look or specific tests, similar to a medical check-up. These special surveys or tests add another layer to the cost structure.

The success of the construction project depends on the Developed and Technical Design phase, despite the fact that it requires a sizable financial investment.

6.1.2. Construction Phase Costs

Effective management of costs and schedule is crucial during the construction phase of the project.

Several types of costs come into play at this phase, including:

1. The Contract Sum: This includes all agreed-upon costs for the project, with provisions for valuation and payment. It's important to understand what's included and what's excluded and to account for Provisional and Prime Cost Sums and fluctuation clauses.
2. Professional Fees: These include architect and consultant fees, which require calculated adjustments, and costs for other professional services.
3. Risk Allowances: Contingency sums must be agreed upon, with procedures in place for reporting and managing drawdowns.
4. Other Project Costs: These can include statutory fees, planning obligations, enabling works, and finance costs.

6.1.2.1. The Contract Sum

The Contract Sum plays a crucial part in the Construction Phase Costs. A crucial component of any construction project's dynamics is this entire financial commitment. Let's dissect the Contract Sum and understand its significance:

1. What's Included and Excluded: The Contract Sum includes all agreed-upon works and is comprehensive. But typically, it doesn't include things like client modifications, unanticipated circumstances, and risk allowances.
2. The Basis for Valuation and Payment: The starting point for figuring out value and payment terms is the Contract Sum. It is an essential component of cost management.
3. Provisional and Prime Cost Sums: These sums are usually stipulated in the Contract Sum, covering specific items or services whose details are not determined at the contract's inception.
4. Fluctuations Clauses: These clauses allow for adjustments to the Contract Sum due to changes in costs.

This construction cost component controls spending and provides financial transparency, which has a positive effect on project delivery.

6.1.2.2. Professional Fees

During the construction phase of your project, be aware of the additional costs that are often overlooked – professional fees.

1. Contract Administration: It costs are central to a project's financial success. Effective management in this area ensures that the project stays within budget and adheres to contractual and regulatory requirements, ultimately contributing to a successful project outcome.
2. Architect and Consultant Fees: These are integral to the project's design and management. Fees depend on the project's size and

complexity. The architect's and consultants' expertise plays a crucial role in cost management and adherence to building regulations.

3. Calculating Fee Adjustments: Adjustments may arise due to changes in the project's scope. An agile construction approach can help manage these changes effectively, ensuring the project stays within budget.
4. Other Professional Services: These may include services like soil testing, landscape design, and project management. The costs of these services vary based on the project's needs.

While hiring professionals might appear as an extra cost, their participation is crucial in preventing expensive errors and guaranteeing a positive project result.

6.1.2.3. Risk Allowances

Since there is some degree of uncertainty in every construction project, risk allowances are an essential part of the budgeting process. These are amounts put aside to cover expenses that might occur from unforeseen circumstances during the project.

1. Design Errors or Omissions: Contingencies can absorb the financial impact of having to modify or rework aspects of the project due to design mistakes.
2. Weather-Related Delays: Including a contingency for weather-related issues safeguards the client against the financial strain caused by unexpected project delays.
3. Contractor Default: A contingency can help offset the costs of replacing the contractor, ensuring the project stays on track without additional financial burden to the client.
4. Unforeseen Site Conditions: Contingencies can cushion the client from the financial blow of dealing with unexpected site conditions like contaminated soil or archaeological finds.
5. Scope Creep: A contingency allows the client some leeway in accommodating new requirements without destabilizing the initial budget.

By facilitating quick responses to unforeseen events, taking risk allowances into account improves project resilience and ensures a more efficient workflow.

6.1.2.3.1. Other Project Costs

Throughout the course of a construction project, there are supplementary expenses referred to as Other Project Costs that must be considered. While these costs may not be directly related to the construction itself, they are crucial for ensuring the project is carried out successfully.

1. Statutory Fees: These are government-imposed charges related to planning permission and building regulations approval. They could also include any environmental taxes if applicable.
2. Planning Obligations: Also known as Section 106 payments, these are agreements made with the local authority to mitigate the impact of your project on the local community and infrastructure.
3. Enabling Works: These are preparatory works to make the site ready for construction and may include demolition, site clearance, and setting up temporary facilities.
4. Finance Costs: Interest payments on a construction loan or any other

financial charges associated with funding the project should also be factored in.

Remember that implementing Lean Construction techniques and adhering to ISO 41001 standards can aid in the effective and efficient management of these expenses.

6.1.3. Handover and Closeout

As the construction project reaches its final phase, we step into the realm of Handover and Closeout. This last lap isn't without its own collection of costs that contribute towards the final tally. These are:

1. Snagging Costs: This includes expenses related to fixing minor defects and completing unfinished tasks – the little bits and bobs that weren't addressed in the construction phase.
2. Commissioning Costs: These costs are associated with testing systems and equipment to ensure they are operating correctly and safely before the building is handed over.
3. Documentation Costs: This involves collating all relevant paperwork, like warranties and manuals, into a comprehensive handover pack.
4. Legal and Administrative Costs: Any legal fees or administrative charges related to the handover process are included here.
5. Training Costs: These are the expenses related to training the client's team to operate and maintain the new systems and equipment.
6. Move-in Costs: These costs encompass packing, transportation, and unpacking of the client's possessions into the new building.
7. Fit-Out Costs: Any additional costs related to customizing the space to the client's specific needs after the handover.
8. Initial Maintenance Costs: The first round of costs related to maintaining the new facility fall under this category.

Prior knowledge of these expenses can help prevent unpleasant surprises and guarantee a seamless transition. Every one of these expenses helps ensure that the construction project is managed successfully overall.

6.2. COST PLANNING

Project management involves three key dimensions: costs, quality, and deadlines. In practice, costs in particular are frequently problematic. Budget overruns are not unheard of. For a company to succeed, sound and realistic cost planning is therefore crucial. However, there are a few requirements that must be satisfied in order to understand the anticipated total costs of a project or order as precisely as possible.

For the management of any project, quotation, production process, or work package, thorough cost planning is necessary. It demonstrates which variables have an impact on the overall costs as well as how to determine each item specifically. Furthermore, we tackle the issue of how to obtain an ideal quotation that enables stable pricing for clients. Effective cost planning can greatly reduce the likelihood of unfeasible projects and cost overruns (Hwang & Zhao, 2014).

6.2.1. What Is Cost Planning?

In general, an instrument for projecting future costs is referred to as "cost planning." Within the context of business, this can apply to the organization as a whole, specific business units, goods, and services. Cost planning is also used to estimate project- or order-related expenses in quotation costing and project management.

The main objective is to ascertain and provide as accurate a forecast as possible for the costs of individual work packages or processes when planning customer-related projects or individual orders. This serves as the foundation for calculating quotes and guarantees to businesses that actual costs do not significantly differ from the initial estimate, which could lead to orders that are not cost-effective. Another function of cost planning is that it makes it easier to compare the target and actual amounts spent on the project, which makes it a crucial tool for project management.

In the event that inadequate project details make cost planning unfeasible, a minimum approximate estimation of projected project expenses needs to be established. This makes it clear whether or not a project is feasible. Over the course of the project, this preliminary cost estimate is improved upon and replaced by a thorough calculation.

6.2.2. Resource Planning as a Basis for the Cost Plan

A project must first be broken up into separate sections in order to estimate its overall costs. The costs of each of these subsections are totaled and each section is assessed separately. Planning for project costs can start as soon as you have an accurate list of all project tasks and a resource schedule. On the other hand, cost planning rarely makes use of other planning data, such as sequence and milestone planning. The following resources and cost types affect the project costs:

- Material: Direct material (raw materials, parts, components), auxiliary materials, consumables, operating materials (valued at acquisition prices or production costs).
- Personnel: Costs for in-house employees involved in the project (valued at daily or hourly rates).
- Plant, machinery, and equipment provided (evaluated at cost price).
- External services: Costs for project-related external services by service providers (evaluated by obtaining quotations).
- Incidental Costs: For example, equipment, travel costs, costs of seminar rooms, insurance, etc.
- Capital and interest costs: Relevant for the external pre-financing of project costs.

6.2.3. Unit Cost Planning

Costs that are clearly and directly attributable to a particular cost unit are known as direct costs. For this reason, they are sometimes called direct costs or direct cost unit costs. The following types of direct costs must be considered in project management:

- Direct Material Costs: Material needed for the product as the main ingredient.
- Direct Labor Costs: Wages that can be directly attributed to the cost unit.
- Special Direct Costs: Are not allocated to individual products but to orders with similar products (e.g., packaging, transport, commissions).

The "expected factor quantities" (e.g., costs, labor hours, material quantities) are the basis for individual cost planning. Standards are the term used to describe this "factor quantity." In order to determine the standards, there are additional possibilities to choose from:

- Parts lists;
- Construction drawings;
- Estimates based on experience;
- Statistical historical values;
- External standard values (e.g., consumption average of a certain industry).

To reach the required degree of precision in planning, a few presumptions must generally be met. For instance, set planned prices for labor and materials shouldn't deviate significantly from the going rate. When determining cost prices, the most appealing offer should be taken into consideration. Furthermore, it is important to consider other pricing factors like freight charges and transportation insurance.

6.2.4. Total Cost Planning

Naturally, during cost planning, an organization will also incur overhead costs in addition to directly attributable direct costs. In essence, these include overheads related to materials, production, administration, and sales. Transport expenses or warehouse operations costs are two examples of material overheads. Typically, an overhead rate that was established in the cost accounting sheet is used to assign the overhead costs to a project, a product, or an order.

Allocating the incurred overhead costs to the total costs is essential if a quotation is to be prepared for a customer. It is only possible to obtain a fully cost-covering calculation by using this method.

A financial buffer should be budgeted for as well, as this lowers the possibility of budget overruns. It should be remembered, though, that this directly affects the bidding process and distorts the project's profitability from what it may actually be.

6.2.5. Quotation Costing

Project costing is expanded upon by quotation costing. The pricing is eventually calculated for the customer by adding various surcharges. These are typically legally binding offers, so in order to calculate the order, a precise clarification and explanation of the order must be provided beforehand. Therefore, in order to be able to create a trustworthy calculation, the requested service needs to be specified as precisely as possible.

Quotation costing adheres to a predetermined framework. It consists of the project proposal price, the cost of goods sold, and the production costs. We go into more detail about the subjects that are typically covered in an overhead cost estimate in the sections that follow.

6.2.5.1. Material Costs

The price of components and raw materials is considered when manufacturing a product. Included are the material overhead surcharges.

6.2.5.2. Production Costs

This category contains an overview of all production-related expenses. Surcharges for overhead are also included here.

6.2.5.3. Third-Party Costs

This includes the price of services rendered by subcontractors, the price of product refinement, and the price of other services (like technical drawings or event conception and execution). The subcontractors' offer typically serves as the foundation for this kind of computation.

6.2.5.4. Costs for the Companies Own Contributions and Services

This role includes the expenses incurred by the company's own employees in person-days or hours. Typically, the scope is computed using the company's hourly rates and is estimated based on the project plan.

6.2.5.5. Other Costs

These are additional project-related costs, like travel, training, and insurance costs, that are required for the provision of services.

6.2.5.6. Overhead Surcharges

The manufacturing costs are the total of all the topics that have been listed thus far. Overhead cost surcharges are now included. Distribution, shipping, management, and administrative costs are classic examples. Standard surcharges may also include risk or warranty charges, depending on the industry.

6.2.5.7. Profit Mark-Up

The "cost of goods sold" is the total of the production costs plus the overhead surcharges. Next, the targeted profit margin is included.

6.2.5.8. Discounts

Surcharges for anticipated profit reductions (discounts, rebates) are applied to the price if needed. This gives room for some wiggle room in the price.

6.2.5.9. Quoted Price

The final quotation price is the outcome if the profit markup and markups for price reductions have been added to the prime costs.

6.3. COST CONTROL, MONITORING, AND ACCOUNTING

In construction, cost control refers to the methods managers use to control labor, material, and overhead costs in order to keep project costs within reasonable bounds and complete the project on schedule.

Effective cost management depends on accurate estimates and ongoing monitoring throughout the duration of a project. Without proper cost control, project expenses can escalate rapidly, depleting the budget and potentially resulting in minimal profits or financial loss for the construction company. Therefore, solid cost control practices are essential for ensuring the success of any construction business.

6.3.1. The Cost Control Problem

Project managers and other construction process participants find that procedures for record-keeping and project control become indispensable tools during the project's execution. These tools record the financial transactions that take place and also provide managers with information about the issues and status of a project. An old definition of a project as any collection of loosely related activities that are 90% complete, over budget, and late sums up the issues with project control quite nicely. Project control systems' job is to provide a reasonable indication of the presence and severity of these issues.

During a project, we take into account the issues related to resource utilization, accounting, monitoring, and control. We focus on the applications of accounting information in project management in this discussion. It is usually difficult to interpret project accounts after a project is finished because by then, it is too late to make changes to the project management process. Confusion can persist in the accounting results even after a project is finished. Thus, in order to manage projects, managers must be able to interpret accounting data. Though the goal is not to present a thorough analysis of accounting practices, we will talk about some common accounting systems and conventions while we examine management problems.

It is important to emphasize project control's narrow goal. The main goal of project control procedures is not to recommend potential areas for cost savings, but rather to find deviations from the project plan. The advanced stage at which project control becomes significant is reflected in this characteristic. The project's planning and design phase is when significant cost savings can be realized. Modifications during the construction phase have the potential to cause delays and excessive cost increases. Therefore, rather than looking for major improvements and cost savings, the focus of project control is on completing the original design plans or indicating deviations from these plans. The construction schedule typically only undergoes significant modifications in the event that a rescue operation is necessary.

Lastly, there will need to be some discussion of the problems related to information integration. Although functional concerns and project management activities are closely related, many times the methods employed do not allow for a thorough or integrated analysis of project activities. For instance, schedule data and cost accounts are typically maintained apart. Therefore, in addition to their own field observations, project managers themselves must synthesize a comprehensive view from the various reports on the project. Specifically, managers are frequently left to estimate the financial effects of schedule modifications without the assistance of tools. Integration or communication of different kinds of information can be beneficial in many ways, but it does need special consideration when establishing project control protocols (Hussein & Razzaq, 2021).

6.3.2. The Project Budget

The construction plan and related cash flow estimates can serve as a baseline for future project monitoring and control in terms of cost control. Schedules allow for the monitoring of activity progress by comparing the project schedule with individual activity progress and milestone completions. The standards by which to evaluate and guarantee the necessary level of construction quality are provided by the contract and job specifications. A baseline for evaluating the project's financial performance is provided by the final or detailed cost estimate. The project is regarded as being under financial control to the extent that expenses are within the comprehensive cost estimate. Overruns in specific cost categories indicate which specific problems are being encountered as well as the likelihood of problems. The categories that are included in the final cost estimation are the focus of expense-oriented construction planning and control. This emphasis is especially pertinent to tasks like grading and paving roads that require a lot of repetition and few activities.

To facilitate control and monitoring, the detailed cost estimate is usually transformed into a project budget, which acts as a reference for management. The detailed cost estimate's specific items are converted into job cost elements. Costs accrued throughout the project are documented in job cost accounts and compared with the initial estimates in each category. Job cost accounts typically serve as the primary unit for cost control. Alternatively, these accounts can be further broken down into work elements that correspond to specific scheduled activities and cost accounts.

The project budget usually includes information on labor inputs and material quantities within each job account in addition to cost amounts. This data allows for a comparison between the actual and anticipated use of labor and materials. Because of this, cost overruns or savings on specific items can be attributed to variations in unit prices, labor productivity, or material consumption.

The quantity of cost accounts linked to a specific project can differ significantly. For builders, a modest project could require the use of about 400 different cost accounts. All of the transactions connected to a project are documented in these accounts. As a result, distinct accounts may be needed for various material kinds, equipment usage, payroll, project offices, etc.

Resources, both tangible and intangible, are included, including overhead costs like interest and computer use. A typical set of cost accounts that could be utilized in the construction of a building is summarized in Table 6.1. It should be noted that this group of accounts is arranged in a hierarchical manner, with multiple sub-divisions under each of the seven main divisions (accounts 201–207). The costs associated with the superstructure (account 204), for instance, would be the total of the underlying subdivisions (i.e., 204.1, 204.2, et cetera) or finer levels of detail (204.61, 204.62, etc.) thanks to this hierarchical structure. For financial accounting purposes, the sub-division accounts in Table 6.1 could be further subdivided into personnel, material, and other resource costs.

Table 6.1. Illustrative Set of Project Cost Accounts

201	Clearing and preparing site
202	Substructure
202.1	Excavation and shoring
202.2	Piling
202.3	Concrete masonry
202.31	Mixing and placing
202.32	Formwork
202.33	Reinforcing
203	Outside utilities (water, gas, sewer, etc.)

204	Superstructure
204.1	Masonry construction
204.2	Structural steel
204.3	Wood framing, partitions, etc.
204.4	Exterior finishes (brickwork, terra cotta, cut stone, etc.)
204.5	Roofing, drains, gutters, flashing, etc.
204.6	Interior finish and trim
204.61	Finish flooring, stairs, doors, trim
204.62	Glass, windows, glazing
204.63	Marble, tile, terrazzo
204.64	Lathing and plastering
204.65	Soundproofing and insulation
204.66	Finish hardware
204.67	Painting and decorating
204.68	Waterproofing
204.69	Sprinklers and fire protection
204.7	Service work
204.71	Electrical work
204.72	Heating and ventilating
204.73	Plumbing and sewage
204.74	Air conditioning
204.72	Fire alarm, telephone, security, miscellaneous
205	Paving, curbs, walks
206	Installed equipment (elevators, revolving doors, mailchutes, etc.)
207	Fencing

Source: https://www.cmu.edu/cee/projects/PMbook/12_Cost_Control,_Monitoring,_and_Accounting.html.

An appropriate coding or numbering scheme is necessary when creating or implementing a cost accounting system in order to enable information sharing and correct cost information aggregation. Specific cost accounts are used to show the costs related to particular projects as well as the costs incurred by the entire organization on specific items. These are some instances of various viewpoints on the same material, which can be summed up in various ways depending on the objective. A specific cost account can therefore be used by multiple application programs and multiple cost information aggregations. When it comes to project cost accounts or financial transactions, distinct identifiers for the kind of cost account and the particular project must be given.

A challenge when creating a project budget involving cost accounts is determining how to handle contingency amounts. These allowances are added to project cost estimates to cover unexpected events and costs. However, it can be difficult to accurately predict where these contingency expenses will come from before the project is finished. It is advisable to create a specific line item in the budget for contingency allowances whenever they are included in the final cost estimate.

Another issue when creating a project budget is dealing with inflation. Usually, the final cost estimates are calculated in real dollars, and an amount for inflation is added on as a percentage or lump sum. This inflation buffer is then distributed to specific cost items based on the projected inflation during the timeframe in which expenses will occur.

6.3.3. Forecasting for Activity Cost Control

It is insufficient to take into account a project's past expenses and revenues for the purposes of project management and control. Effective managers ought to concentrate on upcoming income, upcoming expenses, and technological issues. Traditional financial accounting schemes are insufficient to accurately reflect a project's dynamic nature for this reason. Typically, accounts are used to track regular expenses and previous outlays related to activities. Generally speaking, past expenses are sunk costs, meaning they cannot be changed in the future and may or may not be significant. For instance, after finishing a task, it might be found that a quality issue makes the effort pointless. Regrettably, the resources used on the defective construction will usually be sunk and cannot be recovered for re-construction (though it might be possible to shift who is responsible for paying for these resources through charges or financial withholding; owners usually try to hold designers or constructors accountable for changes resulting from defects in quality). Since financial accounts are by their very nature historical, management control requires a way to project or forecast a project's future course.

An example of forecasting used to assess the project status is shown in Table 6.2. In this example, costs are reported in five categories, representing the sum of all the various cost accounts associated with each category:

1. Budgeted Cost: The comprehensive cost estimate created at the beginning of the project is the source of the budgeted cost. Both a prose description and a cost account would make reference to the cost factors.

2. Estimated Total Cost: The current best estimate of costs in each category is based on progress and any changes since the budget was created. Estimated total costs include cost to date, commitments, and exposure.
3. Cost Committed and Cost Exposure: Each category's estimated cost to completion is broken down into estimated additional costs or exposure and firm commitments. Material orders or subcontracts with fixed dollar amounts committed can be represented by commitments.
4. Cost to Date: Column 6 records the actual cost incurred thus far, which can be obtained from the financial record-keeping accounts. Over or (Under): Table 6.2's final column shows the amount that each category has spent above or below the budget. This column shows how much the project budget has deviated; items with abnormally high overruns would raise special managerial concerns. Keep in mind that the term "variance" in project control refers to a discrepancy between planned and actual spending. The definition and usage of the term in mathematical analysis or statistics differ significantly. Table 6.2 shows that while subcontracts are less than anticipated, labor costs are running higher than anticipated.

As of right now, 23% of the project's planned costs have been incurred, resulting in a $5,950 projected budget overrun.

Table 6.2. Illustration of a Job Status Report

Factor	Budgeted Cost	Estimated Total Cost	Cost Committed	Cost Exposure	Cost To Date	Over or (Under)
Labor	$99,406	$1,02,342	$49,596		$52,746	$2,936
Material	88,499	88,499	42,506	45,993	15,139	0
Subcontracts	1,98,458	1,96,323	83,352	97,832	13,920	(2,135)
Equipment	37,543	37,543	23,623	–	32,076	0
Other	72,693	81,432	49,356	1,43,825	1,13,881	8,739
Total	4,96,509	5,06,139	2,48,433			5,950

Source: https://www.cmu.edu/cee/projects/Pmbook/12_Cost_Control,_Monitoring,_and_Accounting.html.

In order to effectively manage a project, managers need to closely monitor any significant deviations from the budgeted amounts. Specifically, project managers should pay close attention to cost overruns in labor and other expenses as highlighted in Table 6.2. It is important to then analyze the individual components within these categories to determine the root causes of any cost overruns. Issues such as lower productivity, higher wage rates, material costs, or other factors could be contributing to these deviations. Additionally, factors like inadequate training, lack of resources, or excessive rework may also be impacting productivity levels. Simply reviewing a job status report is just the beginning of project control – further analysis and action are necessary to effectively manage the project.

The job status report, as shown in Table 6.2, uses precise estimates for the final cost for every expense category. These approximations are employed to determine the true trajectory and condition of a cost category. Simple linear extrapolations of the productivity or cost of the work completed thus far on each project item can be used to generate estimates. A linear estimation formula typically takes one of two algebraic forms. Using a linear extrapolation of costs, the forecast total cost, C_f, is:

$$C_f = \frac{Ct}{p_t} \qquad (1)$$

where; C_t is the cost incurred to time 't' and p_t is the proportion of the activity completed at time 't.' For instance, the estimated total cost of an activity that is half completed and costs \$40,000 is \$40,000/0.5 = \$80,000. More complex cost forecasting techniques would break down costs into several categories, with the total cost being the sum of the estimated costs for each category.

Alternatively, the use of measured unit cost amounts can be used for forecasting total cost. The basic formula for forecasting cost from unit costs is:

$$C_f = W\ c_t \qquad (2)$$

where; C_f is the forecast total cost; W is the total units of work, and c_t is the average cost per unit of work experienced up to time 't.' If the average unit cost is \$50 per unit of work on a particular activity and 1,600 units of work exist, then the expected cost is (1,600) (50) = \$80,000 for completion.

The unit cost in Eqn. (2) may be replaced with the hourly productivity and the unit cost per hour (or other appropriate time period), resulting in the equation:

$$C_f = Wh_t u_t \qquad (3)$$

where; the cost per work unit (c_t) is replaced by the time per unit; h_t, divided by the cost per unit of time, u_t.

More complex forecasting systems may be able to identify specific issues related to work on specific items and adjust these straightforward proportional cost estimates. For example, the estimate of total costs for an item may be revised lower if productivity increases as employees and managers gain a better understanding of the project activities. In this case, the estimating equation would become:

$$C_f = C_t + (W\text{-}Wt)C_t \qquad (4)$$

where; forecast total cost, C_f, is the sum of cost incurred to date, C_t, and the cost resulting from the remaining work (W – W_t) multiplied by the expected cost per unit time period for the remainder of the activity, c_t.

Let's take a numerical example where the most recent cost per unit of work during a project is \$45 and the average unit cost has been \$50. The cost estimate for completing the activity would be (50) (800) + (45)(800) = \$76,000 if the project manager was given the assurance that the increased productivity could be sustained for the remaining 800 units of work out of a total of 1,600 units of work. Take note that the productivity forecast for the remaining work is based on the actual average productivity of the first 800 units. This kind of non-linear changes in work productivity on specific activities over time may also be represented by historical changes in productivity.

Productivities can be changed, but so can other elements of the estimating formula or the replacement of more precise estimates. When projecting future expenses, one may take into account factors like the shift in unit prices brought about by new labor contracts or the cost of materials from suppliers. Though the quantity and degree of uncertainties in the project environment decrease as work goes on, the process of creating exposure estimates essentially involves the same issues that were encountered when creating the detailed cost estimate. The sole exception to this rule is if there is a chance that the finished work will have quality issues and needs to be rebuilt. All the estimating techniques mentioned above necessitate up-to-date data regarding the level of work completion for specific tasks. There are several possible methods to develop such estimates, including:

1. Units of Work Completed: The actual percentage of finished work amounts can be measured for quantities that are easily measured. To estimate the percentage of piping work completed, for instance, the installed linear feet of piping can be compared to the required amount of piping.
2. Incremental Milestones: Specific tasks can be broken down into smaller milestones, which can then be used to track progress and determine the estimated percentage of work completed using past performance data. For example, the work effort involved with installation of standard piping might be divided into four milestones:
    - Spool in Place: 20% of work and 20% of cumulative work.
    - Ends Welded: 40% of work and 60% of cumulative work.
    - Hangars and Trim Complete: 30% of work and 90% of cumulative work.
    - Hydrotested and Complete: 10% of work and 100% of cumulative work.

Thus, a pipe section for which the ends have been welded would be reported as 60% complete.

1. Opinion: Inspectors, managers, or supervisors themselves may prepare subjective assessments of the percentage completed. Clearly, optimism, pessimism, or erroneous observations can all bias this estimated technique. To achieve a sufficient level of accuracy with this method, competent estimators and sufficient field observations are needed.
2. Cost Ratio: Using the expenses accrued up to this point can also be utilized to gauge the progress of the work. For instance, if a task was originally projected to cost $20,000 and by a specific date only $10,000 had been spent, then according to the cost ratio method, the estimated completion percentage would be 10,000/20,000 = 0.5 or 50%. However, this method does not provide any additional insight into the actual completion percentage or any potential discrepancies in the budgeted cost. The projected cost will always align with the original budgeted amount. Therefore, managers should tread carefully when using the estimated remaining costs calculated through the cost ratio method.

By applying various estimating methods systematically to each project activity, the percentage complete and productivity estimates can be calculated for the preparation of job status reports.

Automated data acquisition may be implemented in certain instances to track work accomplishments. For instance, transponders could be relocated to new work areas at the end of each day to calculate their new positions and compare them with project plans. These measurements of progress should be recorded in a centralized database and analyzed to update the project schedule.

- Example 1: Estimated Total Cost to Complete an Activity: Let's say we want to calculate how much it will cost overall to finish a project's piping construction. For the convenience of management, the 1,000 linear feet of piping that make up the construction are divided into 50 sections. As of right now, $40,000 and 500 man-hours of labor have been used to install 400 linear feet of piping. With a productivity of one foot per man-hour, a unit cost of $60 per man-hour, and a total material cost of $30,000, the initial budget estimate was $90,000. We have received firm material delivery commitments for the estimated cost of $30,000.

The initial step is to determine the proportion of work that has been completed. There are two estimates available. The first estimate is based on the amount of pipe that has been installed, with 400 linear feet out of a total of 1,000 linear feet completed, making the proportion 0.4 or 40%. This method is known as the "units of work completed" estimation. The second method, known as the cost ratio method, calculates the work completed by dividing the cost-to-date by the total cost estimate, resulting in $40,000/$90,000 = 0.44 or 44% complete. Alternatively, the "incremental milestones" method can be used by assessing the completion percentage of each pipe section and then combining them to determine the overall completion percentage.

Complete (100%)	380 ft
Hangars and trim complete (90%)	20 ft
Ends welded (60%)	5 ft
Spool in place (20%)	0 ft

Then using the incremental milestones shown above, the estimate of completed work would be 380 + (20) (0.9) + (5) (0.6) + 0 = 401 ft and the proportion complete would be 401 ft/1,000 ft = 0.401 or 40% after rounding.

The estimated cost to finish the task can be computed once an estimate of the work completed is available. Using the 40% estimate of work completed, a straightforward linear extrapolation of costs yields an estimate for the piping construction of $40,000/0.4 = $100,000. The cost overrun predicted by this estimate is $1,00,000 – $90,000 = $10,000.

Second, a linear extrapolation of productivity results in an estimate of (1,000 ft.) (500 hrs/400 ft.) ($60/hr) + $30,000 = $1,05,000 for completion of the piping construction. This estimate suggests a variance of $1,05,000 – $90,000 = $15,000 above the activity estimate. In making this estimate, labor and material costs entered separately, whereas the two were implicitly combined in the simple linear cost forecast above. The source of the variance can also be identified in this calculation: compared to the original estimate, the labor productivity is 1.25 hours per foot or 25% higher than the original estimate.

- Example 2: Estimated Total Cost for Completion: The forecasting processes made the assumption that future costs would increase linearly, either from recent experience or from the activity's whole history. It may be possible to estimate a generally non-linear profile of cost expenditures and completion proportions for activities with good historical records. One potential non-linear relationship based on experience in a specific activity is shown in Figure 6.1. One can compare the progress made on a new job to this historical record. Point A in Figure 6.1, for instance, indicates a larger expenditure than is typical for the completion proportion. This point indicates that 40% of the work was finished at a cost of 60% of the budget. Even if work efficiency can be raised to historical averages, a 60 – 50 = 10% cost overrun is anticipated because the historical record indicates that only 50% of the budget should be spent at the time of 40% completion. The cost to finish will be much higher if similar cost overruns keep adding up.

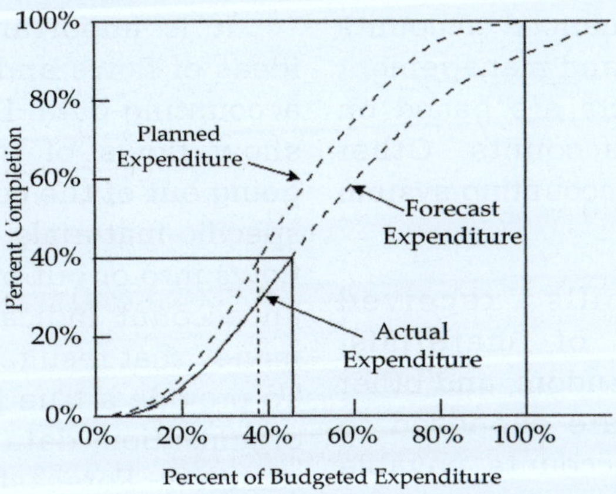

Figure 6.1. Illustration of proportion completion versus expenditure for an activity.

Source: https://www.cmu.edu/cee/projects/PMbook/12_Cost_Control,_Monitoring,_and_Accounting.html.

6.3.4. Financial Accounting Systems and Cost Accounts

A financial accounting system consists of multiple components, of which the cost accounts are just one. Prior to delving deeper into the application of cost accounts in project control, it is pertinent to acknowledge the connection between project and financial accounting. Accounting information is generally used for three distinct purposes:

- Internal reporting to project managers for day-to-day planning, monitoring, and control;

- Internal reporting to managers for aiding strategic planning; and
- External reporting to owners, government, regulators, and other outside parties.

Contractual reporting requirements or widely accepted accounting practices limit external reports to specific formats and processes. Financial accounting refers to the process of creating these kinds of external reports. On the other hand, the goal of cost or managerial accounting is to support internal managers in their duties related to control, monitoring, and planning.

Project expenses are always accounted for in the organization's financial accounting system. All expense transactions are centrally stored in a general ledger within this system. Both the financial accounts for an entire organization and management reports on specific projects are based on the general ledger of accounts. Other components of a financial accounting system include:

- Records of bills received from suppliers of materials, subcontractors, vendors, and other external parties are supposed to be kept in the accounts payable journal. This system keeps track of both charge invoices and checks that are written off as payment. Charges to specific cost accounts are sent to the General Ledger via relay or posting.
- The accounts receivable journal serves as the counterpart to the accounts payable journal. It records billings to clients and receipts, while also transferring revenues received to the general ledger.
- Job cost ledgers, which are organized in the different cost accounts used for the project budget, provide an overview of the expenses related to specific projects.
- To determine the amount of materials available at any given time, inventory records are kept.

In conventional bookkeeping methods, daily transactions are initially documented in journals. In double-entry bookkeeping, each transaction is logged as a debit and a credit to specific accounts in the ledger. For instance, paying a supplier's bill results in a debit to a project cost account and a credit to the company's cash account. Periodically, transaction details are compiled and moved to ledger accounts. This procedure, known as posting, can be done immediately or on a daily basis in computerized systems.

It is important to keep in mind the ideas of flows and stocks when examining accounting data. Daily transactions usually show flows of money coming into and going out of the company. In a similar vein, specific materials' use or receipt represent flows into or out of inventory. A balance on an account indicates the total amount of money that results from these daily inflows. To provide a true picture of the state of an organization, data on stocks and flows are required. Forecasts of upcoming changes are also necessary for efficient management.

The organization's financial reports, which include balance sheets and income statements for each period, are put together using data from the general ledger. These reports, which are fundamental outputs of the financial accounting process, are frequently used to evaluate an organization's performance. A typical income statement for a small construction company is presented in Table 6.3, with a net profit of $3,30,000 after taxes. The transaction flows over the course of a year are summed up in this statement. The comparable balance sheet is displayed in Table 6.4, which shows a net increase in retained earnings equal to the net profit. The impact of income flows throughout the year on the organization's total value is reflected in the balance sheet.

Table 6.3. Illustration of an Accounting Statement of Income

Income Statement	
Gross project revenues	$7,200,000
Direct project costs on contracts	5,500,000
Depreciation of equipment	2,00,000
Estimating	1,50,000
Administrative and other expenses	6,50,000
Subtotal of cost and expenses	6,500,000
Operating income	7,00,000
Interest expense, net	1,50,000
Income before taxes	5,50,000
Income tax	2,20,000
Net income after tax	3,30,000
Cash dividends	1,00,000
Retained earnings, current year	2,30,000
Retention at beginning of year	6,50,000
Retained earnings at end of year	$8,80,000

Source: https://www.cmu.edu/cee/projects/PMbook/12_Cost_Control,_Monitoring,_and_Accounting.html#12.4%20Financial%20Accounting%20Systems%20and%20Cost%20Accounts.

Table 6.4. Illustration of an Accounting Balance Sheet

Balance Sheet	
Assets	**Amount**
Cash	$1,50,000
Payments receivable	7,50,000
Work in progress, not claimed	7,00,000
Work in progress, retention	2,00,000
Equipment at cost less accumulated depreciation	1,400,000
Total assets	$3,200,000
Liabilities and Equity	
Liabilities	
Accounts payable	$9,50,000
Other items payable (taxes, wages, etc.)	50,000
Long term debts	5,00,000
Subtotal	1,500,000
Shareholders' funds	8,20,000
40,000 shares of common stock	8,80,000
(Including paid-in capital)	1,700,000
Retained earnings	$3,200,000
Subtotal	
Total Liabilities and Equity	

Source: https://www.cmu.edu/cee/projects/PMbook/12_Cost_Control,_Monitoring,_and_Accounting.html#12.4%20Financial%20Accounting%20Systems%20and%20Cost%20Account.

The way incomplete contracts are handled in financial reports presents unique challenges for private construction companies. Income is only reported for projects that are finished under the completed-contract method. Only work completed on ongoing projects is shown on the balance sheet as an asset in the event that contract billings exceed costs, or as a liability in the opposite scenario. When a project is finished, the entire net profit—or loss—is recorded as income in the last period. A percentage of all project revenues (or billings) equal to the percentage of work completed during the period is added to actual costs on the income statement under the percentage-of-completion method. The ratio of costs incurred thus far to the project's total estimated cost determines the percentage of work completed. Therefore, the contract revenues earned would be $1,000,000(0.2) = $2,00,000 if 20% of a project was completed in a given period at a direct cost of $1,80,000 on a project with expected revenues of $1,000,000. This amount shows the profit and overhead contribution for the period: $2,00,000 – $1,80,000 = $20,000. Be aware that actual receipts and billings may differ from the estimated revenues of $2,00,000. When an organization uses the percentage-of-completion method to calculate its balance sheet, an asset is typically reported to show billings and estimated or calculated earnings that exceed actual billings.

One benefit of reporting period earnings using the percentage-of-completion method is that it shows actual estimated earnings for each period. Because of this, the income stream and ensuing profits are less vulnerable to sudden fluctuations upon project completion than they might be when using the completed contract method of income calculation. The percentage-of-completion, however, has the drawback of depending on estimates that are hard for outside observers to replicate or that can be altered to hide a company's true situation. Subtleties also exist, like deferring all computed project income until a certain project completion threshold is met. Because of this, it's not always easy to understand a private organization's income statement and balance sheet. The percentage-of-completion method has additional tax drawbacks in that corporation taxes on anticipated profits may become payable during the project rather than being postponed until its completion. The General Accounting Office's study of 47 construction companies revealed that the completed-contract method resulted in $280 million in taxes being postponed between 1980 and 1984. This serves as an example of the tax implications of the two reporting methods.

It is clear that the percentage-of-completion accounting method offers only a rough estimate of a project's actual profit or status. On the other hand, the completed contract method of accounting is solely retrospective and does not provide useful guidance for management. These are just a few examples of accounting practices introduced to align with industry standards, but they may not necessarily benefit effective project management. For instance, equipment depreciation schedules are often used to allocate equipment purchase costs, which can result in significant fluctuations in financial indicators within a project. However, these types of allocations may not have any real significance for overall project management or profit in the long run. As Johnson and Kaplan argue:

- The management accounting information available today is outdated, overly summarized, and unreliable for managers to make effective planning and control decisions, as it is driven by the organization's financial reporting system procedures and cycles.
- Operating managers find management accounting reports to be unhelpful when trying to cut costs and boost productivity. In fact, these reports often hinder productivity by making managers spend valuable time deciphering and justifying variances that bear little resemblance to the actual economic and technological situation of their operations.
- The management accounting system does not accurately determine product costs. Costs are allocated to products using simple and arbitrary methods, typically based on direct labor, which do not accurately reflect the resources required by each product.

Therefore, as discussed in the sections before and after, in order to achieve effective project control, complementary procedures to those utilized in traditional financial accounting are needed. Even though financial statements offer reliable and crucial information about the state of a company as a whole, project managers must interpret and supplement them extensively.

6.3.5. Control of Project Cash Flows

Project managers are responsible for evaluating the project's progress, monitoring activities, managing finances, and tracking payments and receipts. This entails overseeing costs, invoices, payments to contractors, amounts owed to suppliers, and cash flows related to financing plans.

The elements in Table 6.5 include:

1. Costs: The job cost accounts show that as of July 2, total costs were \$8,754,516 compared to the original estimate of \$65,863,092, making the project approximately 13.292% complete. The project manager now anticipates a total cost of \$66,545,263, which is \$6,82,171 higher than the original estimate. This new estimate considers the actual progress of the project and any changes in labor or material costs. The increase in expected costs is concerning to the project manager.
2. Billings: This row, which is not included in reports to owners, summarizes the cash flow situation with regard to the facility owner. 9,276,621-, or 13.741%, of the \$67,511,602 contract amount has been billed. Under the terms of the agreement between an owner and an engineer, architect, or constructor, the amount of allowable billing is specified. In this instance, the total amount billed has surpassed the projected percentage of project completion. The final column shows the \$9,66,339 currently estimated net earnings. The contract amount less anticipated costs is used to calculate this amount: 67,511,602 – 66,545,263 = \$9,66,339. Keep in mind that discounting and time value of money are not reflected in this profit figure.
3. Payables: The Payables section provides a summary of the outstanding debts owed by the contractor to material suppliers, labor, and subcontractors. As of the date of this report, a total of \$6,719,103 has been paid to subcontractors, material suppliers, and other parties. There are currently outstanding invoices totaling \$1,300,089 that have not yet been settled. Additionally, a retention amount of \$3,91,671 is being withheld from subcontractors, and direct labor expenses amounting to \$3,43,653 have been incurred. The total amount of payables matches the total project expenses as indicated in the first row of costs.
4. Receivables: The cash flow of the owner's receipts is shown in this row. Keep in mind that due to late payments or retainage on the owner's part, the actual receipts from the owner may differ from the amounts billed. The net billed is the same as the gross billed minus the owner's retention. In this instance, the billings row displays the gross billed amount of \$9,276,621, the net billed amount is \$8,761,673,

and the retention is $5,14,948. The owner still owes us $2,067,277, which is a substantial amount considering that only $7,209,344 has been received.

5. Cash Position: This row presents the project's cash position as though all project-related costs and revenues were consolidated into a single account. The owner has paid $7,209,344 for the actual expenses, which total $7,062,756 (calculated as the total costs of $8,754,516 less subcontractor retentions of $3,91,671 and unpaid bills of $1,300,089). Consequently, there is a $1,46,588 net cash balance that can be used to cover shortfalls on other projects or placed in a bank account that yields interest.

The rows in Table 6.5 are based on separate financial accounts. Additional reports can also be generated to analyze financing cash flows related to bonds or interest expenses in an overdraft account.

Table 6.5. An Example of a Cash Flow Status Report

Costs 7/02	Charges 8,754,516	Estimated 65,863,092	Percentage Complete 13.292	Projected 66,545,263	Change 6,82,171
Billings 7/01	Contract 67,511,602	Gross Bill 9,276,621	% Billed 13.741	Profit 9,66,339	
Payables 7/01	Paid 6,719,103	Open 1,300,089	Retention 3,91,671	Labor 3,43,653	Total 8,754,516
Receivable 7/02	Net bill 8,761,673	Received 7,209,344	Retention 5,14,948	Open 2,067,277	
Cash position	Paid 7,062,756	Received 7,209,344	Position 1,46,588		

Source: https://www.cmu.edu/cee/projects/PMbook/12_Cost_Control,_Monitoring,_and_Accounting.html#12.4%20Financial%20Accounting%20Systems%20and%20Cost%20Account.

The project's status needs to be assessed by combining the varied information presented in Table 6.5. Each accounting system represented in the table offers a unique perspective on the project's progress. For instance, while the budget data shows higher costs than anticipated, there is still projected profitability. However, there is concern about the owner's outstanding payment, as delays could pose a problem. On a positive note, the project currently has a favorable cash balance, which is advantageous in terms of avoiding financing charges.

The main instrument for project cost control is the job status reports that are discussed in this and the preceding sections. For each person involved in a project, a different report would be prepared, with varying levels of detail and item reports. Summaries would be sent to upper management, specific staff members' reports would highlight their duties (payroll, purchasing, etc.), and project managers would receive detailed reports.

6.3.6. Schedule Control

The primary tool for controlling project costs is the job status report, discussed in this section and the previous one. Each project member receives a tailored report with varying levels of detail. Summaries are sent to upper management, while specific staff reports highlight their responsibilities such as payroll and purchasing. Project managers receive detailed reportGs.

The methods used for forecasting completion times of activities are directly analogous to those used for cost forecasting. For example, a typical estimating formula might be:

$$D_f = Wh_t \qquad (5)$$

where; D_f is the forecast duration; W is the amount of work; and h_t is the observed productivity to time 't.' As with cost control, it is important to devise efficient and cost-effective methods for gathering information on actual project accomplishments.

Figure 6.2, for instance, compares the project's actual progress to its initial schedule. This figure is created by adding up the percentage of each activity that is finished at various times. The amount of effort that goes into each activity can be used to weight this total. The project in Figure 6.2 was behind schedule for the portion of time that included point A, but it is currently behind schedule at point B by the amount equal to the horizontal difference between the expected and actual progress that has been observed thus far.

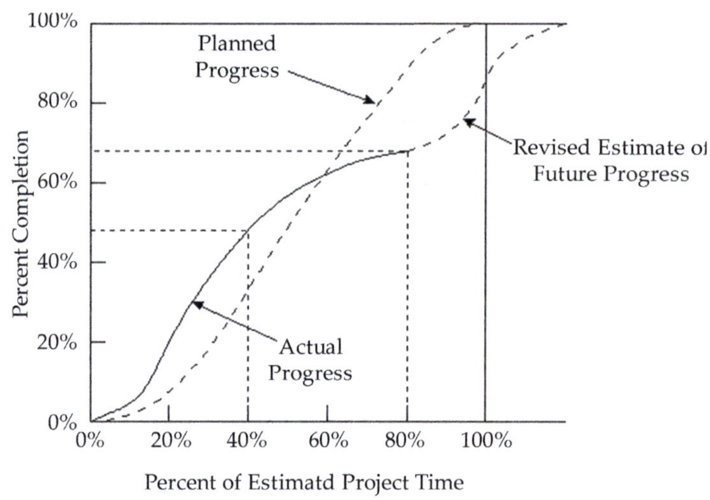

Figure 6.2. Illustration of planned versus actual progress over time on a project.

Source: https://www.cmu.edu/cee/projects/PMbook/12_Cost_Control,_Monitoring,_and_Accounting.html#12.4%20Financial%20Accounting%20Systems%20and%20Cost%20Accounts.

Geometric models of a facility can also be used to show schedule adherence and the state of a project. On a computer screen, for instance, an animation of the building sequence may be displayed, with various colors or other coding schemes designating the various types of activities taking place on each facility component. Another way to illustrate deviations from the schedule is through color coding. As a result, a system for identifying ongoing work as well as schedule adherence unique to certain facility components is created.

When assessing the progress of the schedule, it is crucial to consider that certain activities have scheduling leeway, or float, while delays in activities on the critical path will result in project delays. Specifically, a project delay may result from the planned progress delay at time t, or it may be absorbed by the float of activities, resulting in no overall delay in project completion. Because of this uncertainty, it is best to revise the project schedule in order to create a precise representation of schedule adherence. A new project schedule can be obtained by running it through a scheduling algorithm. A report or graph akin to Figure 6.3's can be created for cash flow planning purposes in order to compare actual and planned spending at any given point in time. There are numerous situations in which updating the schedule and budget may be necessary, and this process of rescheduling to reflect schedule adherence is just one of them.

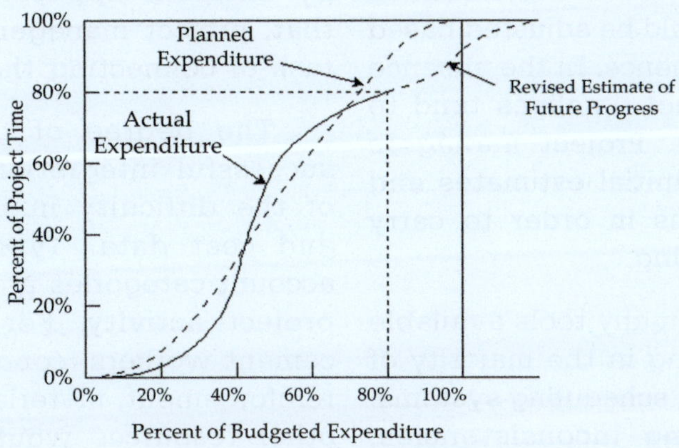

Figure 6.3. Illustration of planned versus actual expenditures on a project.

Source: https://www.cmu.edu/cee/projects/PMbook/12_Cost_Control,_Monitoring,_and_Accounting.html#12.4%20Financial%20Accounting%20Systems%20and%20Cost%20Accounts.

6.3.7. Schedule and Budget Updates

Project planning and scheduling are ongoing activities that last the whole duration of the project. New schedules should be created and the project schedule and cost estimates adjusted as adjustments or differences arise between the realization and the plan. All too frequently, a planner in the central office creates the schedule just once, and only occasionally or incompletely makes revisions or modifications afterward. As a result, there may not be enough efficient project monitoring, which could lead to eventual chaos on the project site.

On fast-track projects, the building process starts before the facility design is complete. The coordinated scheduling of design and construction activities needs to be given extra consideration in this situation. Change orders reflecting final design modifications are frequently issued to incorporate changes requested by the owner, even in projects where the design is completed before construction starts.

It is particularly crucial to update future activity durations and budgets on a regular basis to prevent overly optimistic outcomes in struggling projects. In the event that management adjustments are not made, linked activities that encounter delays on a project are also likely to experience delays. When similar materials, tools, laborers, or site features are used, a variety of closely related activities are typically involved in construction projects. A project plan should also take anticipated cost changes into account. Essentially, future activity duration and cost estimates should be adjusted based on real-world job experience. In the absence of this updating, project timelines tend to deteriorate over time. Project managers require access to the initial estimates and estimating assumptions in order to carry out this kind of updating.

Sadly, there aren't many tools available for this kind of updating in the majority of project cost control and scheduling systems. A method for locating inconsistencies, determining the cause, projecting the outcome, and spreading this outcome to all associated tasks is needed. Although these steps can be completed manually, it would be beneficial to have computers support interactive or even automatic updating.

6.3.8. Relating Cost and Schedule Information

Real projects have intricate relationships between budget and schedule. Project delays have an impact on costs, and financial issues may necessitate modifying activity schedules. Time and cost trade-offs are inherent in project planning; allocating additional resources to a project activity may shorten its duration but will likely increase its overall cost. Unexpected circumstances may cause an activity's completion time and cost to increase. For instance, issues with excavation could easily result in significantly lower productivity than expected for tasks requiring digging.

Although time and cost have an implicit relationship in projects, effective project control systems that incorporate both factors are uncommon. Project schedules and costs are typically tracked and reported by different application programs. After that, project managers have the laborious task of connecting the two sets of data.

The degree of detail necessary for successful integration is the main cause of the difficulty in integrating schedule and cost data. Typically, multiple cost account categories are involved in a single project activity. For instance, laborers, cement workers, concrete forms, concrete, reinforcement, material transportation, and other resources would all be needed for a foundation preparation project. Even a more simplified definition of an activity, like building foundation forms, would require a wide range of resources, including laborers, carpenters, forms, nails, and material transportation. Once more, these different resources would typically be recorded under separate cost accounts. Similarly, costs related to specific cost accounts may be incurred in a variety of activities. Standard piping, for instance, is a material that can be used for a variety of schedule activities. Analysis must begin with the disaggregated charges for particular activities and cost accounts in order to integrate cost and schedule data.

Determining individual work elements that correspond to the resources in a given cost category connected to a specific project activity is an easy way to relate time and cost data. As shown in Table 6.6, work elements would stand for an element in a two-dimensional matrix of activities and cost accounts. For work elements, a system of numbering or identification would comprise the relevant cost account and the related activity. In some circumstances, it could also be preferable to list the work components according to the accountable company or person. In this instance, a three-dimensional depiction of the work components is needed, with the responsible individuals in the third dimension. In general, contemporary computerized databases can accommodate a flexible data representation structure to facilitate aggregation from multiple angles.

This information organization could result in the creation of several management reports or views. Specifically, the total of the work elements that show up in any row in Table 6.6 could be used to determine the costs related to particular activities. These expenses could be applied to determine the expected project cash flow over time as schedule modifications occur, or they could be used to assess alternative technologies for carrying out specific tasks. As costs would be accumulated at such a disaggregated level, issues arising from specific activities could be quickly identified from a management standpoint. Project control consequently becomes more exact and comprehensive.

Table 6.6. Illustration of a Cost Account and Project Activity Matrix

Project Activity Group	Cost Amount for Superstructure					
	204.1	204.2	204.3	204.4	204.5	204.6
First floor	x	x		x		x
Second floor		x		x		x
Third floor		x	x	x		x
Fourth floor		x	x			x
Fifth floor		x	x		x	x

Source: https://www.cmu.edu/cee/projects/PMbook/12_Cost_Control,_Monitoring,_and_Accounting.html#12.4%20Financial%20Accounting%20Systems%20and%20Cost%20Accounts.

Regrettably, creating and managing a work element database may require a significant amount of organizing and data gathering. As was previously mentioned, 400 distinct cost accounts and 400 activities would not be out of the ordinary for a building project. Up to 400 × 400 = 1,60,000 distinct work elements would be the outcome. Not every activity is associated with every cost account. However, even with a density of just 2%, which would mean each activity has eight cost accounts and each account has eight associated activities on average, there would still be nearly 13,000 work elements to consider. Setting up this database initially is a significant task, and project bookkeepers must also record project events for each of these work elements. Implementing work element project control systems has often struggled with the challenges of collecting, storing, and managing data.

The use of work elements to control activities in large projects is likely to be challenging to implement until data collection is better automated. On the other hand, some project activity segments can greatly benefit from this kind of organization. This makes it possible to track material requirements in particular. The burden of data collection and control for materials is significantly less than that of an entire system because materials only involve a portion of all cost accounts and project activities. Furthermore, since delivery schedules are directly impacted and bulk order discounts may be found, the advantages of integrating schedule and cost information are especially evident in materials control. As a result, it makes sense for materials control systems to include a work element accounting system.

Costs related to specific activities are typically estimated in the absence of a work element accounting system by adding up all costs directly related to the activity and a percentage of costs in cost accounts shared by two or more activities. Usually, the amount of work or resources needed for each activity would serve as the basis for allocating costs. For instance, depending on the quantity of work (measured in cubic yards of concrete) involved in each concreting activity, the costs of supervision may be distributed among the various activities. Cost estimates for specific work activities can be obtained with these allocations.

6.4. COST RECONCILIATION IN CONSTRUCTION PROJECTS

Millions of dollars are spent on personnel, materials, and equipment during construction projects. While cost reconciliation is a routine task, we find that many construction project finance, accounting, and commercial teams waste valuable time trying to finish their reconciliation tasks each month.It is evident that the problems are universal and not specific to any construction project. For decades, project teams have been reconciling expenses in the same way, so it's hard to even think of a better method.

6.4.1. Cost/Expense Reconciliation

Cost reconciliation involves verifying that the expenses of a project align with the ongoing activities. Typically, a three-way matching process is used to validate the contractor's costs by comparing them against two other sources of evidence, such as receipts, invoices, time cards, and transaction history (Javad, Nafiseh, & Paria, 2019) (Figure 6.4).

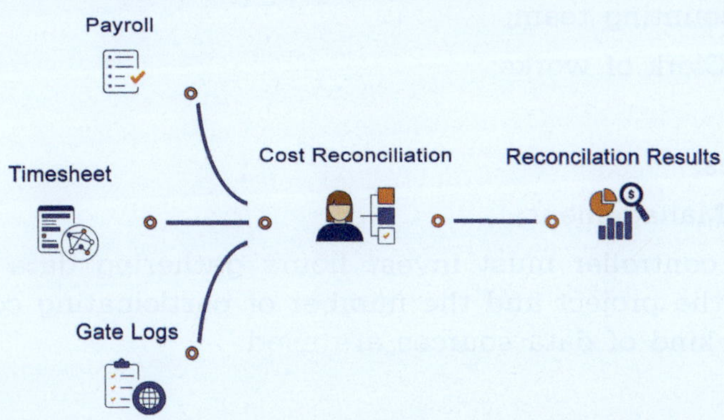

Figure 6.4. Cost reconciliation.

Source: https://blog.morta.io/cost-reconciliation

It would be relatively simple to feel confident in the spending if you were the only one overseeing the project. However, in actual projects, this is almost never the case because numerous items are purchased for the project. Organizations can feel secure knowing that their spending was authorized, necessary, and supported by the required documentation thanks to the reconciliation process.

The process of reconciling costs can differ depending on the item being analyzed, as well as the project or organization in question. When it comes to labor costs, the focus is on comparing the actual cost with the work completed, usually documented in a timesheet, and verifying the worker's physical presence through an independent source, such as a timekeeper or gate data. In some cases, wearable technology may also be used for verification. Ultimately, every cost incurred is cross-checked with some form of evidence to ensure accuracy.

6.4.2. Challenges in Reconciliation Processes

In actuality, reconciliation should be relatively easy work, but it turns into a nightmare very fast for everyone involved. The problems are usually the same whether the company uses an ERP, accounting system, project management system, or even an Excel spreadsheet.

6.4.2.1. Number and Variety of Data Sources

Cost reconciliation on a construction project entail sharing cost-related information among various parties involved. This usually includes different types of costs, like equipment rentals, material purchases, and labor costs, and multiple stakeholders, such as the owner, consultant, main contractors, and subcontractors.

Consequently, one would expect the presence of numerous sources of information with multiple unique formats like:

- Timecards – Site team;
- Payroll – Accounting team;
- Gate data – Clerk of works;
- Invoices;
- Delivery notes;
- Approvals – Management.

As a result, the controller must invest hours gathering data from these various teams. The scope of the project and the number of participating companies determine how many and what kind of data sources are used.

6.4.2.2. Can Only Be Reconciled by People (PDF)

The cost data contained in those files can only be reconciled by people due to the construction industry's widespread acceptance of the PDF format as a means of transferring cost-related information. It typically takes a team several days to reconcile the costs of a large-scale construction project from those PDF files due to the volume and variety of information involved in typical construction projects. This adds up to a significant amount of additional expenses that the owner or contractor must pay in order to balance the budget for the project or projects.

6.4.2.3. Poor Feedback/Visibility

We've talked about the difficulties posed by the quantity and variety of data sources that are used in cost reconciliation, the fact that only humans can reconcile cost reconciliation documents, and the possibility of inaccurate information quality and format. We have now reached two outcomes or after-effects of cost reconciliation: visibility and feedback. It's a common saying that goes, "Garbage in, garbage out."

CASE STUDY

FACTORS EFFECTING THE COST MANAGEMENT IN CONSTRUCTION PROJECTS:

Construction sector is an important sector in developed and developing countries and affects other sectors in direct and indirect ways. Therefore, special attention should be given to control and manage it wisely particularly in the aspect of cost management. Cost management is one of the important factors determining project completion successfully within limited time with reasonable profit. Cost management can be defined as a strategic process that stresses the optimization of efficiency and focuses on the customer and on profitability. It is a philosophy, an attitude, and a set of techniques to create more value at lower cost, consisting of six steps including: Understanding what causes the cost and revenue structure of the business, understanding and reducing inter-functional complexity, Providing the tools to manage costs, Involving employees in decisions, Increasing effectiveness and continuously improving costs, Measuring decisions against the strategic business plan. Burke et al suggested that there are two types of construction costs: Direct costs such as material, labor, and equipment, and Indirect costs such as general overhead and job overhead. These costs need to be controlled and managed. Halepota recommended eight techniques in construction projects, namely: budget planning and control, estimating, budgeting, cash flow forecasting, financial reporting and cost reporting, cost code systems, value management, and judgment.

Cost planning should be applied in order to reduce problems throughout the life cycle of construction projects. The six overall cost planning advantages proposed by Ashworth and Perera are as follows: (i) the tender amount is more likely to match the approved budget estimate; (ii) there is less possibility of addendum in bills of quantities or re-costing being required; (iii) it is more likely to achieve cost-effectiveness and value for money design; (iv) a more rational design is likely to result in a balanced distribution of expenditure; (v) the amount of pre-tender analysis by the architect and quantity surveyor should make it possible to make more decisions earlier, resulting in a smoother running of the project on site; and (vi) a sound basis for comparing different projects is cost planning.

- *Problems of Cost Management:* Construction projects are a huge sector in all countries because they involve different types of work that require various materials, experiences, equipment, techniques, and specializations. These need to be managed and planned, so cost management is important in construction projects. Many researchers study the cost performance in different stages throughout the life cycle of construction projects. Life cycle costing is an important contribution to feasibility studies in construction projects, providing a decision-making scheme for investors that allows evaluation not only of investment costs but also of life cycle savings and benefits. This is crucial as it shifts the objective from decision-making based solely on construction costs to life cycle costs. Cost overrun is a major problem in project development and is a common issue in the construction industry. One of the reasons for cost overruns in construction projects is rework. The cost impact value represents the cost effect of rework, if it occurs, and consists of two parts. The first part is the total daily rate of successor activity, which includes daily salaries and daily overheads. The second part is related to any other costs (besides the daily rate of successor activity). Many problems impact cost overruns as determined by various studies. For example, Ahmed et al. stated that minor delays are often neglected because they develop slowly during the construction process, but their cumulative effect impacts the project financially. Other studies concluded that insufficient basic project data and a delay in the process of preparing, submitting, and approving shop drawings could adversely affect the schedule and cost performance of the project. According to Sriprasert, inefficient construction management and poorly established cost control systems cause cost overruns. Koushki et al revealed that material-related issues contribute to cost overruns. In addition to that, a project might face different problems because of inaccurate, incomplete, or untimely information.

- *Factors Affecting Cost Management:* Cost management is one of the fundamental criteria of success in construction projects; thus, it should be managed and controlled along the lifecycle of the project. The study focused on the evaluation of cost management practices in large construction projects in South Malaysia. The results indicated that the most effective cost management techniques were: Cash Flow Forecasting, Tender Budgeting/Estimating, and Elemental Cost Plan. The most important factors that affected cost management of construction projects in South-western Nigeria were: poor leadership and inappropriate management, inefficient deployment of resources, excessive wastage of materials on sites, complex payment mechanisms, theft of materials on sites, and variation during construction works.

Iyer & Jha indicated the success and failure factors from literature review to find the factors affecting cost performance of Indian construction projects. The results of the questionnaire survey concluded that the critical success factors obtained by the analysis were: project managers' competence; top management support; project managers' coordinating and leadership skills; monitoring and feedback by the participants; coordination among project participants; and owners' competence and favorable climatic conditions. On the other hand, the factors affecting the cost performance of projects were: conflict among project participants; ignorance and lack of knowledge; presence of poor project-specific attributes and lack of cooperation; hostile socio-economic and climatic conditions; reluctance in timely decision-making; aggressive competition at the tender stage; and short bid preparation time.

Cost estimation is an important stage in the cost management of construction projects. As mentioned by Hatamleh et al., to estimate the cost of a construction project, there are important elements that should be considered: comprehensive information, expanded knowledge, considerable expertise, and continuous improvement to ensure accurate estimation. This study identified the factors that affect cost estimation to increase accuracy in the cost estimation of projects. The most important factors affecting the accuracy in cost estimation were: clear and detailed drawings and specifications, pricing experience of construction projects, perception of the importance of estimation, equipment (cost/availability/performance), project complexity, clear scope definition, accuracy and reliability of cost information, site constraints (access, storage, services), material availability, financial capabilities of the client, and availability of a database of bids on similar projects (historical data).

The study by Sinesilassie et al. aimed to indicate and evaluate the factors affecting the success of cost performance in Ethiopian public construction projects. The research method followed the steps below: identification of attributes affecting project performance, development of a questionnaire, selection of respondents, and data analysis method. The results revealed seven success factors and six failure factors that affect the cost performance of Ethiopian public construction projects.

These factors were:

1. **Success Factors:** Project manager's competence, scope clarity, owner's competency, monitoring and feedback, coordination and communication among project participants, top management support, quality control and assurance.
2. **Failure Factors:** Conflict among project participants, Project-specific factors, Indecisiveness of project participants, project manager's ignorance and lack of knowledge, socio-economic and climatic conditions, owner's incompetence.

The key factors affecting construction performance are: (i) time; (ii) cost; (iii) quality; (iv) safety & health; (v) internal stakeholders; (vi) external stakeholders; (vii) client satisfaction; (viii) financial performance; (ix) environment; and (x) information, technology, and innovation. These are indicated through the development of a multidimensional performance evaluation framework for a construction company.

Cost overrun is observed as the most frequent problem found in construction projects, which impacts the cost management of projects in different countries. This is in agreement with Shehu et al., who stated that the majority of construction projects in Malaysia suffer from cost overruns. For example, in Afghanistan, construction cost overruns were reviewed by Niazi through a literature review, where 69 causes of project cost overruns were classified into three categories. These causes were initiated by the client, contractor, and consultant. The most important factors were corruption, delays in progress payments by clients, difficulties in financing projects by contractors, security issues, and change orders by clients during the construction phase. India has limited studies on cost overruns of construction projects, according to Wanjari & Dobariya. Their study revealed that there are few studies on cost overruns in India. The researchers identified 15 factors to be ranked through a questionnaire survey distributed to construction professionals across India (northern region, southern region, western region, eastern region, and central regions), with only three factors indicated as the most important in causing cost overruns in construction projects in India: price escalation of raw materials, delays in planned activities, and lack of coordination between construction parties. In order to reduce the impact of these factors on cost overruns, a framework for each factor is suggested.

The main factors of cost overrun in construction projects in the Gaza Strip are: price fluctuations of construction materials, contractor delays in material and equipment delivery, and inflation, while the important factors of delay were as follows: strikes and border closures, material-related factors, lack of materials in markets, and delays in material delivery to the site (Enshassi, Al-Najjar, & Kumaraswamy, 2009).

The findings of the study revealed the different factors affecting cost management in construction projects which were found in the literature review of different researchers in different countries. The following chart indicates the most important factors affecting cost management in different countries.

- *Conclusion:* Cost is one of the important key factors affecting construction performance. It needs to be managed and controlled throughout the life cycle of a construction project. Therefore, cost management is important in construction projects to ensure project success. The most common factors affecting cost management, as indicated in the literature review of this study, are: poor scope definition, inaccurate activity cost estimates, poor work breakdown structure definition, changes in schedule, unrealistic time schedules imposed in contracts, ineffective frequency of project budget updates, lack of proper training and experience of project managers, and not implementing project management tools like Primavera and other management software. The majority of construction projects in different countries have suffered from cost overruns. Cost management helps reduce these overruns and assists project managers in achieving success in their projects. This study encourages engineers and other stakeholders interested in construction projects to actively implement cost management. Future studies can be conducted to identify other factors or to conduct in-depth studies on the same factors, to help stakeholders in construction projects understand cost management and how to apply it throughout the project life cycle.

CLASS ACTIVITY

BUILDING ON A BUDGET: CONSTRUCTION COST MANAGEMENT:

- **Objective:**
  - To understand the key concepts and processes involved in cost management in construction.
  - To apply cost estimation, budgeting, and cost control techniques in a simulated construction project.
  - To enhance teamwork, problem-solving, and decision-making skills.
- **Materials:**
  - Printed project briefs for different construction projects (e.g., residential house, commercial building, bridge);
  - Cost estimation worksheets;
  - Budget templates;
  - Calculators;
  - Flip charts or whiteboards;
  - Markers;
  - Internet access for research (optional);
  - Cost management software (optional).

SUMMARY

- Cost management is a form of management accounting that helps a business reduce the chance of going over budget with more accurate forecasts of impending expenditures.
- Embarking on a construction project is a complex endeavor, with many costs to consider across each phase. The efficiency and success of the project hinge on a comprehensive understanding of these types of construction costs.
- In construction phase costs, the contract sum holds a critical role. This total monetary commitment is a pivotal aspect in the dynamics of any construction project.
- During the lifecycle of a construction project, there are additional costs known as Other Project Costs that need to be accounted for.
- The term "cost planning" generally describes an instrument for estimating costs that will be incurred in the future. In a business sense, this can refer to the company as a whole, individual company divisions, products, and services.
- Cost control in construction is the process by which managers keep expenses under control by managing labor, material, and overhead costs to ensure that the project finishes on budget.
- During the execution of a project, procedures for project control and record-keeping become indispensable tools for managers and other participants in the construction process.
- For cost control on a project, the construction plan and the associated cash flow estimates can provide the baseline reference for subsequent project monitoring and control.
- The forecasting procedures assume linear extrapolations of future costs, based either on the complete experience of the activity or the recent experience.
- When evaluating schedule progress, it is important to bear in mind that some activities possess float or scheduling leeway, whereas delays in activities on the critical path will cause project delays.
- Cost reconciliation is the process of checking records to ensure that the project's expenses match ongoing activity in the project.

REVIEW QUESTIONS

1. What is included in construction costs? Discuss.
2. Explain the importance of pre-construction costs.
3. Focus on resource planning as a basis for the cost plan.
4. What do you understand by the cost control problem?
5. What do you understand by financial accounting systems and cost accounts?
6. How to control of project cash flows?
7. What do you understand by cost reconciliation in construction projects?

REFERENCES

1. Aguinaldo, S., Andrew, P. J., & Carlos, T. F. (2000). Setting stretch targets for driving continuous improvement in construction: Analysis of Brazilian and UK practices. Work Study, 49(2), 50–58.
2. Chadee, A. A., Chadee, X. T., Chadee, C., & Otuloge, F. (2022). Violations at the reference point of discontinuity: Limitations of prospect theory and an alternative model of risk choices. Emerging Science Journal, 6, 37–52.
3. Chadee, A. A., Ray, I., & Chadee, X. T. (2021). Systemic issues influencing technical certainty in social housing programmes in a small island developing state. Buildings, 11(2), 65.
4. Chadee, A., Ali, H., Gallage, S., & Rathnayake, U. (2023). Modelling the implications of delayed payments on contractors' cashflows on infrastructure projects. Civil Engineering Journal, 9, 52–71.
5. Ellram, L. M. (2006). The implementation of target costing in the United States: Theory versus practice. The Journal of Supply Chain Management, 42(1), 13–26.
6. Enshassi, A., Al-najjar, J., & Kumaraswamy, M. (2009). Delays and cost overruns in the construction projects in the Gaza Strip. Journal of Financial Management of Property and Construction, 14, 126–151.
7. Fayek, A. R. (2013). Factors causing cost variation for constructing wastewater projects in Egypt. Alexandria Engineering Journal, 52(1), 51–66.
8. Gouett, M. C., Goodrum, P. M., & Caldes, C. H. (2011). Activity analysis for direct-work rate improvement in construction. Journal of Construction Engineering and Management, 137(12), 1117–1124.
9. Hang, X., & Ni, Z. (2005). Cost management of construction project on BOQ price basis. Building Construction, 27, 67–71.
10. Hussein, H. A., & Razzaq, Z. (2021). Strengthening prestressed concrete bridge girders and building beams with carbon fiber reinforced polymer sheets. European Journal of Engineering and Technology Research, 6(1), 55–57.
11. Hwang, B. G., & Zhao, X. (2014). Investigating the client-related rework in building projects: The case of Singapore. International Journal of Project Management, 32(4), 698–708.
12. Javad, E. M., Nafiseh, N., & Paria, S. (2019). Measuring the impact of soft and hard total quality management factors on customer behavior based on the role of innovation and continuous improvement. The TQM Journal, 31(6), 1093–1115. https://doi.org/10.1108/TQM-11-2018-0182.
13. Jimoh, R., Oyewobi, L., Isa, R., & Waziri, I. (2019). Total quality management practices and organizational performance: The mediating roles of strategies for continuous improvement. International Journal of Construction Management, 19(2), 162–177.
14. Kazaz, A., Birgonul, M., & Uiuberyli, S. (2005). Cost-based analysis of quality in developing countries: A case study of building projects. Building and Environment, 40, 1356–1365.

15. Love, P. E. D. (2002). Influence of project type and procurement method on rework costs in building construction projects. Journal of Construction Engineering and Management, 128(1), 18–29.
16. Ma, Z. (2008). Construction cost control under engineering quantity list valuation. Construction Economy, 313, 116–118.
17. Mezher, T. M., & Tawil, W. (1998). Causes of delays in the construction industry in Lebanon. Engineering, Construction and Architectural Management, 5, 252–260.
18. Peter, E. D., & Lovea, Z. (2003). A project management quality cost information system for the construction industry. Information & Management, 40(7), 649–661.

CHAPTER

7 Quality Management

LEARNING OBJECTIVES

After studying this chapter, you will be able to:

- Understand basics of quality management;
- Identify quality management standards;
- Focus on managing quality production on site: planning and control;
- Know steps to improve construction quality management.

KEY TERMS FROM THIS CHAPTER

Quality management
Customers
Employees
Quality control
Leadership
Decision making
Documentation
Stakeholders
Investors
Suppliers
Quality assurance
Improvement
Relationship management

INTRODUCTION

Construction projects encompass a variety of tasks spanning different sectors and activities. However, they all share a common goal – to provide high-quality work within the designated timeframe and budget. Quality management serves as the cornerstone of construction projects and is crucial in determining their success or failure. Every construction company strives to deliver top-notch work, as it is a key factor in shaping its reputation and standing within the industry. Implementing quality management practices involves setting policies, processes, and procedures to enhance the project approach and ensure that the desired quality standards are met. Adhering to both internal and external principles and guidelines is essential for achieving quality in construction projects.

7.1. BASICS OF QUALITY MANAGEMENT

Construction projects are intricate endeavors with numerous stakeholders and frequently constrained timelines. For these kinds of projects to succeed and satisfy clients, quality control is essential.

7.1.1. Understanding Quality

Making organizations function for their stakeholders is the goal of quality management. This includes ensuring that the entire organization is efficient and fit, as well as enhancing systems, procedures, goods, and services. Managing quality entails pursuing excellence on a continuous basis, ensuring that the work your organization does is appropriate for its intended use and not only remains so but also continues to advance.

Although those things are undoubtedly important, managing quality involves much more than just producing flawless widgets or ensuring that trains depart on time. Your stakeholders will ultimately determine what constitutes an acceptable level of quality for your organization. And by stakeholders, we mean everyone who has a stake in the accomplishments of your company.

For most businesses, customers will be the most significant stakeholder group; however, other stakeholders include investors, employees, suppliers, and members of the general public. In order to provide your organization with acceptable quality, you must identify your stakeholders, comprehend their needs, and meet—or better yet, surpass—their expectations—both now and in the future. Strong governance to clarify the organization's goals and put them into practice, reliable assurance systems to ensure that things proceed as planned, and an improvement culture to continuously improve are the three main components that the CQI believes are essential to this.

7.1.2. Defining Quality Management System (QMS)

A formalized system that records duties, processes, and procedures for accomplishing quality policies and objectives is known as a quality management system, or QMS. A QMS aids in organizing and directing an organization's operations to satisfy legal and customer requirements and continuously enhance its efficacy and efficiency.

The most well-known method for quality management systems is ISO 9001:2015, an international standard that outlines requirements for these systems. Although the ISO 9001 standard or the collection of documents outlining the QMS are sometimes referred to as the "QMS," the term truly refers to the complete system. The system is only described in the documents.

7.1.3. Defining Quality Control in Construction

A system of management known as construction quality control makes sure that deliverables adhere to the requirements and specifications that the client establishes at the start of the construction process. This can involve meeting a variety of requirements, like finishing the project within the allocated time. Quality is ultimately determined by the customer, authorities, and EPA regulations.

Two equally crucial facets of construction quality are quality assurance and control. Quality control refers to the strategies and processes that produce high-quality results, whereas assurance refers to establishing standards for quality management. Problems with construction quality control occur when the elements influencing the quality are not correctly recognized and dealt with. In construction, quality control aims to address issues head-on, deliver superior outcomes, and stop problems from arising later.

7.1.4. Principles of Quality Management System in Construction

The process of quality management is divided into two parts in order to meet its two goals. Each part of the process is specifically responsible for enhancing the overall quality. Quality Assurance (QA) and Quality Control (QC) are the two subsets of quality management. A well-known example of the quality umbrella highlights the elements and their significance. For effective quality management on a construction project, a few fundamentals of a construction quality management system are required. The following are the critical seven principles of a construction quality management system:

- Client focus;
- Leadership;
- Engagement of people;
- The process approach;
- Improvement;
- Evidence-based decision making;
- Relationship management;
- Several activities converge to make a quality management plan. These activities vary from monitoring to auditing.

7.1.5. The Importance of Quality Management in Construction

The most important factor is probably how construction quality impacts your revenue. Finishing tasks at a high standard will keep clients coming back to you and shield you from expensive errors that need to be fixed. On the other hand, subpar construction projects can harm your reputation and cause you to lose future business from both prospective return customers and others who learn about the subpar quality of your work.

Quality can negatively affect your ability to finish closeout and guarantee that all quality issues are resolved throughout a project. According to an Autodesk and Dodge Data and Analytics report, 69% of general contractors say they have trouble "getting off the job" on at least 25% of their projects, demonstrating the industry-wide nature of this issue. In addition, almost 75% of them claim that this has a "high" to "very high" detrimental effect on their profitability.

In addition to having a negative financial impact, subpar construction can waste materials, time, and resources. Improved construction quality leads to increased project efficiency; that is, the more efficient your workflows and processes are, the easier it will be to complete a job on time and without needing to rework anything. Another important issue is project safety, which can be negatively impacted by subpar project execution. Stated differently, superior construction endeavors enhance security and yield safer final goods as a consequence. These project outcomes also frequently last longer and are more effective.

7.1.6. Elements of a Quality Management Process

Following are the elements involved in the development of a quality management process:

1. Quality Standards: Understanding and controlling the final project quality requires first understanding the specifications for each quality requirement. Clarifications of any of the quality requirements and agreements reached with the project owner and/or designer are documented and added to the quality standards. The contractor's quality management program (CQMP) is based on this. The project staff must be given this information.
2. Contractor Quality Management Program: A written document outlining the contractor's methods, practices, and procedures to guarantee that the project's quality requirements are met or surpassed is called the quality management program. The quality assurance procedures and the quality control plan are the two components of the program.

i. QC Plan: The contractor's plan for managing the project's quality during construction is known as quality control, or QC. Every special project quality requirement needs to be outlined in a document dedicated to the project. It outlines who is in charge of meeting the quality standards as well as how this is to be done. It creates a structure with clear guidelines and practices to guarantee that the finished product satisfies or surpasses the quality standards set forth in the project.

ii. QA Process: Quality assurance (QA) refers to the method or technique that the contractor will use to guarantee that the project meets the necessary quality standards. This procedure outlines the requirements for inspections, when they should be conducted, the written report that will be produced, who will receive and review it, and who will correct any errors, along with the properly defined follow-up.

7.1.7. Key Challenges in Construction Quality Management

Despite having numerous contractors and teams working on projects, not every project is able to meet compliance standards. Without quality management systems or policies in place, it becomes exceedingly challenging to ensure quality assurance and control. Superintendents often find themselves overburdened with the responsibility of maintaining quality standards on construction sites. Additionally, the quality and standards of work may vary due to changes in crew members and suppliers. Without standardized systems and processes for conducting quality control, achieving quality assurance becomes a daunting task. Implementing an integrated construction management approach can help improve quality on sites and minimize rework. Quality management in construction begins with recognizing the various factors that can impact safety and quality on projects.

7.1.7.1. Defective and Low-Standard Materials

Improperly graded steel, timber cut from stunted trees, and too much water or sand in concrete mixes can all lead to problems with construction quality. These materials not only have an early failure rate, but they also present a safety risk while being built. For instance, worker injuries frequently occur when sparks fly during cutting or when a poorly supported structure collapses. The sole method to ensure that a project is supplied accurately is to solely purchase from reliable vendors and assign a quality control officer to supervise the shipments of materials.

7.1.7.2. Supplier Quality Issues

Even when the materials themselves are not the cause of the problem, issues with vendors and suppliers can drive up costs and decrease quality standards. For instance, substituting a different brand or material for building supplies that does not match the same standards could result in disgruntled clients and time-consuming requests for rework. Verify that all suppliers are aware of the specifications, and carry out frequent audits to make sure they are adhering to them. Finding new suppliers during a construction project's infancy may seem like a distraction, but it can improve the quality of the work.

7.1.7.3. Poor Subcontractor Management

According to a study, more than half of all construction defects are the result of human error. Subcontractors run the risk of making mistakes that go undiscovered for years when they hire workers without the necessary training or qualifications. It is advisable to screen labor providers, including subcontractors, to make sure they are capable of self-correction and have the necessary skills. For contractors and project managers to identify and address issues early on, audits of subcontractors are still required.

7.1.7.4. Lack of Documentation of Changes and Practices

Sometimes the absence of documentation is the root cause of quality issues rather than mistakes or design modifications. For instance, if all final documentation is not updated, the maintenance team may handle materials improperly if they have different maintenance and replacement intervals. Updates to project documents ought to be simple to make with a Document Change and Control Management System. There is no justification for holding off on updating related files and drawings.

7.1.7.5. Managing Unexpected Changes

Last-minute modifications frequently result in major quality problems during the building process. For instance, a fatal collapse occurred at the Kansas City Hyatt Regency in July 1981 as a result of a modification to the tie rod supports' design. Make sure you schedule extensions or establish deadlines by which designs must be finalized. Allow enough time to test and validate any modifications to the current design.

7.1.7.6. Increased Scope

Most building projects are considerably smaller and easier to start than they become. What causes an elementary bridge to become a multilane highway? Scope creep is the term for this kind of unplanned expansion. Every project's scope varies over time as new expenses, deadlines, and restrictions on a specific location are identified. But the trouble starts when these adjustments lead to compromising in order to fit everything into a constrained budget and time frame. A clearly defined project scope guarantees that the contractors will continue to provide the same level of quality for the duration of the project.

7.1.7.7. Lack of Communication Between Teams

A study found that the majority of quality problems stem from poor communication. Ineffective use of new methods, faulty materials, and a lack of secondary and tertiary testing to identify present issues are all consequences of poor communication. A single construction project can benefit greatly from improved communication between team members when using an integrated eQMS solution, such as Qualityze, which has chatter functionality.

7.1.7.8. Intricacy of Designs

If a high-quality work is too complicated, it will not be affordable. Therefore, in projects involving state-of-the-art infrastructure and commercial construction, complex techniques and unusual features should be minimized whenever possible. Simplified designs also give the construction company a higher profit margin without sacrificing quality.

7.1.7.8. Lack of Proper Process Management

Finding the ideal intervals for reviewing the work for mistakes and omissions can be done with the help of an appropriate construction process management system. Unfortunately, without a quality control and assurance management system, the majority of construction companies neglect to perform necessary on-time inspections of their work. A construction quality management system can make it simple and quick to stay on top of ongoing projects.

7.1.7.9. Failure to Perform Regular Audits

Certain contractors persist in utilizing defective designs or disregarding quality issues, irrespective of the outcomes of their external testing and auditing schemes. There is no clear workflow for managing the process, so reports are passed back and forth between the lead engineer and the project manager. To prevent any disastrous situations, designate the appropriate individual to oversee audits, the corresponding follow-ups, and the timely sharing of improvement recommendations.

7.1.8. Construction Quality Control Approaches

Depending on the requirements and size of a company's projects, there are several approaches to quality control management in the construction industry, each with pros and cons of their own. The ISO 9001 quality standards were created by the International Organization for Standardization. The engagement of people, customer focus, leadership, process approach, improvement, evidence-based decision making, and relationship management form the foundation of these seven standards. There are also four additional primary methods for managing quality control in the construction industry:

- Continuous Improvement: Emphasizes making gradual, ongoing improvements to processes over time. Enhancements are found via internal analytics procedures and customer feedback.
- Kaizen: Kaizen, a Japanese word meaning "change for the better," describes an approach to quality control management that involves constantly seeking out methods to improve. Over time, the adoption of Kaizen by all members of an organization leads to gradual improvements in their daily practices.
- Six Sigma: This framework for problem solving emphasizes the proactive identification and resolution of emerging problems. This quality control management approach consists of the following main steps: define, measure, analyze, improve, and control.
- A Lean Management Approach: The two main components of this strategy are waste reduction and elimination. Waste is defined as unnecessary procedures and materials that don't benefit clients or building firms.

7.2. QUALITY MANAGEMENT STANDARDS

Building companies can become certified by groups like the International Organization for Standardization (ISO) to show that they are dedicated to quality management. The most widely accepted standard for managing quality systems is ISO 9001. A business must prove that it has a quality management system (QMS) in place and that it complies with all standards in order to receive ISO 9001 certification.

7.2.1. Purpose of Quality Management Standards

Companies use standards for a variety of purposes, including meeting the quality requirements of their clients and for the following reasons:

- Ensuring the safety and reliability of their products and services;
- Complying with regulations, often at a lower cost;
- Defining and controlling internal processes;
- Meeting environmental objectives.

Businesses committed to following quality management standards are often more able to:

- Increase their profits;
- Reduce losses or costs across the business;
- Improve their competitiveness;
- Gain market access across the world;
- Increase consumer loyalty.

7.2.2. Benefits of Achieving Construction Industry Standards for Your Business

All of the aforementioned standards were created by ISO to be widely applicable to any type of organization, regardless of sector. Every one of them has a set of standardized procedures and performance standards for tracking results and starting a continuous improvement cycle. Standardization can assist with a number of issues that are common to the construction industry, even though every organization is unique. Some of them are discussed in subsections.

7.2.2.1. Regulatory Compliance

The majority of economic sectors rely heavily on the construction industry. To ensure that they are trustworthy to other businesses, the governmental bodies establish a number of regulatory requirements. In addition to quality standards, the authorities imposed particular rules on construction companies to control energy consumption and guarantee sustainability. The primary reason for this was the growing awareness of the need to protect the environment and adopt greener practices. Achieving the ISO 14001 and ISO 50001 standards for your construction company proves that you follow the rules as they stand. As a result, the standards give your company the ability to comply with legal requirements and offer certifications that assist in avoiding legal issues or fines.

7.2.2.2. Operational Efficiency and Cost Reductions

Reaching quality standards will also enable your construction company to operate more profitably and efficiently on a daily basis. This allows you to adhere to the estimated budget while completing projects on time and to customers of any size or complexity. To achieve compliance with any quality standard, one essential step is to implement a Quality Management System (QMS). A Quality Management System (QMS) aids in the systematic coordination of business processes, performance maintenance, and goal achievement. Therefore, when it is put into practice, it leads to improved communication, visibility, and predictability across all construction operations, which aids in completing tasks on schedule without sacrificing quality. As a result, you can work swiftly and effectively, recognize errors, cut down on reworks, save money, and produce projects with greater accuracy.

7.2.2.3. Workers' Safety

Construction workers are particularly vulnerable to workplace mishaps, injuries, and even fatalities. As a result, ISO 45001, a standard that deals with occupational health and safety, is also regarded as essential for the construction sector. It is a framework for certification that guarantees the proactive application of best practices to stop incidents or hazards related to work. Thus, accomplishing it would show that your construction business is dedicated to producing high-quality projects while fully considering the health and safety of your employees.

7.2.3. Examples of Quality Management Standards

Globally, ISO international standards are the most extensively recognized collection of quality standards. Businesses can use ISO 9001:2015's specifications for a quality management system to create their own quality agendas.

Other standards related to quality include:

- The remaining ISO 9000 family, including ISO 9000 and ISO 9004;
- The ISO 14000 family for environmental management systems;
- ISO 13485 for medical devices;
- ISO 19011 for auditing management systems;
- ISO/TS 16949 for automotive-related products.

7.2.3.1. ISO 9001 Quality Management Standard

An internationally recognized set of written regulations and guidelines known as the ISO 9001 standard offers precise standards for putting in place and upholding a quality system in your company. ISO 9001:2015 is the most recent version of the standard. Seven quality management concepts serve as the foundation for its requirements:

- Customer focus: to recognize and satisfy the needs of the consumer.
- Leadership: to boost workers' motivation.
- Involvement of people: to identify, strengthen, and celebrate the skills of individuals within the organization and their contribution to the provision of high-quality goods and services.
- Process approach, which entails determining your key business operations and treating each one as a separate process.
- Continuous improvement is the process of making progress a long-term goal for the company.
- Evidence-based decision making refers to the process of making decisions based on data analysis.
- Relationship management: acknowledging the mutual reliance between your company and its suppliers.

The principles outline the conditions that must be satisfied by your company, its goods, and its services in order for them to adhere to the standard.

7.2.3.2. Key Requirements of ISO 9001

The 10 sections of the ISO 9001:2015 standard are referred to as ISO clauses. The standard's definitions, terms, normative references, and scope are all covered in clauses 1, 2, and 3. The requirements for each section of the quality management system are outlined in clauses 4 through 10:

- Clause 4 lays out the organization's background and asks you to assess both internal and external problems, interested parties' requirements and expectations, the scope of the quality management system, and its procedures.
- Clause 5 calls for senior management to set and convey a quality policy, assign duties, and exhibit leadership and customer commitment.
- Clause 6 emphasizes the value of planning, calls for you to handle opportunities and risks within the organization, implement changes, and set quality goals.
- Clause 7 mandates that you document data to support the quality management system and provide the infrastructure and personnel that are required, among other resources and assistance.
- In order to guarantee that you provide what customers want, Clause 8 mandates planning for the supply of goods or services. This planning must take into account factors like sales procedures, design and development, purchasing, production, or service activities.

- Clause 9 mandates that you monitor, measure, analyze, and assess your quality management system in addition to conducting performance evaluations.
- Clause 10 calls for constant enhancements to your quality management system, such as resolving non-conformities, implementing preventive and corrective measures, etc.

These specifications can be used by any company in any sector of the economy and serve as a roadmap for any organization looking to implement quality management.

7.2.3.2.1. ISO 9001:2015 Documentation Requirement

To comply with the ISO 9001:2015 standard, you must maintain the following documentation:

- Your quality policy;
- Your quality objectives;
- The scope of the quality management system.

Instead of requiring documented procedures or a quality manual, ISO 9001:2015 refers to the information that an organization must keep, control, and maintain as "documented information." According to the standard, you must keep records of information "to the extent necessary to support the operation of processes." You are free to decide what you require and how you wish to document it. Remember that recorded information serves as proof of compliance, promotes process uniformity, aids in employee training, and aids in knowledge preservation.

7.2.3.2.2. ISO 9004:2018 Standard

If your company has already adopted ISO 9001 and you want to take it a step further, think about implementing the ISO 9004:2018 standard. It uses a quality management strategy built on continuous improvement to assist companies in achieving long-term success. You cannot certify that you have met the voluntary guidelines and recommendations outlined in the standard. Standards can be very general, encompassing a variety of management techniques, or very specific, pertaining to a specific product type. Although using standards is optional, some stakeholders or customer groups might demand it. Additionally, as a condition of doing business, some organizations and governmental bodies may demand that partners and suppliers adhere to standards.

7.2.4. Other Types of Best Practice Standards

Beyond quality management, standards can assist you in achieving best practices in a variety of business activities. For example:

- Accessibility Standards – can help make services or premises accessible to disabled users.
- Occupational Health and Safety Standards – can help reduce accidents in the workplace.

- Information Security Standards – can help keep sensitive information secure.
- Food Safety Standards – can help prevent food from being contaminated.
- Environmental Management Standards – can reduce environmental impact and waste.
- Energy Management Standards – can help cut energy consumption.

7.2.5. Tips to Get Your Construction Business Accredited with Quality Standards

- Inform and train your staff—Involving your staff is essential to achieving and sustaining compliance with internationally recognized standards. Inform them of the advantages of standards for their work and the importance of standards for the expansion of businesses. Regular training sessions are necessary to make sure your staff members possess the necessary skills to adhere to standards and maintain your company's compliance.
- Assure overall accountability: Everyone in a company is accountable for compliance; it is never the responsibility of a single employee, manager, or team. Therefore, by creating a quality policy, hold everyone in your construction company accountable for adhering to the quality processes and putting the standards' best practices into practice.
- Choose your suppliers and partners wisely. The equipment, materials, and inputs utilized in construction projects greatly affect their quality. Therefore, be sure to only collaborate with reputable suppliers and subcontractors if you want your company to be eligible for the quality standards. It is your duty to acquire premium materials and inputs to ensure that projects are completed precisely and in accordance with client expectations.

7.3. MANAGING QUALITY PRODUCTION ON SITE: PLANNING AND CONTROL

The goal of all construction projects is to produce high-quality work within budget and on schedule, even though they involve a variety of functions and specialty parties. A company's success or failure is frequently determined by its quality management, which is a fundamental component of construction project management as a whole. Although every construction company in the world aspires to provide quality work at every stage of the project, the real secret to quality is the establishment of these internal and external principles and guidelines. The following are the main goals of quality management:

- To minimize the defects on asset delivery or handover.
- To identify and solve defects and issues before your customers do – safeguarding your reputation.

Reaching these goals will clearly benefit you in many ways, but nothing will help you more than getting more work and developing a solid reputation.

7.3.1. Creating a Quality Management Plan for Construction Projects

Since preventing errors saves a lot of time and money compared to fixing them, putting in place a robust quality management plan is a smart way to raise standards. Throughout the course of a project, the initial outlay for developing a clear and thorough quality management plan frequently yields significant returns.

Outlined in your quality plan will be four main sections which establish your:

- Quality policies;
- Quality objectives (clear and measurable);
- Requirement standards (ISO accreditations, etc.); and
- Other statutory and legal requirements.

As you can see, construction companies can use a quality management plan as an internal and external tool. It guarantees that you are doing work in a way that complies with legal requirements and practical quality standards. It also makes sure that your internal quality control policies and objectives are set up in a way that promotes continuous improvement and, eventually, high performance.

7.3.2. Establishing Quality Procedures

The creation of procedures that allow you to accomplish your goals is the next logical step in quality management, after you have established a quality plan and your "goal posts." Because there are more moving parts in quality management at this stage of the design process, it can be more challenging for businesses. Planning takes more time than creating procedures for dozens or hundreds of people, each of whom needs to be supervised and trained.

That being said, there are four major process focus areas that, when addressed separately or in conjunction with an all-inclusive quality management system or software, combine to create this crucial process alignment. These are the main areas:

- Control of documents and records;
- Internal quality audits;
- Control of non-conformances; and
- CAPA (corrective and preventative actions).

The way your work progresses can be observed as a stack of these quality procedures: from site capture (documents and records) to corrective and preventative actions (acting upon what is captured on site).

7.3.3. Quality Management Forms and Documentation

The quality management documents that outline and summarize quality management at a high level are usually the first step in the quality management process. These documents are usually comprehensive and plan-oriented. Excellent project quality management processes place more emphasis on how quality is recorded, arranged, and monitored at the site, where it matters most. Many different types of quality management forms exist, such as basic punch lists, witness points, hold points, inspection and test plans (ITPs), and specialized forms for welding quality control.

These forms are distributed to engineers, project managers, and other employees to ensure that quality standards are being met. They also notify management or other relevant parties when quality standards are not being met, allowing the appropriate party to take corrective action.

While some businesses and teams use paper-based forms, Word documents, and PDFs to manage these forms and documentation, others use digital forms that allow for more control, standardization, and insight into the process.

Apart from these forms and documents, images, videos, and other records are also necessary to ensure that the quality records are up to par for audits and other eventualities.

7.3.4. Construction Quality Control

A management system called construction quality control works to make sure the finished product satisfies the requirements and specifications set forth by the client. This entails finishing the project within the allocated time frame and preventing disagreements at every stage of the undertaking. The client, governmental organizations, and policy and environmental guidelines all define quality. A construction quality management plan contains documentation for each of these procedures and requirements for quality assurance.

7.3.4.1. Quality Assurance vs. Quality Control

Although they both fall under the larger umbrella of quality management in the construction industry, quality assurance (QA) and quality control (QC) do differ in their respective functions and final products. In general, quality assurance encompasses design, development, production, and installation tasks, with the goal of making sure your clients "know" that the work you do will be of the highest caliber. In contrast, quality control is primarily concerned with using quantitative measurements and objective data to track the real quality of completed goods.

Crucial components of quality management are both quality assurance and quality control. Building trusting working relationships with other parties is difficult without quality assurance because they base their decisions about whether or not to work with you on your quality assurance and quality plans. Furthermore, without quality control, it is difficult to assess how you are doing in terms of quality and impossible to decide what needs to be improved based on data.

To promote improvement and the feedback loop, you also require this quality control data to support your quality assurance goals and plans.

7.3.4.2. Quality Control Procedures in the Construction Industry

To ensure clarity in the work, quality control procedures are necessary when working on construction quality control. Use these five steps to accomplish this:

- Define What Done Means: This includes meeting code requirements, finishing the project flawlessly, and adhering to the client's specifications. To make sure everyone is in agreement, once done is defined, communicate the outcome to your superintendents, employees, subcontractors, and other parties involved in the construction project plan.
- Have an Inspection Plan: Examine the finished work carefully to determine whether it satisfies your quality acceptance standards. Decide what needs to be inspected and notify the team when it will be inspected by creating an inspection plan. Additionally, someone needs to be in charge of conducting the inspection, whether they are members of your team or an outsider.
- Make a Quality Control Checklist: To ensure that the inspection process is comprehensive and that no items are missed, create a quality control checklist. Give details. The crew should be given a punch list so they can discuss pre-task discussions

regarding how to complete the work. A checklist not only communicates to the crew what is critical to their tasks but also guarantees that quality control is met.

- Correct Work: Work must be corrected when it fails to meet quality management inspection standards. Mark the work that needs to be fixed (take a picture if needed), fix the issue, and then record the fix to make sure it now satisfies your project's quality requirements.
- Review and Edit: The work doesn't end when you find and address a flaw. Examine the reasons behind it and talk with your team about how to prevent it from happening again. To ensure that the problem doesn't resurface elsewhere, the entire construction team should be informed of the outcome. If necessary, update your inspection checklist with the newly discovered information.

7.3.4.3. The Different Levels of Quality Control

It is beneficial to divide quality management into two tiers. It facilitates management and individual project teams' ability to understand and handle it better. Organizational level quality control is the level of quality management that managers need to concentrate on.

The levers that your organization possesses, such as its resources and established processes and procedures, are the sources of the keys to organizational quality control. You'll need to supervise the methodical deployment of tools and systems that can be used by teams and projects. You'll also need to manage, train, and resource projects and activities appropriately to make sure they have the manpower and other resources they require to complete the work within the established quality standards.

The following stage in the process is project-level quality management, which typically involves a collaboration between management personnel and project workers, specifically project managers who possess a deep understanding of how quality is maintained and delivered on-site. During this phase, quality assurance priorities and tools are implemented to ensure that project-related information and observations are properly recorded, organized, and monitored. Subsequently, this data is transmitted to the appropriate individuals so that both the company and project managers have a clear understanding of the situation and can make informed decisions that lead to successful projects.

The team and technical layer of quality management are the last ones. Teams and businesses are concerned with the actual output of work at this level, as well as how personnel, tools, and equipment are used to guarantee that the tasks are completed within reasonable bounds.

7.3.5. Overcoming Common Quality Control Challenges

Quality control in the construction industry faces many challenges, some common to many sectors and some specific to the industry itself, even with the best of intentions and meticulous plans.

7.3.5.1. Addressing Skill Shortages

One issue that the construction industry frequently faces is a lack of skilled labor. This has an impact on both the final product's quality and the craftsmanship's quality. Incentives to keep skilled workers and training programs go a long way toward addressing this.

7.3.5.2. Managing Budget and Time Constraints

Two of the most frequent limitations in any construction project are money and schedule. If not handled carefully, they may cause one to prioritize quantity over quality.

7.3.5.3. Technology Integration and Adaptation

Historically, the construction sector has been hesitant to embrace new technologies. Nonetheless, contemporary instruments such as sensors, BIM, and drones can greatly improve quality control procedures.

7.3.5.4. Ensuring Clear Communication

Miscommunication can result in a misreading of the specifications, which can cause problems with quality. It is crucial that all levels of communication are frequent, clear, and consistent.

7.3.5.5. Dealing with Environmental Factors

Construction sites are dynamic environments where unanticipated site conditions and weather can have an impact on the final product's quality. Having flexibility and backup plans is essential to addressing these issues.

7.3.6. Tips for Creating a Construction Quality Control Plan

Before you start working together to either develop a quality control plan or modify your current construction industry quality program, consider these last few pointers.

7.3.6.1. Communicate Clearly and Effectively

Quality control and communication need to work together. Quality control will be a never-ending process in the absence of an effective communication strategy for policy, compliance, safety standards, and building expectations. Every communication and discussion regarding project specifications should include quality control, and all contractors and other involved parties should be made fully aware of their expectations.

It is imperative for project managers to ascertain the type of communications, their frequency, and the mode of message dissemination throughout the organization. Your quality control plan needs to specify all monitoring and surveillance efforts as well as the standards you have for suppliers and subcontractors. Excellent communication is also necessary with builders and clients, as it is up to them to determine whether a project was completed to their standards. There should be a strategy in place for construction project managers to share client specifications with the crews and individuals involved when they are gathered.

7.3.6.2. Have a Backup Plan

It's common to overlook the importance of having a backup plan—or several backup plans—especially when workflows, suppliers, and processes have been well-established over time. However, as everyone knows, nothing in construction ever goes exactly as planned. Avoiding expensive errors and difficult client conversations can be achieved by having a backup plan, or a series of backup plans, in place and shared with the relevant teams. Maintain a record of all backup plans in a management system, along with information on when and how they were carried out. Quality construction is ensured by ensuring that all stakeholders have clear backup plans and are in agreement.

7.3.6.3. Use Technology to Streamline Processes

Professionals in the construction sector need to look towards tech solutions that can help streamline quality control management in light of the current supply chain environment, which is beset by disruption and slow communications. Many quality control management procedures call for precise planning, carrying out, and adhering to unmatched communication protocols. Through features like automation, accessibility, and visibility, technology can assist construction companies in improving quality control management.

Examining quality control procedures and making necessary adjustments along the way require data analysis. Systems that offer a comprehensive perspective of data through a unified interface open up new possibilities for information gathered from various projects, suppliers, and vendors to be transformed into useful knowledge.

7.4. STEPS TO IMPROVE CONSTRUCTION QUALITY MANAGEMENT

For the best construction quality management, an organized process should be put in place. Any process should start with clearly defined goals and accountability in a structured framework. Still, that's only the start. A construction quality management system should identify and address possible problems at every turn, validate any results, and evaluate how well the quality objectives are being met. Even though it might seem like a lot of work, creating a process only requires a few steps.

7.4.1. Define Quality Requirements

Conducting research on the quality requirements and specifications for every project step is imperative for the effectiveness of a construction quality management process. It goes beyond simply understanding the standards because documentation is required for distribution and clarification. Clarifying specifications through additional documentation may be part of the process. Following the establishment and documentation of quality standards, you can design your training program.

The foundation of the quality management process will be a quality assurance procedure and a quality management plan. A quality control plan establishes quality standards and specifies how to meet them, as well as any special project requirements. Additionally, it offers advice on how to meet and surpass quality standards using the best practices. The quality assurance procedure will also specify who is in charge of reporting, when, how, and what kinds of inspections are required.

7.4.2. Fulfil Training Requirements

While setting standards is important, training is even more so. The individuals in charge of each process must be identified in order to adhere to the construction quality management processes. They must also be kept informed about the necessary standards, requirements, and procedures. They should let their subordinates know the precise guidelines that their team needs to follow because they will be in charge of the project's quality.

7.4.3. Managing and Monitoring Construction Quality Processes

The final step is to evaluate the management's quality after the training is finished and the procedures are set. The quality assurance representative should check that all plans—including those pertaining to work procedures, supplies, and final products—comply with company standards. They are accountable for fixing any defective labor or supplies. In order to guarantee that the project either meets or surpasses standards, they lastly put quality control procedures into place.

CASE STUDY

APPLICATION OF QUALITY TOOLS TO IMPROVE QUALITY: A CASE STUDY OF A BUILDING REMEDIATION CONTRACTOR:

- **Company Background:** The organization under investigation is a U.S.-founded Environmental and Construction Company with a focus on general contracting, restoration, and mold removal. E&C Inc. was founded in 1987 (real name changed) is made up of seasoned business experts, such as project managers, supervisors, customer service representatives, and principals who have received training and certification. Equipped with engineering controls, the newest technology, and knowledge of the indoor air quality sector, E&C Inc. became an excellent option for a thorough and all-inclusive construction, reconstruction, and mold remediation project. The E&C Inc. team has effectively finished hundreds of mold removal and restoration projects for a wide range of clients from its operation centers in Florida, New York, and Virginia, all on schedule and within budget. With more than 16 years of experience, the company has collaborated with a wide range of organizations, including developers and builders, insurance companies, co-ops and condominium associations, homeowners, owners of commercial buildings, and institutions of higher learning and healthcare. Given that E&C Inc. has consistently raised the bar for environmental remediation, restoration, and construction since its founding by providing thorough, dependable, and committed service. It still strives to uphold its commitment to quality.

E&C Inc. was first established as an asbestos removal company, which is a federally and state-regulated required service before any remodelling or demolition. It is a family-run company. A process of declining prices, increased insurance and operating costs, competition, and a contracting market over the first 10 years had E&C Inc. faced with the difficulty of creating new offerings. The company's reliance on asbestos removal was evident, as the use of asbestos in the construction industry has decreased over time.

- **Identification of Challenges:** An interview with the company president was conducted to gain a deeper understanding of the history and challenges faced by the organization. The president disclosed that the primary obstacle facing the company was adjusting to the changing market conditions. The business has always needed to identify new targets in the remediation market. "E&C Inc. Acknowledged the need to explore for alternative services, and the Ballard v. Farmers drew our attention," the company president said, citing a well-known mold incident that happened in Texas. After discovering that their 22-room Texas home was infested with "black mold" shortly after they moved in, the Ballard family filed a lawsuit against Farmers Insurance Group. After Farmers was found guilty of a $32 million verdict, this case gained national attention.

As other companies specializing in asbestos removal entered the demolition sector, E&C Inc. modified operations to satisfy the need for mold removal. The primary clients of this service were insurance companies and policyholders. Without any federal or state regulations, lawsuits and health concerns drove the mold market. The change brought E&C Inc. to compete with contractors for insurance restoration, which presented a significant obstacle for them. Strong marketing, industry reputation, and technical expertise were the primary requirements for project awards.

Mold remediation replaced asbestos abatement as the primary cause of E&C Inc. to encounter a series of obstacles pertaining to field operations, corporate objectives, and employee training. E&C Inc. employees needed to undergo retraining to acquire the principles and guidelines of remediation. Rather than engaging in the demolition process, they began working with personal contents and the residential market. Regretfully, a few seasoned workers who were unwilling to adapt were let go. Meanwhile, E&C Inc. had to pay equal attention to price and quality control. Services related to asbestos are subject to federal regulation and are classified as "life-safety" concerns. The owner has more responsibility because there are no federal regulations in place for the mold industry.

In remediation projects, adapting to external conditions is a constant challenge. Major damage was caused by the intense hurricane seasons of 2004 and 2005, which offered E&C Inc. both an opportunity and a challenge. E&C Inc. effectively rose to the challenges posed by extensive drying and restoration operations. The company's success has been largely dependent on providing this kind of customer service in the face of a slowing mold market brought on by an insurance cap on mold damage payouts, a lack of regulations, and unrestrained competition. The Florida Insurance Commission approved a $10,000 cap per occurrence per policy in 2005, which led to E&C Inc. expecting its revenue to drop by about 50% from the previous year. E&C Inc. changed its emphasis from homeowner's insurance work, which generated 65% of total revenue, to commercial work, which is less reliant on insurance coverage and more driven by liability. Only hurricane-related projects contributed to the 25% increase in revenues in 2005 over the previous year.

At E&C Inc., internal issues also arise on a daily basis. Estimating and sales may disagree due to changes in market trends. As competition drives down prices, estimators and project managers are under pressure to work more efficiently and cut costs, while sales is under pressure to better serve the current clientele. The business made the decision to invest in marketing and technology in order to meet this challenging objective. The company targets clients who value E&C Inc. and are prepared to pay more for its caliber. Since Florida's summers and active hurricane seasons necessitate E&C Inc., the president of the company identifies the main challenge as being able to employ a large number of workers during slow winter months, when less than half of the labor force is needed to maintain quality. Keeping a capable group of project managers and supervisors was the answer in this case. An investment in technology resulted in the creation and deployment of a biometric worker registration system along with an electronic database of workers' names, contact details, and employment histories.

- **Methodology:** The study team employed a variety of data collection techniques to identify current issues and gain an understanding of the business. These techniques included, among others, company information, interviews, and questionnaires. The team used a variety of tools, including a Pareto chart, cause-and-effect diagrams, check sheets, and brainstorming, to identify problems and potential solutions as a result of this information. Following the application of these strategies, potential fixes were arranged and given to the management group.

The research team was able to gain a deeper understanding of the company and pinpoint the areas where this quick study would be most helpful thanks to the questionnaires and interviews that were performed in 2007. First, quality programs should focus on small problems with a high chance of success. This can be achieved by selecting operations that appear to be running smoothly. In order to gather information for creating a Pareto chart, the study team met with the heads of the financial and environmental departments. Every employee in the office received a questionnaire from the study team. The purpose of the questionnaire was to gather feedback from employees about the company and to gauge their level of job satisfaction without disclosing specific answers to managers. Surveys are useful because, as experience has shown, employees frequently perceive the world and particular circumstances more clearly and know more about them than does upper management. When using questionnaires, it must be made clear that their purpose is to aid in improving job performance rather than to punish or discipline employees.

Data Analysis:

- **Pareto Analysis:** A Pareto chart was created using the data collected from various sources, and the key factors contributing to the quality-related issues were examined.
- The top five cause factors are as shown in Table 1.

Table 1. *Top Cause Factors and Number of Occurrence*

Cause Factor	Unit cost	# of occurrences per year
Violations	5000	2
Communication errors	185	80
Over Staffing	240	60
Failed Clearance	1500	36
Cross contamination	12000	1

Source: http://ascpro0.ascweb.org/archives/cd/2008/paper/CPRT240002008.pdf.

The final test for the mold remediation activity must pass or fail. The manager of the company refers to it as a "failed clearance cause." "Cross-contamination cause" refers to the situation where products intended to remove mold are accidentally dispersed over an area that shouldn't be contaminated. Because mold removal has such a high turnover rate, managers would rather hold onto some of their best workers even when there is no work for them. They believe that the cost of keeping those workers is less than the loss the business would suffer if it had to turn down a project because of a staffing shortage. "Overstaffing" was identified as the root cause. "Violations cause" happens when certain workers don't adhere to the terms of the contract. Finally, "communication errors cause" arises when a worker does not fully comprehend the task assigned to him or her; as a result, the worker completes the task correctly but not in the prescribed manner, leading to work duplication. Table 2 displays the total cost of each cause factor.

Table 2. *Cause Factors and Their Cost*

Cause Factor	Total Cost
Violations	10000
Communication errors	14800
Over Staffing	14400
Failed Clearance	54000
Cross contamination	12000

Source: http://ascpro0.ascweb.org/archives/cd/2008/paper/CPRT240002008.pdf.

Looking at the above figure, one could conclude that the company should focus its efforts on fixing the communication errors, since those are the most common. Three environmental department staff members and the manager were shown Table 1. They were all in agreement that the communication errors factor needed to be the main focus. However, the project team's focus shifted to the failed clearance factor after the study team produced Table 3 using financial data that was obtained from the company following a meeting with the director of the financial department. The creation of a Pareto chart, as seen in Figure 1, makes this very clear.

Table 3. *Cause Factors with Their Cost Contribution*

Cause Factor	# of occurrences per year	Unit Cost	Total Cost	% of the Cost	Cumulative %
Failed Clearance	36	1,500	54,000	51.33	51.33
Communication errors	80	185	14,800	14.07	65.40
Over Staffing	60	240	14,400	13.69	79.09
Cross contamination	1	12,000	12,000	11.41	90.49
Violations	2	5,000	10,000	9.51	100.00
Total # of occurrences	179		105,200		

Source: http://ascpro0.ascweb.org/archives/cd/2008/paper/CPRT240002008.pdf.

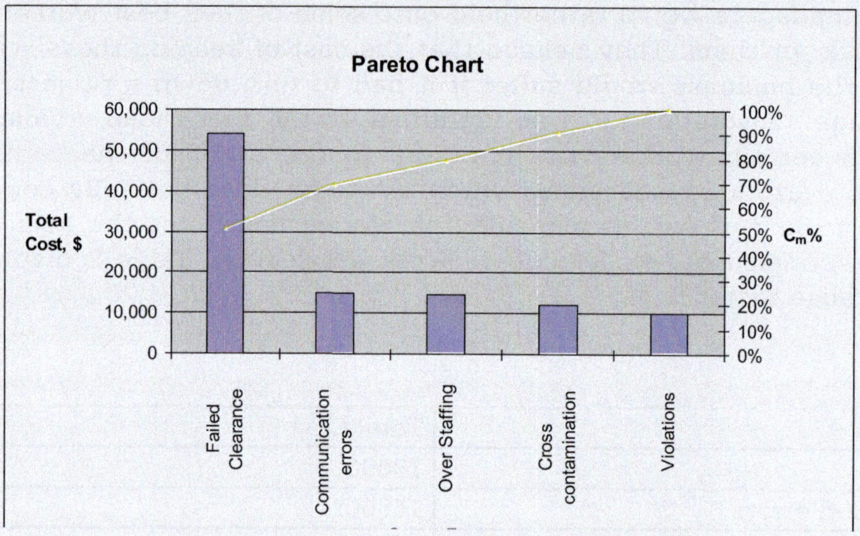

Figure 1. *Pareto chart.*

Source: http://ascpro0.ascweb.org/archives/cd/2008/paper/CPRT240002008.pdf.

According to the graph analysis, 20% of occurrences and 51% of costs are associated with failed clearance. Therefore, the business ought to concentrate on this subject.

- **Cause-and-Effect Diagrams:** Determining the causes that led to the "failure clearance" was crucial. After that, the research team employed cause-and-effect diagrams to determine every potential cause. The method is what makes the diagram known as an Ishikawa diagram; it was created by the late Dr. Kaoru Ishikawa, a renowned Japanese quality expert.

In order to apply the cause-and-effect diagram, the study team met with the following employees of the company: the environmental department president, an engineer, a supervisor, a worker in demolition, a worker in final cleaning, and a worker in contents. The group was informed that the failed clearance was the topic of discussion, and that the goal was to enumerate every procedure-related factor that might have an effect on the denied clearance. The group came up with a list of potential contributing factors through brainstorming. The machine, operator, materials, methods, measurement, and environment were identified as the six main factors, or causes, that could affect the effect based on the list that was generated. The next step was to assign each of the other causes to the major causes that they affect after assigning the major causes and the list.

A fishbone diagram as shown in Figure 2 is produced. It represents a picture of the major factors that can cause failure clearance and in turn the smaller factors that affect the major factors.

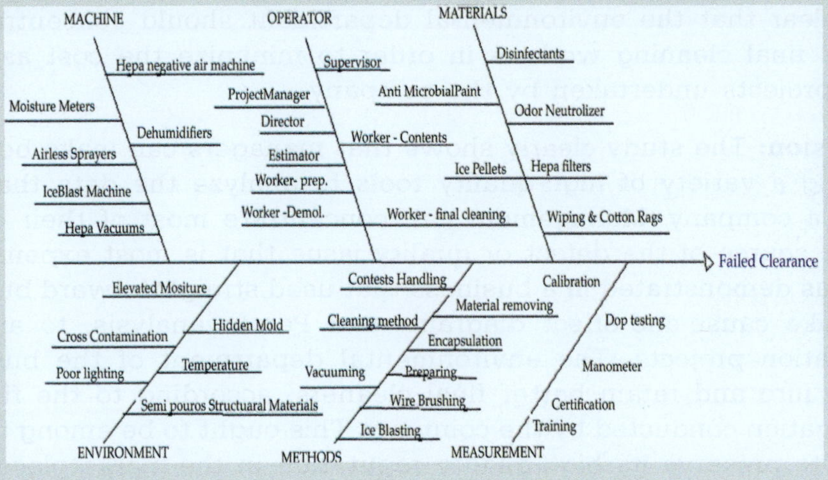

Figure 2. *Fishbone diagram.*

Source: http://ascpro0.ascweb.org/archives/cd/2008/paper/CPRT240002008.pdf.

The fact that it's highly likely that no one person possessed all of that knowledge and information is what makes the diagram so helpful. Because of this, teams of individuals with greatly different areas of expertise typically create cause-and-effect diagrams. The cause-and-effect diagram is a great way to remind the business of the things that need to be addressed if it wants to keep improving the process. The most crucial role for any team or employee is ongoing improvement, even in well-functioning processes.

- **General Analysis:** The failure clearance was the most significant factor, according to the first Pareto chart. The environmental department was shown a list of potential causes for this outcome by the cause-and-effect diagrams. After discussing financial and historical data, the director and the engineers chose the top five causes from that list and set up a meeting. Following the consensus that "failed clearance" was the most crucial component, Table 4 further deduces the contributing factors.

Table 4. *Cause Factors for Failed Clearance*

Cause Factor	# of occurrences per year	# of occurrences	Cumulative %
Final cleaning	15	41.67	41.67
Supervisor	5	13.89	55.56
Hidden Mold	4	11.11	66.67
Calibration	2	5.56	72.22
Semi pouros materials	2	5.56	77.78
Other	8	22.22	100.00
TOTAL	36	100.00	

Source: http://ascpro0.ascweb.org/archives/cd/2008/paper/CPRT240002008.pdf.

Now it is clear that the environmental department should concentrate its efforts in hiring better final cleaning workers in order to minimize the cost associated with defects in the projects undertaken by the company.

- **Conclusion:** The study clearly shows that managers can make better decisions by using a variety of high-quality tools to analyze the data that is available within a company. Management can concentrate most of their efforts on the specific source of the defect or quality issue that is most expensive for them. This was demonstrated in a business that used straightforward but high-quality tools, like cause-and-effect diagrams and Pareto analysis, to aid in building remediation projects. The environmental department of the business should look to hire and retain better final cleaners, according to the findings of the investigation conducted by the company. This ought to be among the company's tactics to preserve its high-quality reputation in the marketplace. To optimize its profit without sacrificing quality, the company needs to conduct additional research to determine how much money it should spend on employing better final cleaners.

CLASS ACTIVITY

LIFTING AN OBJECT:

To inject some fun into your construction team's routine, you can do this strenuous exercise with them during their breaks.

All you need is a rope, a heavy object (e.g., a tire), and construction materials (duct tape, nails, and hammers). Next, split your team up into four-person groups. Lifting the bulky object to a higher level—the second story, for instance—is their aim. However, the rope's length prevents it from reaching the required height. The group needs to think carefully about the tools at hand and how best to raise the object using them or their own bodies. The winning team is the one that finishes the goal first. This activity could take a few minutes or the duration of the break, depending on how imaginative they are.

SUMMARY

- Construction projects are complex undertakings that involve many stakeholders and often have tight deadlines. Ensuring quality in such projects is crucial to their success and the satisfaction of clients.
- Quality management is about making organizations perform for their stakeholders – from improving products, services, systems, and processes, to making sure that the whole organization is fit and effective.
- A quality management system (QMS) is defined as a formalized system that documents processes, procedures, and responsibilities for achieving quality policies and objectives. A QMS helps coordinate and direct an organization's activities to meet customer and regulatory requirements and improve its effectiveness and efficiency on a continuous basis.
- Construction quality control is a system of management that ensures that deliverables meet the standards and guidelines set by the client at the beginning of the construction process. This can include a number of criteria, such as completing the project within the scope of work. In the end, quality is decided by the client, regulatory bodies, and EPA guidelines.
- There are a number of ways to approach quality control management in construction, with each having its own pros and cons depending on the needs and scope of a company's projects. The International Organization for Standardization established a set of quality standards called ISO 9001.
- Construction projects span a number of functions across a number of specialty parties and trades, but the one thing they all share in common is their desire to deliver quality work – on time and on budget. Quality management is a key pillar of overall construction project management, and is often the difference between company's success and failure.
- Preventing mistakes is much more time and cost effective than correcting them – which is why establishing a strong quality management plan is a good way to improve quality. The upfront investment of creating a coherent and comprehensive quality management plan often pays big dividends throughout the life of a project.
- The quality management process typically starts with those quality management documents – which tend to be plan oriented and comprehensive in nature – outlining and summarizing quality management at a high level. Great quality management process on projects focuses more on how quality is captured, organized and tracked from the point where quality matters most – on site.
- Construction quality control is a management system aimed at ensuring the final deliverable meets the standards and guidelines set by the client. This includes completing the project within the scope of work and avoiding disputes throughout the life cycle of the project.
- The Construction Quality Management System should detect and mitigate potential issues at every stage, verify any outcomes, and assess the effectiveness of quality objectives.

REVIEW QUESTIONS

1. What is construction quality management?
2. What are the key challenges in construction quality management?
3. How to improve construction quality management?
4. What are the elements of a quality management process?
5. Why is quality important in construction?
6. What creates quality issues?
7. What are quality processes in construction?
8. Why should organizations care about quality?
9. What is ISO standard for construction quality?

REFERENCES

1. Akhund, M. A., Memon, A. H., Imad, H. U., Siddiqui, F. H., & Khoso, A. R. (2018). Motivational factors for the implementation of ISO-9001 in construction firms of Pakistan. Civil Engineering Journal, 4(9), 2023. http://dx.doi.org/10.28991/cej-03091135.
2. Alshehri, A. R. (2016). Quality management system for building maintenance (Thesis). Heriot-Watt University. http://hdl.handle.net/10399/3092 (accessed on 1 July 2024).
3. Androwis, N., Sweis, R. J., Tarhini, A., Moarefi, A., & Amiri, M. H. (2018). Total quality management practices and organizational performance in the construction chemicals companies in Jordan. Benchmarking: An International Journal, 25(8), 3180–3205. http://dx.doi.org/10.1108/bij-05-2017-0094.
4. Coffey, V., Willar, D., & Trigunarsyah, B. (2011). Quality management system and construction performance. In 2011 IEEE International Conference on Quality and Reliability (ICQR). IEEE. http://dx.doi.org/10.1109/icqr.2011.6031750.
5. Dahmas, S., Li, Z., & Ahmad, M. (2020). Evaluation of implementation preparation for CE based on BEACON model—Taking construction enterprises in Yemen as a case of illustration. Frontiers Research of Architecture and Engineering, 3(1), 7. http://dx.doi.org/10.30564/frae.v3i1.1723.
6. Khan, M. W. J., Alam, M. B., Sulaiman, G., & Ahmed, Z. (2020). Nexus between total quality management practices and construction firms' performance in Pakistan. Review of Economics and Development Studies, 6(4), 787–798. http://dx.doi.org/10.47067/reads.v6i4.278.
7. Low, S. P., & Ong, J. (2014). Construction quality assessment system. In Project Quality Management (pp. 11–27). Singapore: Springer Singapore. http://dx.doi.org/10.1007/978-981-287-074-2_2.
8. O'Brien, J. J. (1997). Total quality management. In Construction Inspection Handbook (pp. 110–121). Boston, MA: Springer US. http://dx.doi.org/10.1007/978-1-4615-6017-3_9.

9. Oakland, J. S., & Marosszeky, M. (2017). Quality management systems. In Total Construction Management (pp. 285–313). Abingdon, Oxon; New York, NY: Routledge. http://dx.doi.org/10.4324/9781315694351-14.

10. Rumane, A. R. (2017). Quality management system. In Quality Management in Construction Projects (pp. 121–168). Boca Raton: Taylor & Francis, CRC Press. http://dx.doi.org/10.1201/9781315098425-2.

11. Shaikh, F. A., & Sohu, S. (2020). Implementation, advantages and management of ISO 9001 in the construction industry. Civil Engineering Journal, 6(6), 1136–1142. http://dx.doi.org/10.28991/cej-2020-03091535.

12. Wesonga Oruma, B. (2014). Top management commitment towards implementation of total quality management (TQM) in construction companies in Nakuru County-Kenya. International Journal of Economics, Finance and Management Sciences, 2(6), 332. http://dx.doi.org/10.11648/j.ijefm.20140206.15.

CHAPTER

8 Health and Safety Management

LEARNING OBJECTIVES

After studying this chapter, you will be able to:

- Examine the health and safety management in construction;
- Determine the crucial role of construction safety management;
- Discuss on risk assessment and hazard identification;
- Learn about the safety training and education;
- Evaluate the emergency preparedness and response;
- Give an overview on construction site traffic management.

KEY TERMS FROM THIS CHAPTER

Workplace health and safety (WHS)	Digital solution
Security management	Supervision
Pandemic	Digital twins
Remote assistance tools	Employee stress
Construction site	Safety management system

INTRODUCTION

Health and Safety Management (HSM) is a systematic approach to identifying, evaluating, and controlling risks to ensure the well-being of employees, contractors, and other stakeholders within an organization. It encompasses policies, procedures, and practices designed to prevent accidents, injuries, and illnesses in the workplace. Effective HSM is rooted in compliance with legal and regulatory requirements, but it extends beyond mere adherence to laws; it fosters a culture of safety where every member of the organization is committed to maintaining a safe working environment.

Central to HSM is the development and implementation of a comprehensive health and safety policy, which serves as a framework for setting and achieving safety objectives. This policy is supported by detailed procedures and guidelines that outline specific responsibilities and actions required to manage risks. Risk assessment is a critical component of HSM, involving the systematic identification of hazards, evaluation of their potential impact, and the implementation of measures to mitigate these risks. This process requires continuous monitoring and review to ensure that controls remain effective and relevant as the workplace evolves.

Employee involvement is another cornerstone of effective HSM. Engaging employees at all levels in safety programs fosters a sense of ownership and accountability. Training and education are essential to equip employees with the knowledge and skills necessary to recognize hazards and respond appropriately. Regular safety meetings, workshops, and drills reinforce the importance of safety and keep it at the forefront of employees' minds. Leadership commitment is crucial for the success of HSM. Senior management must demonstrate a genuine commitment to safety by allocating necessary resources, supporting safety initiatives, and leading by example. This top-down approach ensures that safety is prioritized alongside other business objectives and that it is integrated into the organization's overall strategy.

Incident reporting and investigation are key elements of HSM. Establishing a clear protocol for reporting accidents, near misses, and unsafe conditions allows for timely intervention and prevents recurrence. Thorough investigations help identify root causes and contribute to the continuous improvement of safety practices. Learning from incidents and implementing corrective actions is vital for building a safer work environment.

Moreover, HSM involves maintaining detailed records and documentation to track safety performance, compliance with regulations, and the effectiveness of implemented measures. These records provide valuable data for analysis, helping organizations to identify trends, measure progress, and make informed decisions.

Health and Safety Management is not a static process but a dynamic one that requires ongoing attention and adaptation. Regular audits and inspections ensure that safety practices are being followed and identify areas for improvement. Staying abreast of new regulations, emerging risks, and advancements in safety technology is essential for maintaining an effective HSM system.

8.1. HEALTH AND SAFETY MANAGEMENT IN CONSTRUCTION

Due to frequently occurring occupational hazards, working at heights, heavy machinery, bad weather, and other factors, the construction industry poses serious health and safety risks to site workers. These risks can result in serious injuries, financial losses, and harm to the company's or builder's reputation. To protect the health and safety of employees, the general public, and the environment, construction safety management entails methodically identifying, preventing, and managing risk within construction sites. Ensuring a safe working environment and preventing accidents or injuries to employees involves a range of practices, policies, and procedures.

8.1.1. Workplace Health and Safety

One of the riskiest sectors of the economy to work in is construction, where illnesses and accidents at work claim millions of lives each year. Not only that, but millions also experience stress and psychological risks in addition to non-fatal workplace accidents. Luckily, most of this can be avoided by taking action to improve working conditions. A collection of guidelines and procedures known as workplace health and safety (WHS) teaches us how to control the risks to the health and safety of employees, clients, and members of the general public (such as suppliers and visitors) could run into at work.

In the construction sector, these practices aim to eliminate risks on construction sites such as:

- Fire and explosions;
- Skin hazards;
- Handling heavy objects;
- Handling chemical substances;
- Defective equipment;
- Electric shock;
- Fall from heights;
- Dust inhalation;
- Faulty equipment;
- Elevated noise levels; and
- Many others.

In the high-risk environment of the construction industry, it is essential to create a strong health and safety culture that will ensure safe working conditions for everyone involved.

8.1.2. Who Is Responsible for It?

Construction sites are places of employment with a variety of participants, even though authorities are in charge of keeping an eye on them. They are all accountable for safety, albeit in different ways. Owners of businesses are in charge of ensuring that their workers follow health and safety regulations and laws. Various laws are in effect in various regions of the world. Nevertheless, all employers need to implement safety procedures to guarantee a secure workplace as well as the knowledge and instruction required to keep employees safe. In addition to abiding by the law and the health and safety protocols of the workplace, employees must also behave responsibly and refrain from taking any actions that could put them or other employees in danger (Figure 8.1).

Figure 8.1. Responsible entity for WHS on a construction site.

Source: https://www.resco.net/wp-content/uploads/2022/07/who_is_responsible_for_whs-1024x538.jpg.

Kelly Roofing improves security management on construction sites with the help of a digital solution. The company serves all of Florida with roof replacement and repair services. At a specific location for roofing repair, there is always a crew that oversees the entire process in addition to the installers. The crew is made up of a dispatcher who handles scheduling, an administrator who needs to know the crew's whereabouts and agenda, and a supervisor. The team makes use of a mobile CRM system to provide expert service and foster a secure work environment.

8.1.3. Manage Workplace Health and Safety

1. Risk Assessment and Planning: The first step in managing WHS is risk assessment. Identify and analyze the hazards in your workplace – their likelihood, how severe they are, and what their impact will be. Don't forget to include your workers in the process.
2. Control and Prevention: Implement safety measures to reduce the likelihood of workplace accidents. Put on personal protective equipment (PPE) like safety shoes, hard hats, reflective clothes, earplugs, and eye protection. This helps lower the possibility of common mishaps like falls, electric shocks, and object strikes. In order for your staff to follow safety procedures and safeguard themselves, it is important that they receive regular training. To ensure that your staff members are well-rested and capable of giving their tasks their full attention, make sure they take regular breaks.
3. Health and Safety Management:

Even though employees are aware of safe behavior and how to avoid mishaps and injuries, supervision is still helpful, particularly when dealing with new hires. It helps you identify risky circumstances or behaviors in advance. When using defective or dangerous equipment, even skilled workers are unable to stop incidents. To eliminate risks, take good care of your equipment. Remember to perform preventive maintenance in addition to any required repairs. Establish procedures for employees to report damaged equipment, and provide maintenance staff with training. Even when everything is going well, keep an eye on the safeguards you have put in place. This will show you how well you mitigated the risks, highlight your advantages and disadvantages, and enable you to recognize potential threats before they arise. Mobile field service solutions can support the oversight of your safety precautions and provide workers with the information they require to remain safe.

8.1.4. Current Trends

Safety is of the utmost importance when managing a construction site, as most accidents in this field are caused by bad or neglected safety practices. So, what are the current trends for managing health and safety on construction sites?

1. Focus on Work-Life Balance: Lack of work-life balance and employee stress have long been problems. But the pandemic brought it even more to light. Many people suffered from anxiety, stress, or even depression as a result of the lack of social interactions and the hazy boundaries between work and home. Workers who get enough sleep are more concentrated and make fewer mistakes. Employers are putting their employees' mental health and general well-being first because of this. Businesses are putting in place stress management initiatives, experimenting with 4-day work weeks, or offering their staff access to on-site therapists. In addition, many companies sponsor their employees' participation in competitive sports on a financial basis and encourage them to engage in sports and fitness activities. Very inspiring, don't you think?
2. Wearables: When the Google Glass project was shelved a few years ago, smartwatches were thought of as basic activity trackers, and smart glasses brought back memories. Since then, wearable technology has developed into a desirable product that, because of its many practical features, can nearly completely replace smartphones. Construction workers can obtain a wealth of information, including instruction manuals and client information, by utilizing wearable technology and a suitable mobile solution. They can converse in real time and hands-free by using their wearable technology. Employers can also use wearables to track the health of their employees by gathering data on things like body temperature and level of fatigue. By analyzing this data, they can prevent health problems and establish a secure workplace.

8.1.5. Safety Management in Construction

Construction sites are among the riskiest places for workers to work in terms of their health and safety. The labor-intensive nature of the construction industry, its production processes, and occupational accidents that cause significant financial losses and damage to a company's reputation make it a high-risk industry. In the construction industry, construction site safety is a crucial requirement that is frequently disregarded on project sites. The hazardous and ever-changing nature of the construction industry makes workers vulnerable to both minor and major accidents on a regular basis in the absence of appropriate safety regulations.

Some of the most common hazards in the construction industry are:

- Falling objects;
- Exposure to chemical substances;
- Fall from heights;
- Dust inhalations;
- Vehicular and equipment accidents;
- Working in confined spaces;
- Slips on sites;
- Fire and explosions;
- Electrocutions;
- Defective safety equipments;
- Chemical spills.

Most accidents in the construction industry are caused by a lack of education about construction safety management and training. Negligence and ignorance on the part of any individual or authorities can also result in hazards (Goetsch, 2012).

8.1.5.1. Need for Safety Management

Even though the construction industry has made many efforts to prevent accidents on the job site, provide workers with safe working conditions, and enforce safe practices, required safety is still a long way off. A safety management system is crucial in the construction industry because of the size of the sector and the resulting high rate of accidents. On construction sites, safety management strengthens working conditions against risks associated with tools, materials, and procedures.

8.1.5.2. Safety Management Process

A construction site safety management system can help by methodically identifying hazards and mitigating associated risks, all the while ensuring that these preventative measures are successful in promoting worker safety. The safety management process roughly outlines the following steps:

- Identifying risks associated with the workplace during the planning stage and pre-deciding suitable controls and steps that need to be implemented in order to avoid any accidents or damage to property and ensure construction safety.
- Monitoring work to detect any actions early that may result in mishaps and to check if safe practices are being followed and safe conditions are being maintained.
- Enforcing and promoting effective communication across all levels of the organization.
- Protecting neighboring public and private properties.
- Providing regular safety education and training along with attractive incentive programs for using safe operating methods.
- Abiding by standard occupational health and safety regulations.

8.1.5.3. *Importance of Safety in the Construction Industry*

Public Protection – The public is often seen circulating throughout and around construction sites. Systems for managing construction safety can shield the general public from potential harm caused by construction activities. Thus, public safety is also ensured by construction site safety (Figures 8.2 and 8.3).

Figure 8.2. Useful entity of construction safety.

Source: https://d1gmilvhcqro2r.cloudfront.net/blog/what-is-safety-management-in-construction/what-is-safety-management-in-construction-03.jpg.

1. Efficient Usage of Time and Money: Construction safety aids in saving excess time and money that have to be spent in case of any mishaps. These unwanted situations may take up excessive time in recovering from these downsides. Operating costs are increased due to compensation claims from injured employees. In case of outsiders getting injured, a lawsuit may be filed against the project owner that might ultimately prove to be a huge financial burden. Irregular and improper handling and maintenance of equipment increase time and money expenses. Construction site safety can thus be improved by adopting safety management systems.
2. Promote Compliance: An unsafe workplace environment is always prone to mishaps on some construction sites, even in the absence of a history of site accidents. If employees exhibit a lenient attitude towards adhering to the regulatory compliances outlined in their contracts, then enforcement of those provisions is necessary. The rationale is that the company is at risk if these rules are broken since an unsafe environment could be created on the job site, which could harm the project's quality and financial aspects. Respecting regulatory compliances safeguards the organization as a whole as well as its employees. To keep workers safe from potential risks, safeguard the general public, and maintain project timeliness, construction and process safety management is crucial. Encouraging safety cultures on building sites safeguards all project participants.
3. Safe Site Management: The safety manager needs to establish trust with the staff in order to get them to follow their lead in promoting site safety. The employees need to understand that process safety management is critical to their jobs and that they are subject to safety regulations in order to ensure their safety. The safety manager is responsible for directing and encouraging safe work practices and conditions.

Figure 8.3. Site safety instructions.

Source: https://d1gmilvhcqro2r.cloudfront.net/blog/what-is-safety-management-in-construction/what-is-safety-management-in-construction-04.jpg.

A responsible safety manager has the following responsibilities:

- Worker safety and health must be considered as an occupational value.
- Fully committed to preventing accidents, protecting workers, and improving safety and health culture completely.
- Sufficient resources must be provided to implement and maintain the safety and health of employees.
- Proper medical kits must be installed at various locations based on worker density. Site safety equipment such as fire extinguishers and water cannons can also be placed on site.
- Accessible toilet facilities should also be provided on site to prevent workers from experiencing issues due to sanitation.
- Workers must be properly trained to demonstrate and communicate safety and health management systems.

8.1.5.4. Risk Management for Safety in Construction

Risk management procedures for occupational site health and safety can be roughly categorized as in four simple steps, that are:

- Plan: The safety policies and procedures must be planned as per risk assessment report.
- Do: Enforcements of policies and procedures.
- Check: In order to check for relevance, effectiveness and efficiency of implementation, the safety performance is measured.

- Act: Proper remedial measures must be adopted in order to prevent safety and health issues.

A risk management system for safety on construction sites is more of a method than a product, and every time a method is implemented anywhere on the project, it needs to be documented. This allows the safety in charge to monitor the method's efficacy and use it for auditing and report preparation. Because workplace conditions, negligence, and inadequate safety procedures can all lead to accidents and mishaps, safety in the construction industry is crucial. Since the project management process includes a safety management system, sufficient provisions must be made for this process to feed and document safety procedures that are followed on the job site. Through process analysis and updating, construction project management software can help promote safety systems on various construction sites. Construction project managers can benefit from Powerplay software's all-around capabilities, which help clients, contractors, and other key team members monitor progress and take appropriate action.

8.2. THE CRUCIAL ROLE OF CONSTRUCTION SAFETY MANAGEMENT

Building sites are dynamic centers of activity where talented laborers realize architectural concepts. The industry is essential to the advancement of society, but it also carries risks that must be carefully considered. In addition to being required by law, construction safety management is also morally necessary to protect the lives of those who work on the built environment. We discuss the importance of construction safety management and how it promotes a safe and effective work environment in this blog.

1. The Human Toll: Construction sites are well known for their potential for danger. Workers are exposed to a wide range of risks due to the combination of heavy equipment, tall buildings, and complex tasks. A sizable percentage of workplace fatalities, as reported by the Occupational Safety and Health Administration (OSHA), are related to construction. Around 1,061 of the 5,333 worker fatalities in 2019 happened in the construction industry, underscoring how urgent it is to give safety precautions top priority.
2. Financial Implications: Accidents in the construction sector have serious financial consequences in addition to human costs. Injuries and deaths at work result in higher medical expenses, legal responsibilities, and possible project delays. The average cost of a work-related death in the construction industry was estimated by the National Safety Council to be approximately $1.15 million. Strong safety management procedures are not only life-saving but also safeguard the bottom line of building projects.
3. Regulatory Compliance: To enforce safety standards in the construction industry, governments all over the world have put in place strict regulations. Following these rules is not only required by law, but it is also a cornerstone of moral business conduct. Neglecting safety protocols may lead to harsh fines, the termination of projects, and harm to an organization's image. By ensuring that projects are carried out legally, construction safety management protects businesses and employees from potential legal problems.
4. Productivity and Efficiency: Despite popular belief, safety precautions increase productivity rather than decrease it. An atmosphere of safety at work gives employees a sense of security, which boosts morale and increases job satisfaction.

Workers are more likely to be engaged in their work and produce better work when they feel protected. Furthermore, fewer accidents mean less downtime, which guarantees that building projects move forward without hiccups and on schedule.

5. Changing Perspectives: The construction industry has seen a paradigm shift in its approach to safety over time. It is now seen as an essential component of project management rather than an incidental worry. The industry's commitment to fostering a culture of safety is demonstrated by the implementation of safety management systems and the inclusion of safety procedures in project planning. Both the long-term viability of construction companies and the ongoing well-being of their workforces depend on this cultural shift.
6. Technological Innovations in Safety: Technology has completely changed the way construction safety management is done. Innovative solutions are improving safety on construction sites, from wearables that track vital signs to drones that inspect sites. Another technological development that makes it possible to virtually simulate building projects and identify potential hazards before they materialize on-site is Building Information Modeling (BIM). Adopting these technologies puts construction companies at the forefront of industry innovation while also reducing risks.
7. Global Initiatives: The issue of construction safety transcends national borders and requires international cooperation. International standards for occupational health and safety are being established by organizations such as the World Health Organization (WHO) and the International Labor Organization (ILO). International projects emphasize how intertwined the construction sector is and how important it is to have a common commitment to safety everywhere.

8.2.1. Managing Health and Safety in Construction Efficiently

Putting in place a top-notch health and safety program is the best way to safeguard your most valuable asset, which is your workforce. Avoiding the conventional reactive approach—which involves acting only after an incident occurs—is the key to improving health and safety management in the construction industry. Rather than waiting for hazards to cause illness or injury, project managers and foremen need to identify and address them early on. You can utilize the following 7 suggestions to make your workplaces safer and reap the rewards of improved health and safety procedures, which include increased worker morale, better labor productivity and quality, improved recruiting and retention strategies, and more:

1. Commit to a Safety and Health Program: Efficiency and profitability are not as vital to your company as clearly communicating your commitment to managing health and safety in construction through a policy. You should set clear expectations for everyone—from you to the employees to the program as a whole—and set precise goals and objectives in a written policy. As you work to improve worker safety and health, concentrate on particular actions and set measurable objectives. Next, create a plan to accomplish those goals by identifying and allocating the resources that will be needed, designating personnel with specific

responsibilities, and establishing deadlines for completing the health and safety goals of your program.

2. Involve Workers in Your Safety Practices: Having a solid health and safety program isn't enough—workers and managers must willingly participate in the program.

Show that you value your workers' input and encourage them to participate and report any health and safety concerns by:

- Maintaining an open-door policy: Invite workers of all levels to talk to you and other management about health and safety concerns, questions, or suggestions.
- Establishing simple procedures for reporting injuries, hazards, etc.: Allow anonymous reporting so workers can voice their concerns without fear of reprisal.
- Removing barriers to participation: Ensure everyone has the ability to participate in the program and spend the time to facilitate participation, like holding regular health and safety meetings.

3. Inspect Your Job Site to Identify Health and Safety Hazards: As job site conditions change, new hazards can pop up anytime. It's important to regularly and frequently perform inspections as you move into different building phases, when trades arrive and depart, after inclement weather, etc.

Identifying physical safety hazards is generally easier than health hazards, which are often less obvious and more complex. Check each job site for the following health hazards:

- Chemical Hazards: Low exposure limits, unventilated spaces, etc.
- Biological Hazards: Sources of mold, toxic materials, infectious diseases, etc.
- Physical Hazards: Excessive noise, radiation sources, etc.
- Ergonomic Hazards: Heavy lifting, repetitive motions, etc.

4. Implement a Hazard Control Plan: Creating a hazard control strategy to protect workers on the job site during emergencies or non-routine tasks is another excellent method to enhance your program for managing health and safety in construction industries. In addition to exercising extra caution when performing tasks that employees wouldn't typically perform, you should update your hazard control plan to account for potential events like fires, falls, spills of hazardous materials, natural disasters, etc. Drills are another smart idea to make sure that your employees are aware of the safety protocols you follow and are sufficiently protected.
5. Provide Awareness Education and Training: A huge part of managing health and safety in construction is education and training for employers, managers, and workers. This ensures everyone has the knowledge and skills to do their jobs safely without creating additional hazards, putting themselves or other workers at risk.

Your awareness program should:

- Contain steps for reporting hazards, injuries, etc.

- Detail what to do in an emergency.
- Provide training at literacy levels and in languages all workers can understand.
- List the goals, procedures, and policies of your health and safety program.
- Include contact information for the person heading the training program.

6. Monitor Your Program's Performance and Progress: After establishing new health and safety procedures, it's crucial to evaluate the program frequently to look for opportunities for improvement, track goals and progress, and identify shortcomings.

You should initially and periodically monitor the following indicators:

- Worker participation;
- Safety suggestions;
- Workers' comp data;
- The number of hazards and near misses;
- The number and severity of injuries;
- The number of hazards found during those inspections;
- The frequency of inspections;
- Time taken to respond to health and safety reports;
- Worker feedback;
- Worker exposure monitoring results.

7. Digitize Your QHSE Processes: The management system for Quality, Health, Safety, and Environment (QHSE) guarantees a safe and healthy work environment for all employees. Real-time updates and transparent collaboration are unfeasible, though, as many project managers and organizations still conduct their QHSE procedures manually. One of the most effective ways to manage health and safety in construction is to digitize and automate your QHSE processes. This will enable you to carry out site inspections more quickly, record defects with a complete audit trail, standardize forms and reporting, and maintain employee engagement with your program.

8.2.2. Improve Safety Management at Construction Site

Every construction site faces different difficulties. By assigning specialized safety officers to your company, you can guarantee targeted supervision, customized solutions, and ongoing observation to successfully reduce risks. They pay close attention to detail and take preventative action to improve your worksite's safety, lower risks, foster a safety-conscious culture among employees, and elevate safety management.

8.2.2.1. Role of Safety Officers at Construction Sites

- Safety officers will conduct thorough risk assessments in the workplace to determine potential hazards and create safety regulations and policies to overcome them.
- They will verify that the tools and equipment are of good quality, specify adequate personal protective equipment for workers, and guide them on handling it properly.
- They will find the best way to keep workers safe from fire explosions, chemical spills, and avoid falls and injuries when working at heights.

- They regularly carry out mock drills on managing emergencies and test the effectiveness of site emergency response plans.
- They will investigate accidents, gather evidence to determine their causes, and document their findings to prevent future accidents.
- They will continuously monitor all safety-related documents, reports, and issues and keep them updated.

8.2.2.2. Benefits of Equipping Your Workers with Construction Safety

- It minimizes the risk of accidents, injuries, or fatalities among workers and boosts employee confidence for better job performance.
- They will acquire knowledge to identify workplace hazards, know the safety precautions when working with different hazards, and respond to all industrial warning indicators.
- Training makes them familiar with work at height protection, safety in excavations, scaffolding safety, electrical safety, material handling safety, welding and gas cutting safety, and construction safety management.

8.2.2.3. Golden Safety Tips to Follow in Construction Safety

- PPE: Use appropriate PPE while performing work.
- Work Permit: Obtain the necessary permit before entering into the work.
- Work at Height: Implement fall protection measures while working at heights.
- Confined Space Entry: Follow the safety procedures for safe entry and work in a confined space.
- LOTO Safety: Ensure proper lockout/tagout procedures for machinery maintenance.
- Lifting Operation: Conduct lifting operations with proper precautions and equipment.
- Safety Device: Never bypass or disable safety devices on equipment.
- Safe Driving: Practice safe driving on-site and follow the speed limits.
- Barricades: Install barricades when working under loads or overhead structures.
- Accident Reporting: Immediately report accidents and carry out thorough investigations.

Employee safety is a strong foundation for a thriving construction industry, so empower your employee with our safety training & ensure safety compliance.

8.3. RISK ASSESSMENT AND HAZARD IDENTIFICATION

Risk assessment and hazard identification on construction sites are fundamental processes essential for ensuring the safety and health of workers in the construction industry. These processes involve a systematic approach to identifying potential hazards, evaluating the risks associated with these hazards, and implementing measures to mitigate or eliminate them. The dynamic and complex nature of construction activities, with constantly changing environments and a wide variety of tasks, makes these processes particularly crucial.

The first step in risk assessment is hazard identification, which involves recognizing potential sources of harm that could cause injury or illness to workers. Common hazards on construction sites include falls from heights, slips, trips, and falls, exposure to hazardous substances, electrical hazards, machinery and equipment-related risks, and ergonomic hazards related to manual handling tasks. Identifying these hazards requires thorough site inspections, reviewing past incident reports, and consulting with workers who have firsthand experience and knowledge of site-specific risks.

Once hazards are identified, the next step is to evaluate the risks associated with each hazard. This involves determining the likelihood of an incident occurring and the potential severity of its consequences. Risk evaluation can be conducted using qualitative or quantitative methods, such as risk matrices, which help prioritize risks based on their likelihood and impact. High-priority risks, those with a high likelihood of occurrence and severe potential consequences, are addressed first.

After evaluating the risks, appropriate control measures must be implemented to mitigate or eliminate the identified hazards. The hierarchy of controls is a widely accepted framework used in this process, prioritizing control measures from most effective to least effective: elimination, substitution, engineering controls, administrative controls, and personal protective equipment (PPE). For instance, eliminating a hazard entirely, such as using prefabricated components to reduce the need for working at heights, is the most effective control measure. If elimination is not possible, substituting a less hazardous material or process, implementing engineering controls like guardrails or safety nets, and establishing administrative controls such as safety training and work procedures can significantly reduce risk. PPE should be used as a last resort when other controls are not sufficient to fully mitigate the risk.

Continuous monitoring and review are integral parts of the risk assessment and hazard identification process. Construction sites are dynamic environments where conditions and activities can change rapidly, necessitating regular reassessment of risks and updating control measures accordingly. This ongoing process ensures that emerging hazards are promptly identified and addressed, and that control measures remain effective over time.

Engaging workers in the risk assessment process is crucial for its success. Workers possess valuable insights into the practical aspects of site-specific hazards and can provide feedback on the effectiveness of control measures. Encouraging open communication and involving workers in safety discussions foster a culture of safety and shared responsibility.

8.3.1. Potential Hazards on Construction Sites

Identifying potential hazards on construction sites is a critical task that involves recognizing various sources of risk that can cause harm to workers, property, and the environment. Construction sites are inherently complex and dynamic, with numerous activities occurring simultaneously, each presenting unique hazards. Physical hazards are among the most common and include risks such as falls from heights, which are the leading cause of fatalities in the construction industry. These can occur from ladders, scaffolding, roofs, and unfinished structures. Other physical hazards involve slips, trips, and falls on uneven surfaces, loose materials, or wet conditions. Machinery and equipment-related risks are also significant, including injuries from cranes, forklifts, and other heavy machinery due to improper use or mechanical failure. Struck-by and caught-between hazards are prevalent when workers are in the proximity of moving vehicles, falling objects, or collapsing structures.

Chemical hazards are another major concern on construction sites. Workers may be exposed to hazardous substances such as asbestos, lead, silica dust, solvents, and other chemicals used in paints, adhesives, and cleaning agents. Inhalation of toxic fumes, dust, and vapors can lead to serious health issues, including respiratory problems, skin irritation, and long-term diseases such as cancer. Proper storage, handling, and disposal of these chemicals, along with adequate ventilation and the use of personal protective equipment (PPE), are essential to mitigate these risks. Additionally, workers need to be trained in recognizing chemical hazards and understanding Material Safety Data Sheets (MSDS) to ensure safe practices.

Environmental hazards also pose significant risks on construction sites. These include exposure to extreme weather conditions such as heat, cold, rain, and snow, which can lead to heatstroke, hypothermia, or slippery surfaces. Noise pollution is another environmental risk, with prolonged exposure to high noise levels from machinery and tools potentially causing hearing loss. Vibration from tools and equipment can lead to hand-arm vibration syndrome (HAVS) and other musculoskeletal disorders. Dust and particulate matter present another environmental hazard, contributing to respiratory issues and reducing air quality on site. Natural disasters such as earthquakes, floods, or landslides can pose sudden and severe risks, requiring contingency planning and emergency preparedness.

Moreover, ergonomic hazards should not be overlooked. Manual handling of heavy materials can result in musculoskeletal injuries, including strains and sprains. Repetitive motions, awkward postures, and overexertion are common ergonomic hazards that need to be addressed through proper training, the use of mechanical aids, and the design of work tasks to minimize physical strain.

Identifying these hazards requires a comprehensive approach that includes regular site inspections, risk assessments, and consultations with workers who have firsthand knowledge of the working conditions. Utilizing safety audits and hazard reporting systems helps in the early detection and mitigation of potential risks. Implementing a hierarchy of controls—starting with the elimination of hazards where possible, followed by substitution, engineering controls, administrative controls, and the use of PPE—ensures a systematic approach to managing these risks.

8.3.2. Conducting Regular Risk Assessments

Conducting regular risk assessments to evaluate and prioritize hazards on construction sites is an essential practice to ensure the safety and health of workers. This process involves systematically identifying potential hazards, assessing the risks associated with these hazards, and determining the most effective measures to control or eliminate them. Regular risk assessments are crucial due to the dynamic nature of construction sites, where changes in activities, environments, and personnel can introduce new risks or alter existing ones.

The first step in conducting a risk assessment is hazard identification. This involves a thorough examination of the work environment to identify all possible sources of harm. Physical hazards, such as working at heights, operating heavy machinery, and exposure to noise, are common on construction sites. Chemical hazards include exposure to harmful substances like asbestos, lead, and solvents. Environmental hazards encompass risks such as extreme weather conditions, poor air quality, and natural disasters. Ergonomic hazards related to manual handling and repetitive tasks must also be considered.

Once hazards are identified, the next step is to evaluate the risks associated with each hazard. Risk evaluation involves assessing the likelihood of an incident occurring and the potential severity of its consequences. This can be done using qualitative methods, such as risk matrices, which rank risks based on their probability and impact, or quantitative methods, which involve statistical analysis of past incident data. The goal is to prioritize risks so that the most significant hazards are addressed first. High-priority risks are those with a high likelihood of occurrence and severe potential consequences, while lower-priority risks may pose less immediate danger.

After evaluating the risks, appropriate control measures must be implemented. The hierarchy of controls is a widely accepted framework for this purpose. It prioritizes control measures from most effective to least effective: elimination, substitution, engineering controls, administrative controls, and personal protective equipment (PPE). For example, eliminating a hazard entirely, such as by using prefabricated components to avoid working at heights, is the most effective control measure. If elimination is not possible, substituting less hazardous materials or processes, implementing engineering controls like guardrails and safety nets, and establishing administrative controls such as safety training and work procedures can significantly reduce risk. PPE, such as helmets, gloves, and safety glasses, should be used as a last resort when other controls are not sufficient.

Regular monitoring and review are integral to the risk assessment process. Construction sites are constantly evolving, and new hazards can emerge as work progresses. Continuous monitoring ensures that existing control measures remain effective and that new risks are promptly identified and addressed. This ongoing process involves scheduled site inspections, audits, and consultations with workers, who provide valuable insights into practical safety concerns.

Worker involvement is key in successful risk assessment. They detect hazards first and give feedback on control measures. Open communication and worker participation in safety discussions promote a safety culture and shared responsibility.

8.3.3. Implementing Strategies to Mitigate Identified Risks

Implementing strategies to mitigate identified risks on construction sites is essential for ensuring the safety and well-being of workers. Once hazards are identified and risks are assessed, appropriate measures must be taken to control or eliminate these risks. Effective risk mitigation strategies follow a structured approach based on the hierarchy of controls, which prioritizes measures from most to least effective: elimination, substitution, engineering controls, administrative controls, and personal protective equipment (PPE).

Elimination is the most effective risk mitigation strategy and involves completely removing the hazard from the workplace. For instance, eliminating the need to work at heights by using prefabricated components that can be assembled at ground level significantly reduces the risk of falls. Similarly, automating a process can eliminate the need for manual handling, thereby reducing the risk of musculoskeletal injuries.

When elimination is not feasible, substitution is the next best strategy. This involves replacing a hazardous material or process with a less hazardous one. For example, using water-based paints instead of solvent-based paints can reduce exposure to harmful fumes. Substituting manual lifting with mechanical lifting equipment also minimizes the risk of injury.

Engineering controls are designed to isolate people from hazards. These controls involve modifications to equipment, processes, or the work environment to reduce exposure to risks. Installing guardrails and safety nets to prevent falls from heights, using enclosed cabs for machinery to protect operators from noise and dust, and implementing ventilation systems to control exposure to airborne contaminants are all examples of effective engineering controls.

Administrative controls include changes to work policies, procedures, and practices to reduce risk. These controls are less effective than elimination, substitution, or engineering controls but are still crucial for managing risks. Examples include implementing job rotation to minimize repetitive motion injuries, scheduling regular safety training sessions to ensure workers are aware of hazards and safe practices, and conducting safety meetings to discuss and address ongoing risks. Establishing clear protocols for reporting and investigating incidents also helps in identifying root causes and preventing recurrence.

Personal protective equipment (PPE) is considered the last line of defense and should be used when other control measures are not sufficient to mitigate risks. PPE includes items such as helmets, gloves, safety glasses, high-visibility clothing, and respiratory protection. While PPE can effectively reduce exposure to hazards, it relies on proper use, maintenance, and worker compliance. Providing appropriate PPE, ensuring it is correctly fitted, and training workers on its use and care are essential for its effectiveness.

Regular monitoring and review of implemented control measures are critical to ensure they remain effective over time. Construction sites are dynamic environments where conditions and risks can change rapidly. Continuous monitoring through inspections, audits, and feedback from workers helps identify any shortcomings in control measures and allows for timely adjustments. Engaging workers in the monitoring process encourages ownership and accountability for safety.

Involving workers in the development and implementation of risk mitigation strategies is also crucial. Workers have practical knowledge of site-specific risks and can provide valuable insights into the feasibility and effectiveness of control measures. Encouraging their participation in safety discussions, risk assessments, and decision-making processes fosters a culture of safety and collaboration.

8.4. SAFETY TRAINING AND EDUCATION

Safety training and education in the construction industry are pivotal components for ensuring the health and well-being of workers. Given the high-risk nature of construction work, comprehensive and continuous training programs are essential to equip workers with the knowledge and skills needed to recognize, avoid, and respond to potential hazards. Safety training encompasses a broad range of topics, including general safety practices, specific job-related hazards, proper use of personal protective equipment (PPE), emergency response procedures, and adherence to regulatory requirements. Effective safety training and education programs are designed to foster a culture of safety, where every worker understands the importance of following safety protocols and feels empowered to contribute to a safe working environment.

The foundation of safety training in the construction industry begins with a thorough orientation for all new employees. This initial training provides an overview of the company's safety policies, the specific hazards associated with the worksite, and the procedures in place to mitigate these risks. It sets the tone for the company's commitment to safety and ensures that new hires understand the critical role they play in maintaining a safe work environment. Orientation training typically includes instruction on the proper use of PPE, safe operation of machinery and equipment, fall protection strategies, and the importance of reporting unsafe conditions.

Beyond the initial orientation, ongoing safety training is crucial to keep workers informed about new hazards and changes in safety regulations. Regular refresher courses and toolbox talks are effective ways to reinforce safety knowledge and address specific issues that may arise during the course of a project. These sessions provide opportunities for workers to discuss recent incidents, share best practices, and receive updates on new safety protocols or equipment. Keeping safety training current and relevant helps to maintain a high level of awareness and preparedness among workers.

Specialized training is also necessary for certain high-risk activities within the construction industry. For example, workers involved in tasks such as operating heavy machinery, handling hazardous materials, or working at heights require targeted training to address the unique risks associated with these activities. This specialized training includes both theoretical instruction and practical, hands-on experience to ensure that workers are fully competent and confident in their abilities to perform their duties safely.

Another critical aspect of safety training and education is emergency preparedness. Workers must be trained to respond effectively to emergencies such as fires, chemical spills, or medical incidents. This training includes understanding the emergency procedures specific to their worksite, knowing the locations of emergency exits and equipment, and being able to perform basic first aid. Regular emergency drills help to reinforce this training and ensure that workers can respond quickly and efficiently in a crisis.

Effective safety training programs (Roughton & Mercurio, 2002) also emphasize the importance of communication and collaboration. Workers are encouraged to speak up about safety concerns and to participate in safety committees or meetings. This inclusive approach not only helps to identify potential hazards but also fosters a sense of collective responsibility for safety.

8.4.1. Training Programs for Workers

Providing comprehensive training programs for workers at all levels in the construction industry is essential to ensure safety and efficiency on the job site. These programs must be tailored to address the diverse roles and responsibilities within a construction project, from entry-level laborers to senior management, ensuring that everyone understands and adheres to safety protocols and best practices.

For entry-level workers and new hires, foundational training is crucial. This includes an in-depth orientation covering the basics of workplace safety, company policies, and specific hazards associated with the construction site. Training sessions should introduce the proper use of personal protective equipment (PPE), emergency response procedures, and general safety practices. This foundational training helps new workers integrate safely into the workforce and understand the importance of maintaining a safe work environment from the outset.

Skilled tradespeople, such as electricians, plumbers, and carpenters, require more specialized training to address the unique risks associated with their specific tasks. This training should combine theoretical knowledge with practical, hands-on instruction to ensure that workers are not only aware of the hazards but also competent in using tools and equipment safely. For instance, electricians need to be trained on the dangers of working with live circuits and the protocols for lockout/tagout procedures to prevent accidental energizing of equipment.

Supervisors and site managers need advanced training that includes not only safety protocols but also leadership and communication skills. Their training should focus on hazard identification, risk assessment, and incident investigation. Supervisors play a critical role in enforcing safety standards and must be able to recognize potential hazards before they cause harm, conduct thorough investigations if incidents occur, and effectively communicate safety expectations to their teams. They should also be trained in conflict resolution and motivational techniques to foster a positive safety culture on site.

For senior management, comprehensive training programs should encompass strategic planning and the integration of safety into all business operations. This level of training should highlight the importance of regulatory compliance, risk management, and the financial implications of workplace safety. Managers must understand their role in setting the tone for safety culture within the organization, including allocating appropriate resources for safety initiatives and ensuring that safety is a core value in all decision-making processes. Leadership training for senior management should also focus on the development and implementation of comprehensive safety policies and programs.

Continuous training is also vital for all levels to stay updated with new regulations, technologies, and evolving best practices in construction safety. Regular refresher courses, safety workshops, and ongoing professional development opportunities ensure that all workers remain knowledgeable about the latest safety standards and innovations. Additionally, incorporating real-world scenarios and simulations in training programs can enhance learning by providing practical experience in a controlled environment.

Engagement and feedback from workers are critical components of effective training programs. Encouraging workers to share their experiences and suggestions for improving safety practices helps to identify gaps in training and develop more relevant content. This inclusive approach not only improves the quality of training but also empowers workers to take ownership of their safety and contribute to a culture of continuous improvement.

8.4.2. Employees' Education About Site-Specific Risks and Safety Procedures

Ensuring that employees are educated about site-specific risks and safety procedures is a critical component of construction site safety management. Construction sites are dynamic environments where the specific hazards can vary significantly from one project to another, necessitating tailored safety education for each location. Effective education about site-specific risks begins with a comprehensive site orientation for all workers, including new hires and subcontractors, upon their arrival. This orientation should provide a detailed overview of the particular hazards associated with the site, such as the presence of hazardous materials, high-voltage areas, areas where heavy machinery operates, and any ongoing activities that might pose risks.

Site-specific training should also cover the layout of the site, highlighting safe routes for pedestrian traffic, locations of emergency exit, first aid stations, and fire extinguishers. Workers need to be familiar with the site's emergency response plan, including evacuation procedures, the roles and responsibilities of emergency response teams, and the communication protocols to be followed in case of an incident. This training ensures that all employees know how to act swiftly and effectively during an emergency, minimizing the potential for injury and chaos.

To enhance the relevance and retention of site-specific safety information, training should include hands-on demonstrations and practical exercises. For instance, workers should practice using site-specific safety equipment, such as harnesses for fall protection if working at heights, or respirators if there are airborne contaminants. These practical sessions help reinforce theoretical knowledge and build the confidence of workers in handling actual site conditions.

Regular safety briefings and toolbox talks are essential for keeping site-specific safety education up-to-date. These briefings should address any new hazards introduced by changes in the work environment or construction activities, such as new equipment being brought on site or changes in weather conditions that could affect safety. They also provide an opportunity to discuss recent incidents or near misses, analyze their causes, and emphasize the lessons learned to prevent future occurrences.

Involving workers in identifying site-specific risks and developing safety procedures is another effective strategy. Workers often have valuable insights into the practical aspects of site operations and can help pinpoint areas where safety improvements are needed. Encouraging their active participation fosters a sense of ownership and accountability for safety, leading to a more vigilant and cooperative workforce.

To support continuous learning, safety information should be made readily accessible through various means such as safety posters, digital platforms, and mobile apps. These resources should provide clear, concise information about site-specific hazards and the corresponding safety procedures, enabling workers to refresh their knowledge as needed.

Supervisors and site managers play a crucial role in ensuring that site-specific safety education is effectively communicated and adhered to. They must consistently enforce safety protocols, conduct regular site inspections, and provide immediate feedback to workers. Additionally, they should lead by example, demonstrating a commitment to safety through their actions and decisions.

8.4.3. Update Workers on New Safety Practices and Regulations

Offering ongoing training to keep workers updated on new safety practices and regulations is essential for maintaining a safe and compliant construction site. The construction industry is constantly evolving, with new technologies, methods, and regulations emerging regularly. Continuous training ensures that workers stay informed about these changes and can apply the latest safety practices in their daily activities.

Regularly scheduled refresher courses are a key component of ongoing training. These courses revisit fundamental safety concepts and update workers on any changes to existing protocols or the introduction of new regulations. By periodically reinforcing core safety principles, these sessions help to prevent complacency and ensure that safety remains a top priority for all employees. Furthermore, refresher courses provide an opportunity to address any observed deficiencies in safety practices and to reinforce the importance of compliance with safety standards.

Toolbox talks, or brief, informal meetings focused on specific safety topics, are another effective method of ongoing training. These talks are typically held at the start of a shift and cover relevant issues such as recent incidents, seasonal hazards, or new equipment. By addressing current and site-specific topics, toolbox talks keep safety at the forefront of workers' minds and encourage continuous dialogue about safe practices. They also provide a platform for workers to voice concerns and suggest improvements, fostering a collaborative approach to safety management.

In addition to formal training sessions, leveraging digital tools and resources can enhance ongoing safety education. Online training modules, webinars, and mobile applications offer flexible and accessible options for workers to stay informed about the latest safety practices and regulations. These tools can provide interactive and engaging content, such as videos and quizzes, to reinforce learning. Digital platforms also allow for the tracking of training completion and progress, ensuring that all workers receive the necessary updates.

Ongoing training should also include scenario-based exercises and simulations. These practical exercises help workers to apply their knowledge in realistic settings, enhancing their ability to respond effectively to potential hazards. For example, conducting emergency drills for situations such as fires, chemical spills, or equipment malfunctions helps to ensure that workers are prepared to act swiftly and safely in real emergencies. These simulations can also reveal gaps in current safety protocols and highlight areas for improvement.

Keeping abreast of regulatory changes is critical for compliance and safety. Construction companies must stay informed about updates from regulatory bodies such as the Occupational Safety and Health Administration (OSHA) or equivalent organizations in other countries. This includes changes in safety standards, reporting requirements, and industry best practices. Providing regular updates on these regulatory changes through training sessions ensures that workers understand their responsibilities and the steps needed to remain compliant.

Engagement and feedback from workers play a crucial role in the effectiveness of ongoing training. Encouraging workers to participate actively in training sessions, ask questions, and share their experiences helps to create a culture of continuous learning and improvement. Workers are more likely to embrace new safety practices and regulations if they feel their input is valued and their concerns are addressed.

8.5. EMERGENCY PREPAREDNESS AND RESPONSE

Emergency preparedness and response are critical components of construction site safety management, aimed at ensuring the safety and well-being of workers in the event of unexpected incidents or disasters. Construction sites are inherently dynamic environments where various hazards exist, making it imperative to have robust emergency plans and procedures in place to effectively respond to emergencies and minimize their impact.

The first step in emergency preparedness is the development of a comprehensive emergency response plan (ERP) tailored to the specific risks and conditions of the construction site. This plan should outline procedures for various types of emergencies, including fires, chemical spills, medical emergencies, natural disasters, and other potential incidents. The ERP should designate roles and responsibilities for personnel, establish communication protocols, and identify evacuation routes and assembly points.

Training plays a crucial role in emergency preparedness, ensuring that workers are familiar with the ERP and know how to respond effectively in emergency situations. All workers should receive thorough training on emergency procedures during their site orientation and regular refresher training thereafter. This training should cover topics such as emergency notification procedures, evacuation protocols, first aid techniques, and the proper use of emergency equipment such as fire extinguishers and first aid kits.

Regular emergency drills and exercises are essential for testing the effectiveness of the ERP and familiarizing workers with emergency procedures. These drills should simulate various emergency scenarios and involve all workers, subcontractors, and other stakeholders on the construction site. By conducting drills regularly, potential weaknesses in the ERP can be identified and addressed, and workers can gain confidence in their ability to respond quickly and appropriately in real emergencies.

Effective communication is critical during emergencies to ensure that information is disseminated quickly and accurately to all personnel on the construction site. The ERP should include protocols for communicating emergency alerts, instructions, and updates to workers, including designated communication channels and emergency contact information. Additionally, establishing a system for accounting for all personnel during emergencies helps to ensure that everyone is safely evacuated from the site.

Emergency preparedness also involves ensuring that appropriate emergency response equipment is readily available and properly maintained on the construction site. This includes fire extinguishers, first aid kits, personal protective equipment (PPE), and any specialized equipment needed to address specific hazards. Regular inspections and maintenance of emergency equipment are essential to ensure that it remains in good working condition and ready for use when needed.

Collaboration with local emergency responders, such as fire departments, paramedics, and law enforcement agencies, is vital for effective emergency preparedness and response. Construction companies should establish partnerships with these agencies and coordinate emergency response plans to ensure a seamless and coordinated response in the event of an emergency. This collaboration may involve conducting joint training exercises, sharing information about site-specific hazards, and establishing protocols for requesting assistance during emergencies.

8.5.1. Developing and Communicating Emergency Response Plans

Developing and communicating emergency response plans (ERPs) is a critical aspect of construction site safety management, ensuring that workers are prepared to respond effectively to unexpected incidents or disasters. ERPs provide a structured framework for identifying potential emergencies, outlining response procedures, assigning roles and responsibilities, and communicating vital information to all personnel on the construction site. Developing a comprehensive ERP involves several key steps to ensure that it is tailored to the specific risks and conditions of the site and that it can be implemented swiftly and efficiently in times of crisis.

The first step in developing an ERP is to conduct a thorough risk assessment of the construction site to identify potential hazards and vulnerabilities. This assessment should consider a range of factors, including the types of construction activities being performed, the presence of hazardous materials, environmental risks, and the likelihood of natural disasters such as earthquakes or severe weather events. By understanding the specific risks facing the site, construction companies can develop response plans that address these hazards effectively.

Once potential hazards have been identified, the next step is to develop response procedures for various types of emergencies. These procedures should outline step-by-step instructions for responding to specific incidents, including actions to take to ensure the safety of workers, how to mitigate the effects of the emergency, and protocols for coordinating with emergency responders if necessary. Response procedures should be clear, concise, and easy to understand, with consideration given to the unique challenges of the construction environment.

Assigning roles and responsibilities is another crucial aspect of ERP development. Each member of the construction team should have a clearly defined role in the emergency response process, including responsibilities for initiating the response, coordinating evacuation efforts, providing first aid, securing the site, and communicating with emergency responders. By assigning specific roles and responsibilities, construction companies can ensure that everyone knows what is expected of them during an emergency and can act quickly and decisively to protect themselves and others.

Communication is a key component of emergency response, and effective communication protocols should be established as part of the ERP. This includes establishing designated communication channels for transmitting emergency alerts, instructions, and updates to all personnel on the construction site. Communication protocols should specify how information will be disseminated, who is responsible for issuing alerts, and how workers can confirm receipt of important messages. Additionally, construction companies should establish protocols for communicating with emergency responders and other external stakeholders, such as neighboring businesses or local authorities.

Once the ERP has been developed, it is essential to communicate it effectively to all personnel on the construction site. This can be done through a combination of training sessions, written materials, and regular drills and exercises. Training sessions should cover key aspects of the ERP, including emergency procedures, roles and responsibilities, communication protocols, and evacuation routes. Written materials, such as posters, handouts, and digital resources, can serve as reference guides for workers to consult in an emergency. Regular drills and exercises provide an opportunity to practice implementing the ERP in a controlled setting and help ensure that workers are familiar with their roles and responsibilities.

8.5.2. Emergency Drills and Training Sessions

Conducting emergency drills and training sessions for workers is a crucial aspect of construction site safety management, ensuring that personnel are prepared to respond effectively to unexpected incidents or disasters. These drills and training sessions provide valuable opportunities for workers to practice emergency procedures, familiarize themselves with evacuation routes, and learn how to use emergency equipment. By simulating realistic emergency scenarios in a controlled setting, construction companies can assess the effectiveness of their emergency response plans, identify areas for improvement, and enhance overall preparedness.

Emergency drills should be conducted regularly, with participation from all personnel on the construction site, including workers, supervisors, and subcontractors. The frequency of drills will depend on factors such as the size and complexity of the site, the nature of the work being performed, and any specific regulatory requirements. However, drills should typically be conducted at least annually, with additional drills scheduled as needed to address specific concerns or changes in site conditions.

Before conducting an emergency drill, it is essential to establish clear objectives and scenarios for the exercise. This might involve simulating various types of emergencies, such as fires, chemical spills, medical incidents, or severe weather events, to test different aspects of the emergency response plan. Scenarios should be realistic and relevant to the specific risks and conditions of the construction site, ensuring that workers gain practical experience that can be applied in real-world situations.

During the drill, workers should be expected to follow the same procedures and protocols that they would in an actual emergency. This includes initiating emergency response procedures, evacuating the site according to established routes, assisting with first aid and rescue efforts as needed, and communicating with supervisors and emergency responders. By actively participating in the drill, workers can gain confidence in their ability to respond effectively and identify any areas where additional training or resources may be needed.

After the drill is completed, it is essential to conduct a thorough debriefing session to review the exercise and discuss lessons learned. This debriefing should involve all participants in the drill, including workers, supervisors, and emergency responders if applicable. Key topics for discussion may include the effectiveness of communication protocols, the efficiency of evacuation procedures, any challenges or obstacles encountered during the drill, and recommendations for improving the emergency response plan.

In addition to emergency drills, construction companies should also provide regular training sessions to educate workers about emergency procedures and protocols. These training sessions should cover topics such as recognizing potential hazards, initiating emergency response procedures, evacuating the site safely, providing first aid, and using emergency equipment such as fire extinguishers and first aid kits. Training sessions should be tailored to the specific needs and roles of workers on the construction site, ensuring that everyone receives the information and skills they need to respond effectively in an emergency.

8.5.3. Ensuring Emergency Equipment

Ensuring that emergency equipment, such as fire extinguishers and first aid kits, is accessible and maintained is crucial for construction site safety management. These essential safety devices play a critical role in responding to emergencies and providing immediate assistance to injured workers. Proper maintenance and accessibility of emergency equipment help to minimize response times and ensure that workers have the resources they need to address emergencies effectively.

The first step in ensuring the accessibility of emergency equipment is to strategically place it throughout the construction site in easily accessible locations. Fire extinguishers should be positioned at regular intervals throughout the site, with clear signage indicating their locations. First aid kits should be readily available in multiple areas, such as in tool trailers, work areas, and break rooms, to ensure that workers can access them quickly in the event of an injury.

Emergency equipment should be stored in designated cabinets or containers that protect it from damage and exposure to environmental factors such as moisture, dust, and extreme temperatures. Cabinets should be clearly labeled and kept unlocked or easily accessible in case of an emergency. Additionally, emergency equipment should be positioned away from potential hazards such as machinery or materials that could obstruct access during an emergency.

Regular inspections and maintenance are essential to ensure that emergency equipment is in good working condition and ready for use when needed. Fire extinguishers should be inspected and serviced annually by a qualified technician to ensure that they are fully charged, have no visible damage, and operate correctly. Any fire extinguishers that are found to be defective or expired should be replaced immediately.

Similarly, first aid kits should be inspected regularly to ensure that they are fully stocked with the necessary supplies and that all items are within their expiration dates. Common items found in first aid kits include bandages, gauze pads, adhesive tape, antiseptic wipes, scissors, and disposable gloves. Any depleted or expired items should be replenished promptly to ensure that the first aid kit is fully functional in an emergency.

In addition to regular inspections, construction companies should provide training to workers on the proper use of emergency equipment. This training should cover topics such as how to operate fire extinguishers, administer first aid, and perform basic life-saving techniques. By ensuring that workers are familiar with emergency procedures and equipment, construction companies can increase the likelihood of a swift and effective response in the event of an emergency.

8.6. CONSTRUCTION SITE TRAFFIC MANAGEMENT

For several reasons, it's critical to safely manage traffic flow and pedestrian areas on building sites. Usually, a variety of vehicles, such as loads, cranes, HGV lorries, and diggers, are on sites. These vehicles are specialized commercial work vehicles with moving parts and restricted visibility. Vehicles will usually need to enter and exit your construction site to carry out various tasks, depending on the nature of your work. You need clearly marked entry points, room for large lorries to turn around safely, and walkways so that staff and guests can move around the site without taking unnecessary risks—from delivery trucks to staff cars. Dedicated traffic controllers or supervisors are always beneficial, but given the realities of running a busy site, you also need to think about other fail-safe security measures you can implement to control vehicle flow through your property.

8.6.1. Types of Traffic on Construction Sites

As mentioned, there are many different types of vehicles flowing through a construction site on a daily basis, including:

- Staff, supervisors and tradespeople arriving for work, often in different shift patterns.
- Deliveries of materials, resources and construction components.
- Low-loaders transporting specialist vehicles to and from the site.
- Plant, including dump trucks, HGVs and excavators.

Traffic management strategies are crucial to ensuring site managers can access the relevant facility safely and minimizing crossovers between traffic through routes and pedestrian walkways, as many have a variety of vehicles requiring access to a construction area throughout the day. Not only must the location and flow of vehicles into and out of the construction site be taken into account, but also the areas for turns, parking, and offloading.

8.6.2. Legal Requirements for Construction Site Traffic Management

Construction sites are required to provide for vehicular activity in accordance with the Construction (Design and Management) Regulations of 2015. Site planners must arrange traffic so that, to the extent that it is practical, they accommodate vehicle and pedestrian traffic so that there is no risk to public health or safety.

In essence, the law requires construction site managers to:

- Risk assesses their sites, identify the key areas requiring action, and review those risk assessments periodically or whenever circumstances change.
- Share the outcomes of such an assessment with the relevant personnel, and ensure that safety measures have been correctly implemented.
- Provide training and safe working awareness training to all personnel or site visitors, or appoint staff to accompany visitors and supervise their safety.
- To consult with the workforce when assessing critical areas of risk and establish a dialog about how they manage exposure to potential hazards.
- Appoint appropriately trained and skilled staff members to be responsible for overseeing operations in alignment with risk mitigation policies.

8.6.3. Practical Construction Site Traffic Management Actions

Every construction site is different, so there is no 'one size fits all' solution that will remove every site's hazard. However, some of the common ways to manage traffic safety in the construction industry are:

- Create segregated traffic through points and pedestrian walkways with clear barriers and lighting, and crossings where access points meet.
- Install signage and directions to ensure vehicles only access appointed areas.
- Schedule deliveries and material offloads in designated areas, and in assigned time slots.
- Assign different entry points and exit points for different vehicle types and/or pedestrian and vehicular access.
- Keep roadways clear of blockages, hazards, and potholes.
- Block off unauthorized entrance points to ensure vehicles are prevented from entering pedestrian zones.

If there are any doubts about your core risks and the right way to resolve them, the Clearway team is always available to assist with the risk assessment process and make recommendations for the best solutions.

8.6.4. Construction Site Traffic Management Plan Options

In order to ensure that you reduce the risk factor of all your identified hazards, Clearway UK collaborates with planners and managers of construction sites of all sizes, providing a broad range of solutions. All our vehicle control options are flexible and can be combined with other safety measures to provide a complete solution, or they can be used as a stand-alone service to lessen your critical hazards.

Essential construction site traffic management services include:

- Access Control: Using a sophisticated access control system is one of the most effective ways to control vehicle access to your building site. High-functioning remote passes and intercoms are available as options to guarantee that only authorized individuals are allowed entry. Using an automated gate lifting system, supervisors can ensure that visitors are escorted to the designated area of the site only with the presence of a designated staff member.
- Concrete Barrier Hire: Using a sophisticated access control system is one of the most effective ways to control vehicle access to your building site.
- Perimeter Security Solutions: The weakest aspect of your site's security is always its boundaries. In order to protect your perimeters and make sure that unintentional or criminal access is never a problem, we provide a wide range of options. You can defend your perimeters and make sure that all site activity is safely contained in a number of ways, from anti-climb fences to manned security patrols to mobile inspections and CCTV surveillance on building sites. Even at the busiest construction sites, managers can safely manage all vehicle access and activity around them by implementing one of several effective strategies. However, managers must always take their responsibility for traffic management on construction sites seriously.

8.6.5. Examples of Good Construction Traffic Management

Since workers, deliveries, and equipment all require access, construction sites can be hectic locations. When trying to prevent major and deadly accidents involving your employees, traffic management is crucial. These 10-construction traffic management best practices are examined in this chapter. Construction sites are among the places of employment where coexistence between humans and machines is frequently required. Plants and machinery, however, present a risk even though they can help workers with tasks like excavation and material transportation. The results will not be favorable if a person is struck by a machine weighing more than 10 tons on the spot. More than 30 workers lost their lives and almost 4,000 were hurt in accidents involving moving machinery or being hit by a moving car last year. However, there are other jobs where people and traffic must coexist. Construction is just one example. However, compared to other sectors, the construction industry may have more challenges with transport management. In the construction industry, the challenge lies in the fact that each site is unique, with varying layouts, access points, requirements, tasks, equipment, and user counts, among other factors. Furthermore, at different project stages, the same site requires different layout configurations.

Here are our top 10 examples of good traffic management on construction sites:

1. Keep People Away from Plant and Machinery: Keeping pedestrians and cars as far apart as possible is the simplest approach to prevent people from being hurt by plants and vehicles on your property. Although it may seem easier said than done, take into account the locations on your property where people and plants are moving as they go to and from their work areas. When they head to the canteen, nobody expects to come into contact with an excavator. It's possible that site visitors are unaware of the cars you have listed. Additionally, when people are maneuvering or leaving the site, there is a higher chance that they will come into contact with vehicles. Therefore, it is easier to keep people and machines apart along traffic routes because they are both high-risk areas and places where they are not required to be together. To keep foot traffic and vehicle traffic apart, use barriers, access routes, and entrances that are designated specifically for pedestrians. Make crossing areas clear and safe.
2. Keep Site Plant Away from Hazards: The drivers of machines and vehicles on your site need to be aware of more than just people. Although it is important to ensure that your traffic plan provides adequate, level routes for traffic on the site, construction sites do not have the luxury of a completed road or traffic route until the project is completed. Vehicle overturning can be a risk due to excavations and uneven or unstable ground. However, there are numerous risks on a building site that could harm equipment and machinery in addition to that challenge. When planning your vehicle's access routes, keep in mind that construction sites are busy places. Avoid obstacles like excavations, structures, overhead cables, uneven or unstable ground, and watercourses. Other things to consider include temporary works, new structures, skips, waste materials, and services.
3. Minimize Movements: While construction site traffic is often unavoidable, it should be reduced as much as possible. Do workers' vehicles need to be on the site? Probably not. Would planning the layout of the site so that the storage area is close to the entrance help reduce the amount of traffic crossing the site? Probably yes.

Any vehicles that are on site should need to maneuver as little as possible. Reversing is usually where visibility issues and fatal accidents can occur in construction, so using one-way systems and planning turning circles to avoid reversing can be a great way to improve safety.

Control vehicle entry. Use a one-way system and plan turning areas especially in busy areas. Keep parking, and where possible deliveries, away from the work area.

4. Visibility is Vital: We need to see all the construction site obstacles that we talked about, including people, excavations, uneven surfaces, other machinery and plant, other structures, watercourses, and materials. It takes clear visibility to ensure safe traffic flow on the property. Banksmen, cameras,

alarms, and mirrors can all be useful when navigating the site. People walking on your property ought to be dressed in bright colors. Perhaps more lighting is required.

5. There's No Need for Speed: Drivers have less time to react and pedestrians have less opportunity to move aside as plants and vehicles move more quickly. There will frequently be crossings and other locations where people and cars may come into contact, even though it should be the top priority to keep site traffic and pedestrians apart. Determine a reasonable speed limit by evaluating your website. To convey the speed limit, use signs and inductions; these should be watched over and followed.
6. Train to Gain: Every plant operator must be capable and fit. When people attempt to operate vehicles for which they lack the necessary experience, training, or knowledge, accidents may result. Only after competency has been verified and appropriate training has been provided should authority to operate the plan be granted on site. Furthermore, not just the individuals using the site's machinery and cars for transportation need training. Banksmen and signallers are examples of people who assist and direct movements; they must also be competent to do so. Make sure the necessary training is received by all of your plant operators, bankers, and signalers, and that it is regularly renewed.
7. Provide Information: You have designated routes, a speed limit, and a course of action. It is important to ensure that all users are aware of the traffic plan and site rules, including pedestrians and drivers. You can communicate your traffic rules through toolbox talks, safety briefings, instructions, and inductions. Additional tools for providing clear instructions and serving as a reminder include traffic light systems, road and speed limit signs, and speed checks placed throughout the site. Put up signs, label your routes, display a site plan, and make sure that everyone receives an orientation that addresses traffic management.
8. Get Ready for Change: The layout of a construction site is likely to change as buildings and other structures are built because of the nature of the work involved. Hazardous conditions involving vehicles and vegetation on the property may arise from an outdated traffic management plan or from the absence of one altogether. As the site layout changes over the course of the project, keep your traffic management plans up to date.
9. Keep Routes Clear: Similar to the well-known tale of waiting for several buses to arrive at once, deliveries can also occur, particularly when the site is being set up. To ensure safe management of deliveries and other vehicle movements, plan ahead and schedule them. The location will be less congested and safer the fewer cars there are at any given time. During construction health and safety inspections, one of the main things HSE inspectors look for is well-organized sites with clear walkways and access routes.

Maintain your access routes free of obstructions to deter drivers from finding a different route.

10. Maintain and Inspect: Every vehicle and plant on the property must be secure. For built-in controls like brakes, alarms, and guards to be effective, they must function. Maintaining equipment in good working order is required by law and is crucial for the safe movement and operation of vehicles on the site. Make sure your equipment and cars are maintained and inspected to ensure their safety while on the job. Maintaining standards in traffic management is essential; once other project demands become apparent, don't let them slip. Implement near-miss reporting, solicit worker input, and enforce regulations as needed. To make sure your plan is working, evaluate and monitor it frequently.

CLASS ACTIVITY:

CONSTRUCTION SITE SAFETY SIMULATION

- **Objective:** To understand and practice safety protocols and hazard identification on a construction site.
- **Materials Needed:** Safety vests, hard hats, safety goggles, gloves, cones or markers, printed hazard signs, etc.
- **Assessment:** Begin with a brief presentation on the importance of safety in construction sites. Highlight key safety protocols, common hazards, and the importance of personal protective equipment (PPE). Have students write a brief reflection on what they learned about construction site safety, including specific hazards they identified and how they can apply these lessons in real-world scenarios.

SUMMARY

- Health and Safety Management (HSM) is a systematic approach to identifying, evaluating, and controlling risks to ensure the well-being of employees, contractors, and other stakeholders within an organization.
- Effective HSM is based on legal and regulatory compliance, going beyond mere adherence to laws to cultivate a safety culture where all organization members are dedicated to maintaining a safe work environment.
- Employee involvement is crucial for effective HSM, as engaging employees at all levels in safety programs promotes a sense of ownership and accountability.
- Leadership commitment plays a vital role in the success of HSM. It's important for senior management to show a genuine commitment to safety by allocating the required resources, supporting safety initiatives, and setting a good example.
- The construction industry poses significant health and safety risks to site workers due to their working at heights, heavy machinery, adverse weather conditions, and frequent occupational hazards, resulting in serious injuries, financial losses, and damage to the reputation of the company or builder.
- Workplace health and safety (WHS) is a set of principles and practices that show us how to manage the health and safety risks that workers, customers, and the public (visitors, suppliers, etc.) might face in the workplace.
- Safety is of the utmost importance when managing a construction site, as most accidents in this field are caused by bad or neglected safety practices.
- Construction industry sites are one of the most hazardous workplace environments to work in with regard to health and safety of employees.
- Construction site safety is a very important requirement in the construction industry that is often neglected on project sites.
- The construction industry has taken numerous steps to avoid accidents on site and provide workers with a safe working condition and ensure safe practices but even now requisite safety in the construction industry is still a distant dream.
- A safety management system can assist in systematically identifying hazards and controlling risks associated with it, while assuring that these controlling steps are effective in aiding construction site safety.
- Construction safety aids in saving excess time and money that have to be spent in case of any mishaps. These unwanted situations may take up excessive time in recovering from these downsides.
- Construction and process safety management is essential to protect employees from safety hazards, protect the public in the surroundings, and keep the job on track and schedule. Promoting safety cultures on construction sites protects everybody related to the project.
- A risk management system for safety on construction sites can be considered more of a method rather than a product, and this method has to be documented

each time it is being implemented anywhere on the construction project so that the safety in charge may keep a check on its effectiveness and can also be used for auditing as well as report preparation.

- Construction sites are bustling hubs of activity, where skilled workers bring architectural visions to life. While the industry is pivotal for societal development, it also presents inherent risks that demand meticulous attention.
- Digitizing and automating your QHSE processes is one of the most impactful things you can do for managing health and safety in construction more efficiently—it allows you to perform site inspections faster, capture defects with a full audit trail, standardize your forms and reporting, and keep your workers connected with your health and safety program.
- Physical hazards, like falls from heights, are common in construction and are the leading cause of fatalities. These can happen from ladders, scaffolding, roofs, and unfinished structures.
- Chemical hazards are also a major concern, with workers exposed to substances like asbestos, lead, silica dust, solvents, and other chemicals found in paints, adhesives, and cleaning agents.

REVIEW QUESTIONS

1. How can workplace health and safety be effectively managed?
2. What are the key components of safety management in construction?
3. How can health and safety in construction be managed efficiently?
4. Why is it important to conduct regular risk assessments, and how should they be conducted?
5. What strategies can be implemented to mitigate identified risks on construction sites?
6. How can employees be educated about site-specific risks and safety procedures?
7. What are the best practices for developing and communicating emergency response plans?
8. What are the legal requirements for construction site traffic management?

REFERENCES

1. Goetsch, D. L. (2012). Construction Safety & Health (2nd ed.). Prentice Hall.
2. Griffith, A. (2011). Integrated Management Systems for Construction: Quality, Environment and Safety. Harlow: Prentice Hall.
3. Griffith, A., & Howarth, T. (2001). Construction Health and Safety Management. Pearson Education.
4. Hill, D. C. (Ed.). (2014). Construction Safety Management and Engineering (2nd ed.). ASSE.

5. Howarth, T., & Watson, P. (2009). Construction Safety Management. Wiley-Blackwell.

6. Lingard, H., & Rowlinson, S. (2005). Occupational Health and Safety in Construction Project Management. Spon Press.

7. McAleenan, C., & Oloke, D. (Eds.). (2010). ICE Manual of Health and Safety in Construction. Thomas Telford.

8. Poon, S. W., Tang, S. L., & Wong, F. K. W. (2008). Management and Economics of Construction Safety in Hong Kong. Hong Kong University Press.

9. Roughton, J. E., & Mercurio, J. J. (2002). Developing an Effective Safety Culture: A Leadership Approach. Butterworth-Heinemann.

CHAPTER

9 Environment Management

LEARNING OBJECTIVES

After studying this chapter, you will be able to:

- Give an overview of environment management;
- Discuss the environmental and sustainable development;
- Explain the environmental issues associated with sustainable development;
- Explain the concepts of environmental management;
- Focus on environmental management tools and techniques.

KEY TERMS FROM THIS CHAPTER

Biodiversity
Sustainable construction
Water resources
Social equity
Raw materials
Soil stability
Greenhouse gas
Transparency
Social sustainability

INTRODUCTION

The goal of environmental management in construction is to minimize the negative effects of construction operations on the environment while advancing sustainable development. Fundamentally, it is an all-encompassing strategy that includes a range of tactics, laws, and procedures to reduce environmental risks and support conservation initiatives. Conducting Environmental Impact Assessments (EIAs) before starting a project is one of the main facets of environmental management in the construction industry. Environmental Impact Assessments assess the possible environmental effects of proposed construction projects, encompassing implications on biodiversity, soil stability, air and water quality, and nearby communities. Early in the planning stage, construction teams can identify potential risks and impacts and devise strategies to mitigate negative effects and guarantee adherence to environmental regulations and standards.

The use of sustainable building techniques is essential to environmental management. This entails applying sustainable design and construction principles to every facet of the process, from waste management and construction techniques to material selection and design. By consuming less energy and reducing resources from the environment, using eco-friendly materials—such as recycled or locally sourced materials—helps construction projects have a smaller environmental impact. Using energy-efficient building designs and construction methods can also lower building lifetime energy consumption and greenhouse gas emissions. Another crucial component of environmental management in the construction industry is waste management. Significant volumes of waste, including steel, concrete, and wood, are produced by construction operations. Reducing waste generation, recycling materials whenever possible, and putting plans in place to reduce the amount of waste dumped in landfills are all part of proper waste management. Construction projects can reduce their environmental impact and aid in resource conservation by implementing a circular economy strategy that reuses, repurposes, or recycles materials.

Since construction projects can have a big impact on nearby water resources, water management is also essential in the construction industry. Water quality can be preserved and the sustainable use of water resources can be ensured by putting policies in place to manage stormwater runoff, prevent water pollution, and reduce water consumption. This can entail putting in place erosion and sediment control techniques, employing water-efficient fixtures and equipment, and creating vegetated buffer zones to filter pollutants from runoff. In addition, continuous reporting and monitoring are needed for environmental performance and regulatory compliance in the construction industry as part of environmental management. Construction teams can spot any deviations from environmental management plans through routine monitoring and take necessary corrective action. Notifying project stakeholders of findings encourages accountability and transparency, which builds mutual respect and teamwork. In order to reduce the negative effects of construction on the environment and to support sustainable development, environmental management is crucial in the construction industry. Construction projects can reduce environmental risks, conserve resources, and contribute to a more sustainable and healthy built environment by incorporating sustainability principles into all stages of the process, from planning and design to construction and operation.

9.1. OVERVIEW OF ENVIRONMENTAL MANAGEMENT

Environmental management in construction is a critical aspect aimed at mitigating the environmental impact of construction activities. Here are some key points:

1. Environmental Impact Assessment (EIA): Conducting thorough EIAs before starting construction projects helps identify potential environmental risks and impacts. This assessment considers factors like air and water quality, biodiversity, soil stability, and the well-being of local communities.
2. Compliance with Regulations: Adhering to environmental regulations and standards set by authorities is essential. Obtaining necessary permits and approvals ensures that construction activities meet legal requirements and minimize environmental harm.
3. Sustainable Construction Practices: Implementing sustainable practices reduces the environmental footprint of construction projects. This includes using eco-friendly materials, minimizing waste generation, optimizing energy efficiency, and promoting biodiversity conservation.
4. Waste Management: Proper waste management is crucial to minimize the amount of construction waste sent to landfills. Recycling materials, reusing resources, and adopting waste reduction strategies help mitigate environmental impacts.
5. Water Management: Construction activities can affect water resources through pollution and excessive usage. Implementing water management strategies, such as reducing consumption, preventing pollution, and managing stormwater runoff, helps protect water quality and ensure sustainable water use.
6. Biodiversity Conservation: Construction projects should consider the conservation of biodiversity in surrounding areas. Protecting natural habitats, replanting trees, and creating green spaces contribute to biodiversity conservation efforts.
7. Noise and Air Pollution Control: Construction activities often generate noise and air pollution, impacting the environment and communities. Implementing measures to control noise and air pollution, such as using sound barriers and dust control measures, minimizes adverse effects.
8. Ecological Restoration: Restoring disturbed ecosystems post-construction promotes ecological balance and biodiversity conservation. Rehabilitating affected areas and reintroducing native vegetation help restore natural habitats.
9. Monitoring and Reporting: Regular monitoring of environmental performance ensures compliance with environmental management plans and regulations.

Reporting findings to stakeholders promotes transparency and accountability in environmental management practices.

10. Community Engagement: Engaging with local communities and stakeholders fosters collaboration and addresses concerns related to environmental management. Involving communities in decision-making processes and addressing their feedback enhances project acceptance and promotes sustainable outcomes. Environmental management can effectively mitigate the environmental impact of construction activities and promote sustainable development by incorporating these key points into construction practices. The environmental management plan (EMP) for a project ought to encompass all facets and tackle all noteworthy consequences that may emerge throughout the building stage. CEMP offers a manual for use by management and construction workers involved in the work, as well as the overall strategy to manage all project-related environmental issues on site.
    - Define the environmental objectives and goals to be achieved during the construction phase;
    - Outline the applicable environmental guidelines, including local as well as international best practice standards that must be followed during construction;
    - Identify environmental impact sources from construction of the project, such as chemical and hazardous waste, dust, exhaust gases, and noise emissions;
    - Provide specific mitigation measures to avoid or minimize any impacts on the air, water, soil, and groundwater quality, such as site preparation and clearing activities, control installations, and waste minimization protocols, in order to ensure environmental compliance;
    - Provide emergency procedures including an emergency control and action plan, related corrective measures, adequate equipment, and a training program for personnel to deal with case-specific emergency situations on-site;
    - Provide safeguard and control measures to ensure the occupational health and safety of all personnel working at the construction site;
    - Define the responsibilities and actions required by all parties (project proponent, main contractor, subcontractors, etc.) during construction to maintain compliance with the environmental requirements;
    - Recommend implementation procedures such as the appointment of a full-time senior supervisor based on-site to ensure that compliance with the requirements of the CEMP is met, and that all relevant construction issues are dealt with in a timely and efficient manner;

- Provide a structured reporting system with standard forms for communication, documentation, and review of environmental compliance activities;
- Provide a framework for a monitoring and inspection program to determine whether the applicable standards and protocols are being complied with.

9.1.1. What Is Sustainability?

For humans to coexist on Earth for a very long time, sustainability is a social objective. There is debate over this term's definitions, which change over time, depending on the literature and context. The three pillars of sustainability are typically environmental, economic, and social. The environmental component is emphasized in many definitions. This may entail tackling important environmental issues like biodiversity loss and climate change. At the international, national, and personal levels, decisions can be influenced by the concept of sustainability. Sustainable development is a related idea, and the two phrases are frequently used interchangeably. UNESCO makes the following distinction between the two: sustainability is frequently seen as a long-term objective (i.e., a more sustainable world), whereas sustainable development describes the various methods and approaches used to get there. Sustainability involves meeting current needs without jeopardizing the ability of future generations to meet their own needs. It requires maintaining a balance between environmental, social, and economic factors in order to protect natural resources, promote social equality, and sustain economic prosperity for the long term. Here's a breakdown of the key components of sustainability:

1. Environmental Sustainability: This aspect of sustainability focuses on protecting and preserving natural resources and ecosystems. It involves reducing pollution, minimizing waste, conserving biodiversity, and mitigating the impacts of human activities on the environment. Environmental sustainability aims to maintain ecological balance and ensure the long-term health and resilience of ecosystems.
2. Social Sustainability: This emphasizes the well-being and equity of communities and individuals. It involves promoting social cohesion, justice, and inclusion, ensuring access to essential services such as healthcare, education, and housing, and fostering a sense of belonging and cultural diversity. Social sustainability also encompasses issues such as human rights, labor rights, and community engagement in decision-making processes.
3. Economic Sustainability: It focuses on maintaining long-term economic prosperity and stability while minimizing negative impacts on the environment and society. It involves promoting economic growth that is inclusive, equitable, and environmentally responsible. Economic sustainability encompasses principles of resource efficiency, innovation, fair trade, and responsible investment practices.

At its core, sustainability involves meeting current needs without jeopardizing the ability of future generations to fulfill their own needs. This necessitates a comprehensive approach that takes into account the interdependence of environmental, social, and economic systems, and strives to harmonize conflicting interests in order to promote lasting prosperity for both people and the Earth. Sustainable development strives to establish a world where every individual can prosper while respecting the constraints of natural resources and ecological limits, fostering a sustainable relationship between humanity and the environment.

9.1.2. Sustainability and Construction

In the construction industry, sustainability is the process of creating, erecting, maintaining, and dismantling infrastructure and buildings in a way that satisfies current needs without jeopardizing the ability of future generations to satisfy their own. It entails taking into account how construction operations will affect the environment, society, and economy over the course of a project's whole lifecycle.

Here are some key aspects of sustainability in construction:

1. Environmental Considerations: Sustainable construction aims to minimize the environmental impact of buildings and infrastructure by reducing resource consumption, minimizing waste generation, and mitigating pollution. This includes using environmentally friendly materials, optimizing energy and water efficiency, and preserving natural habitats and ecosystems.
2. Energy Efficiency: Sustainable construction emphasizes energy-efficient building designs and technologies to reduce energy consumption and greenhouse gas emissions. This may involve incorporating passive design strategies, such as orientation for optimal solar gain and natural ventilation, as well as utilizing energy-efficient HVAC systems, lighting, and appliances.
3. Resource Conservation: Sustainable construction promotes the efficient use of resources, including materials, water, and land. This involves using recycled or renewable materials, minimizing construction waste through recycling and reuse, and implementing water-saving technologies and practices.
4. Green Building Certifications: Many sustainability standards and certifications, such as LEED (Leadership in Energy and Environmental Design) and BREEAM (Building Research Establishment Environmental Assessment Method), provide frameworks for evaluating and certifying the sustainability performance of buildings and infrastructure projects. These certifications encourage the adoption of sustainable practices and demonstrate a commitment to environmental responsibility.
5. Life Cycle Assessment (LCA): It is a method used to assess the environmental impacts of a building or infrastructure project throughout its entire lifecycle, from raw material extraction and manufacturing to construction, operation, and end-of-life disposal or recycling. By considering the environmental impacts of materials and construction processes, LCA helps identify opportunities for

improvement and inform decision-making to minimize environmental burdens.

6. Social Equity: Sustainable construction also considers social aspects such as community well-being, health, and equity. This includes designing buildings and infrastructure that promote occupant health and comfort, providing access to essential services and amenities, and fostering inclusive and resilient communities.
7. Economic Viability: Sustainable construction seeks to balance environmental and social objectives with economic considerations. While upfront costs for sustainable materials and technologies may be higher, investments in sustainability often yield long-term savings through reduced operational costs, improved asset value, and enhanced market competitiveness.

In general, building sustainability entails using all-encompassing methods to design, build, and operate that put economic prosperity, social responsibility, and environmental stewardship first. The construction industry can help create a more sustainable, equitable, and resilient built environment for both the current and future generations by incorporating sustainability principles into its practices.

9.1.3. A Sustainable Construction Industry for the Future

Future sustainable construction practices will prioritize social justice, economic resilience, and environmental responsibility. This represents a paradigm shift in the industry. By acknowledging the interdependence of the environmental, social, and economic systems, this transformative approach seeks to develop infrastructure and buildings that satisfy the requirements of both the current and future generations while maintaining the health of the environment and community well-being. Using green building practices and concepts is the cornerstone of a sustainable construction industry. This entails creating infrastructure and buildings with eco-friendly, resource- and energy-efficient designs and construction. Throughout the course of a project, using renewable energy sources, passive design techniques, and incorporating sustainable materials and technologies can help minimize resource consumption, reduce carbon emissions, and mitigate environmental impacts.

Moreover, social sustainability plays a critical role in determining how the construction sector develops in the future. Fostering inclusive and equitable development that puts the needs and goals of all stakeholders—workers, communities, and marginalized populations—at the top of the priority list is necessary to achieve this. Building socially sustainable projects that improve quality of life and foster social cohesion requires embracing fair labor practices, encouraging diversity and inclusion, and involving local communities in decision-making processes.

Another important tenet of the building industry of the future is economic sustainability. This entails implementing business strategies and models that minimize adverse externalities while placing a premium on long-term economic prosperity. While lowering reliance on finite resources and fossil fuels, investing in innovation, technology, and green infrastructure opens up possibilities for economic growth, job creation, and market competitiveness. In addition, adopting circular economy concepts—such as recycling, waste reduction, and resource reuse—helps the building industry maximize resource efficiency and open up new revenue streams.

Facilitating the shift towards a sustainable construction industry requires cross-sector collaboration and partnerships. To create and execute laws, rules, and incentives that support sustainable development goals, governments, corporations, academic institutions, and civil society organizations must collaborate. Putting money into R&D, capacity-building, and knowledge-sharing programs makes it easier for people to work together to solve global issues like urbanization, social inequality, and climate change.

Building a sustainable construction industry for the future calls for a shared dedication to social responsibility, economic prosperity, and environmental stewardship. The building sector can be a key player in developing a more resilient and sustainable built environment for future generations by adopting green building practices, advancing social equity, and stimulating economic resilience.

9.2. ENVIRONMENTAL AND SUSTAINABLE DEVELOPMENT

In the early 1970s, the Cocoyoc Declaration on Environment and Development coined the phrase "sustainable development." Since then, international organizations working to achieve development that is beneficial or environmentally benign have adopted it as a catchphrase. The term has sparked discussions on how economic change interacts with the natural resource base that underpins it (Figure 9.1).

Figure 9.1. Sustainable development represents an opportunity to construct a new approach, and the success of these efforts has powerful implications for issues of peace and security.

Source: https://5.imimg.com/data5/XQ/CD/KR/SELLER-1206617/sustainable-solution-for-environment.jpg.

Scholars are now using the term "sustainable development" more often than just "development." It is now a hotly debated subject among social scientists, leaders, statesmen, and social groups. Since it has led to imbalances in the environment and among nations, the industrial-technological and economic progress that humanity has experienced over the past 100 years has not proven to be a true development. Environmental degradation brought on by the irresponsible and overuse of the resources of our planet has the potential to undermine, if not already negatively impact, the advancements made by different societies. Infrastructure and industrial projects are unquestionably means of development. Previously, the viability and usefulness of these initiatives were exclusively assessed in terms of financial returns. But as environmental and human rights concerns have gained prominence recently, it is thought to be crucial that these projects be designed, planned, and carried out in a way that can guarantee the long-term and actual socioeconomic welfare of the populace without endangering the environment. Gross Domestic Happiness, or GDH, is the concept we must consider in place of GDP and GDP per capita. The protection of development should never come at the expense of people's human rights, especially those whose lives will be directly and physically impacted by such initiatives. This new, optimistic perspective is symbolized by the concept of sustainable development. It is now widely accepted that achieving stable, long-lasting, genuine, and sustainable global peace, security, and development requires sustainable development.

9.2.1. Definition and Meaning of Sustainable Development

In 1987, Brunt-land Commission prescribed: "Sustainable development meets the needs of the present without compromising the ability of future generations to meet their own needs."

It has got certain objectives or goals to achieve for the over-all development of mankind:

1. We are common, having same future.
2. The development process must be participatory, from all sections of people.
3. The objectives are same but ways might be different.
4. The development process must be balanced and conducive with nature.
5. The development process must be continuous and long-term.
6. Equal distribution and social justice should be taken care of.
7. All development process must be integrated and coordinated.

This development process should not be contradictory with nature. Rather, it should be corollary with natural productive process. So, sustainable development should be ecologically conformable, economically viable, and socially acceptable.

9.2.1.1. Characteristics of Sustainable Development

1. It attempts to minimize greenhouse gases, reduce global warming and preserve environmental resources.

2. It emphasizes on green architecture and eco-friendly building practices.
3. Use of natural, biodegradable building material.
4. Use of renewable resources of water.
5. Emphasis on renewable energy sources such as Solar and wind.
6. Protection of natural habitat.
7. Planned replacement for any resources used.
8. Non-polluting construction practices and industries.

9.2.2. Goals of Sustainable Development

There are three primary goals of sustainable development:

1. To minimize the depletion of natural resources when creating new developments.
2. To create development that can be maintained and sustained without causing further harm to the environment.
3. To provide methods for retrofitting existing developments to make them into environmentally friendly facilities and projects.

Global organizations such as the United Nations, NGOs, aid organizations, and even governments are increasingly sponsoring efforts to ensure sustainable development goals are realized for every individual across the board. Some other important sustainable development goals set by these bodies include:

1. Eradication of Poverty Across the World: These organizations primarily focus on the least developed and low-income countries where poverty is prevalent. They aim to eradicate poverty by expanding social protection programs like school feeding, cash transfers, targeted food assistance, social insurance, and labor market programs such as skill training, old age pensions, wage subsidies, unemployment insurance, disability pensions, and so on.
2. Promotion of Good Health and Well-Being: This sustainable development goal seeks to ensure good health and well-being for all at each stage of life. The goal addresses main health priorities such as maternal and child health, reproductive health, environmental, communicable and non-communicable diseases, universal health coverage, and access to quality, safe, effective, and affordable vaccines and medicines. It also advocates for enhanced health financing, increased research and development, and strengthening the capacity of every country engaged in health risk prevention and management.
3. Provision of Quality Education for All: These bodies have realized that the level of child school dropout is at an all-time high. This gap must be closed to ensure sustainable future development even as international communities work to ensure quality and equity in the education sector. In a nutshell, this goal seeks to ensure equitable and inclusive quality education and promotion of lifelong learning opportunities.
4. Provision of Clean Water and Sanitation: Water and sanitation are at the top of the chart regarding sustainable development. They are critical to the survival of

humans and the planet. This goal aims to address aspects relating to sanitation, hygiene, drinking water, and the quality and sustainability of water resources across the globe.

5. Building Up Strong Infrastructure, Supporting Inclusive and Sustainable Industrialization, and Incubating Innovation: This goal takes into account three aspects of sustainable development: industrialization, infrastructure, and innovation. Infrastructure is vital because it offers the basic framework necessary for the smooth running of enterprise and society at large. Industrialization drives economic development, yields job opportunities, hence reducing levels of poverty. Innovation enhances technological abilities of industrial sectors and triggers the development of innovative skills.
6. Enabling Access to Affordable and Clean Energy: Energy is the most critical resource to achieving most of the sustainable development goals. Energy plays a vital role in mitigating poverty through advancements in industrialization, education, water supply, and health, and fighting climate change. This sustainable development goal focuses on developing and expanding renewable energy resources such as sun, wind, hydropower, liquid and solid biofuels, biogas, and geothermal. These renewable sources of energy don't emit greenhouse gases into the atmosphere and are ideal for the environment and human health.
7. Achieving Gender Equality: In the past few decades, gender equality and women's empowerment have been agendas for most governments for long-term sustainable development. Access to education for girls has improved, the percentage of child marriage has plummeted, and huge leaps have been taken in the domain of sexual and reproductive health and rights, such as a dramatic reduction in maternal health. Although there is still a long way to go to reach this milestone, organizations are using every ounce of their energy and resources to ensure the dream is realized.

Other sustainable development goals established by these organizations include the promotion of decent jobs and economic growth, the fostering of sustainable cities and communities, the preservation of sea, ocean, and marine resources, the fight against climate change, the promotion of sustainable consumption and production practices, and more.

9.2.2.1. How Can We Make It Happen?

To make sustainable development the norm, we have to change the vision of the cultures of each country. To change the vision of the culture two things, have to occur.

1. The culture must value a global benefit more than a local one.
2. A responsibility towards providing and sustaining resources for the future must be of more value than profit in the present must be developed.

Both tasks are challenging because they require individuals to prioritize selflessness over personal gain in today's society. The importance of future generations is often overlooked in the pursuit of profit and convenience. While it is natural for each generation to focus on their own needs, history has shown that it is possible to also consider the well-being of future generations, leading to more responsible decision-making in the present.

9.2.2.2. What Prevents Sustainable Development From Happening?

There are two main obstacles standing in the way of sustainable development. The first is that using sustainable materials and methods is costly for many aspects of development. Although sustainability does prove to be less expensive in the long run than traditional development, the initial stages of creating a sustainable project may be far more costly. The lack of widespread acceptance of the necessity of sustainable development is the second significant problem. The solution to this educational problem might take many years.

Since sustainable projects are more expensive to start, contractors and investors are unable to recognize their significance. It can be very challenging to develop the long-term vision needed to comprehend the significance of sustainability when the immediate goal is financial success. In order to achieve this, more rules and incentives have been implemented by the government, which should encourage program and project managers to consider sustainable development.

9.2.3. Importance of Sustainable Development

Sustainable development is a hard topic to nail down because it consists of a wide range of things. Due to the technicality and complexity of this topic, it's best to check out its importance holistically to be able to grasp it easily. Population is the main factor driving up sustainable development campaigns. So, the importance of sustainable development can be viewed from this perspective:

1. Provides Essential Human Needs: The explosion of population means people will have to scramble for the limited life essentials like food, shelter, and water. Adequate provision of these basic needs almost entirely hinges on infrastructure capable of sustaining them for a long time. If governments insist on utilizing fossil fuel-based sources of energy instead of renewable and sustainable options, the cost and environmental effects of supplying these basic needs would become a tall order.
2. Agricultural Requirement: Growing population means agriculture must catch up. Finding ways to feed more than 3 billion people can be staggering. If the same unsustainable cultivation, planting, irrigation, spraying, and harvesting techniques are utilized in the future, they might prove to be financially burdening considering fossil fuel resources are projected to run out. Sustainable development focuses on sustainable agricultural methods such as effective seeding techniques and crop rotation to promote high yields while maintaining the integrity of the soil, which produces food for a large population.
3. Manage Climate Change: Climate change can be mitigated by sustainable development practices. Sustainable development practices seek to reduce the use of fossil-

based sources of fuel like oil, natural gas, and coal. Fossil fuel sources of energy are unsustainable since they will deplete in the future and are responsible for the emission of greenhouse gases.

4. Financial Stability: Sustainable development practices have the ability to create more financially sustainable economies across the globe. Developing countries that can't access fossil fuels can leverage renewable forms of energy to power their economies. From the development of renewable energy technologies, these countries can create sustainable jobs as opposed to finite jobs based on fossil fuel technologies.
5. Sustain Biodiversity: Unsustainable development and overconsumption practices greatly impact biodiversity. Life ecosystem is designed in such a way that species depend on one another for survival. For instance, plants produce oxygen that humans need for respiration. Humans exhale carbon dioxide that plants need for growth and production. Unsustainable development practices like emission of greenhouse gases in the atmosphere kill many plant species resulting in reduction of atmospheric oxygen. This is not good for humans. Sustainable development practices encourage the use of renewable energy resources, and organic farming practices that do not emit any greenhouse gas to the atmosphere.

9.2.3.1. Examples of Sustainable Development

1. Wind Energy: It is energy harnessed from the motion of wind using wind turbines or windmills. Wind energy is renewable, which means it's never-ending and can be used to substitute energy on the grid. This makes it a good sustainable development practice.
2. Solar Energy: This is energy harnessed from the sun using solar panels. It's advantageous since it's absolutely free and its supply is infinite. These factors make it beneficial to consumers and good for Mother Nature because it doesn't contribute to the emission of greenhouse gases (Figure 9.2).

Figure 9.2. This is energy harnessed from the sun using solar panels.

Source: https://i0.wp.com/nenow.in/wp-content/uploads/2022/07/Bright-future-for-Electrical-Engineers.jpg?fit=1200%2C675&ssl=1.

3. Green Space: These are places where wildlife and plants are allowed to thrive. Parks are categorized as green spaces as well. People have amazing opportunities to enjoy outdoor recreation in green spaces, especially in large cities where it can be difficult to find a place to relax. In addition to mitigating the warming effects of paved surfaces, green spaces also protect rivers and streams from contaminated runoff and help control climate and air quality.
4. Crop Rotation: The practice of planting different crops on the same farm in order to improve soil fertility and help control insects and diseases is known as crop rotation. One of crop rotation's many advantages is that it doesn't require chemicals. This makes the farming method you're employing a sustainable development strategy since it preserves the integrity of your soil.

There won't be a discussion about sustainable development in the long run. Only sustainable development can expand our economies without endangering the environment and public health. It has also been shown to be cleaner and possibly more effective. Global governments are placing increasing emphasis on sustainable development, which has led to a shift in consumer preference towards renewable energy sources such as geothermal, hydropower, wind, and solar power. Resources will be depleted more quickly as more people jump on this bandwagon and demand a greater quantity of them. Sustainable development will eventually become unaffordable for people who prefer to lead healthy lives.

9.2.4. Inner and Outer Limits of Sustainable Development

The noted environmentalist and devoted protagonist of sustainable development, Barbara Ward, first coined the words 'inner and outer limits of sustainability' to express the different limiting forces that work against sustainable development. The inner limits are always reflected by the internal socio-economic inequality within the population, while outer limits are social cohesion. Beyond these two limits, sustainable development cannot be sustained; it breaks down suddenly. Social disharmony and inequality among inhabitants lead to environmental disorder, war, social tensions, or ethnic conflict.

To break or broaden the outer and inner limits of growth, certain measures are recommended:

- The present pattern of production and consumption should be in accordance with strict environmental norms so that it could be both socially and environmentally sustainable.
- Inequality among people should be minimized.
- Enhancing creativity and sensitivity so that lifestyles and activities should be diverse, and people would have plenty of consumption choices.
- Ethical and moral values should be preserved, and mass education should be implemented to curb materialistic consumption.
- Participation of the people in the decision-making process.
- Diversification of the economy to enable the population to select their lifestyle.
- Poverty alleviation programs and eradication of economic imbalance.

9.2.3.2. Recent Efforts for Sustainable Development

To achieve the goal of sustainable development, the convention stated: "The use of goods and services that respond to basic needs and bring a better quality of life, while minimizing the use of natural resources, toxic materials, and emissions of waste and pollutants over the life cycle, so as not to jeopardize the needs of future generations."

Regarding the identification of problems of sustainability and recommending the way out, the convention pointed out some key aspects:

- Unsustainable consumption patterns are more visible in physical infrastructures like energy, housing, transport, and waste management. Cultural habits and societal outlook are responsible for that.
- The business and industrialist community has a specific role to manage the environmental impacts of the goods and services it provides.
- Political institutions have a role to change consumption patterns rather than consumption volumes.
- For rectification in the chain of production-consumption and final disposal, the government must provide incentive measures, infrastructural facilities, proper leadership, and legislative regulations.
- Responsible organizations like trade unions must involve themselves in the agenda of consumption and production pattern change.
- Individuals have a great role in different aspects like consumer, householder, worker, decision-maker, and even as voters.

9.2.3.3. Understanding Regulations and Incentives

To make sustainable development the better course of action, there is an absolute need for changes to the laws and incentives that control development. Mandated sustainable development could make it unfeasible to create projects that are needed for the common good, so it is really not feasible. However, more of a project can meet the sustainability requirements by utilizing different tax credits and incentive programs that are assigned to different levels of sustainability in a development program. This is a step in the direction of giving people the option of fully sustainable development, but it also takes implementation costs into account and offers solutions that won't obstruct advancement.

9.2.5. Sustainable Development: Approach

This section throws light upon the four main approaches to sustainable development. The approaches are:

1. Appraisal of the Environment: Any project aimed at sustainable development must first assess the state of the environment. It is important to consider the symbiotic relationship that exists between the local ecology, social structure, and economic circumstances. Every region's biophysical state and economic structure are closely linked, and this has an impact on societal advancement. A comprehensive biophysical investigation facilitates the evaluation process. The evaluation

report suggests an action plan. The project can only be implemented with the help of an "action plan."

2. Estimation of the Environmental Impact: The environment and its wise use are essential components of sustainable development. Unplanned development can have a detrimental effect on the environment and tarnish sustainability as a whole. Only development that achieves perfect harmony between the social and environmental spheres is considered sustainable. Identification of the main advantages, as well as their optimal use and monitoring, are required to meet this goal.

It is important to carefully evaluate and monitor the various ecological inputs as well as the climate, soil, vegetation, and drainage resources. It is also important to critically examine how development processes impact these natural systems. Thus, the analysis of the relationship between the development process and natural systems constitutes the estimation of environmental impact.

3. Natural Resource Accounting: Because of this, it is rare for logging profits to fully account for the future revenue losses brought on by forest degradation. The exploitation of other resources, such as air, water, and soil, which are not capitalized in enterprise or national accounts, results in similar incomplete accounting. Economic development must fully account for the improvement or deterioration in the stock of natural resources when measuring growth in all nations, rich or poor.

However, the depletion of natural resources has a direct impact on national income, productivity, and resource creation. The economy and ecology are closely intertwined. Ecological losses must impede economic growth. For development to be sustainable, natural resource accounting is therefore essential. But in this sense, the conventional accounting system is not very effective. Accounting systems that are more exact and sensitive are needed for sustainable development.

4. Government Policies and Economic Outlook: Sustainability in development is hampered by the growing divide between rich and poor nations, personal inequality, extensive trade, capital flows, and a lack of consistency in governmental policies. As is the case in the majority of developing countries, sustainable development may be hampered by a lack of integrated, cogent development policies, price incentives, awareness, and information campaigns. Taxes, as well as the reduction or elimination of subsidies, should be used to curb the unsustainable use of natural resources. Therefore, government policy can have a significant impact on the sustainable development process.

9.3. ENVIRONMENTAL ISSUES ASSOCIATED WITH SUSTAINABLE DEVELOPMENT

With the aid of science and technology, man has rapidly advanced the fields of forestry, land management, transportation, mining, agriculture, and industry. Regrettably, this has caused significant disruptions to the way the natural environment functions.

Now, people have begun to realize the side-effects of the developmental activities and have started their efforts in minimizing these effects by adopting methods and technology for sustainable development. Some of the important environmental issues associated with development, and the role of ecology in possible solution of these problems, i.e., sustainable development, are as in subsections.

9.3.1. Food Production and Organic Farming

In many parts of the world, particularly in developing nations like India, China, Indonesia, Egypt, Brazil, and many other Afro-Asian countries, the growing population's demand for food has become a serious issue. Scientists studying agriculture have been able to produce food and other crops in greater quantity and with higher quality. This has been made feasible by the use of chemical fertilizers, high-yielding seeds, crop rotation techniques, pesticides, weed control, and agricultural mechanization.

As a result, in an effort to boost food production in developing nations, the Green Revolution got underway in the late 1960s. In comparison to the traditional varieties, new varieties of wheat, rice, and corn responded better to fertilizers, irrigation, and pest and weed control chemicals. Unfortunately, as food production increased, so did our reliance on pricey, energy-based technologies, and an array of ecological issues resulted from our overuse of chemicals as pesticides and fertilizers. Today, it is widely acknowledged that the use of chemical fertilizers and other chemicals degrades soil productivity and contributes to air and water pollution.

In addition to upsetting the environment, the Green Revolution is to blame for the Social Revolution. Farmers in nations such as the USA are currently moving away from the regular application of pesticides, fertilizers, herbicides, etc., and are turning to chemical-free organic farming. Only organic materials are utilized as fertilizers, such as crushed limestone, manure, etc.

This system is more environmentally friendly than the ancient Indian system. Using organic farming is one approach to achieve sustainable development. Even in this system, high-yielding varieties can be developed by using appropriate management techniques. It is possible to develop aquaculture in addition to agriculture. The best course of action is to expand fish farming, which meets the growing demand for food—particularly protein.

1. Climate: Constraints and impacts.
2. Energy and Inputs: Energy resources, fertilizers, plant protection, ecological farming, techniques and technology.
3. Policy and Management: Political dimensions including economical, cultural and social issues, research and development as well as population control policy.
4. Genetic Resources: Identification, evaluation and utilization of genetic resources.
5. Soil and Water: Resources and requirements.

9.3.2. Energy Crisis and the Ecosystem

Within the ecosystem, energy is a crucial component. The planet's initial energy source was the sun. All forms of energy, i.e., nuclear, wind, coal, electric, fuel, water, and so on, come from the solar source. The earth's surface receives a vast amount of energy from the sun, far more than what is needed for human consumption. However, the sources of energy in the modern era are hydroelectricity, nuclear fission, and the use of fossil fuels. The following table indicates the production and consumption of various sources of energy in the world:

Sources	Production	Consumption
Petroleum	1,26,900	1,27,222
Natural gas	72,791	68,290
Coal	1,02,322	1,01,660
Nuclear energy	18,373	18,373
Hydroelectricity	22,493	22,493
Total	3,42,879	3,38,038

An energy crisis has arisen as a result of the quick exploitation of these energy sources. Fossil fuels, including coal, gas, and oil, will become scarce in the future. The effects of large dams, thermal power plants, and nuclear power plants worry environmentalists. Therefore, using unconventional energy sources like solar, wind, tidal, etc., is essential for sustainable development. Solar energy is the best of these, assuming we can harness it using the least expensive technology. The only energy source that can supply electricity in tropical nations like India without harming the environment is solar energy. Photovoltaic cells, solar-powered water pumps, air conditioners, solar water heaters, solar home heating units, solar cookers, etc., have already started to be developed. In this area, research is also ongoing. The creation of less expensive technology is what's required to harness solar energy and other unconventional energy sources.

The most important social issues affecting sustainable development are closely tied to energy. These include issues related to economic and security as well as poverty, employment, income levels, gender inequality, access to social services, population growth, agricultural productivity, climate change, and environmental quality. It is impossible to achieve the global social, economic, and environmental goals of sustainability without giving enough consideration to the vital importance of energy and all of these factors. It is true that a significant, fundamental shift in the way that energy is produced and consumed globally is required.

Energy is clearly and vitally important for food preparation, consumption, and agricultural production (UNDP, 1997). Without properly considering the effects of its uses on society, the environment, the economy, and security, the world still searches for energy to meet its needs. It is now obvious that the energy strategies used today are unsustainable. Political institutions have an obligation to make sure that end users are involved in the research and development of technologies that support sustainable systems. To maintain sustainability and prevent chaos in the natural life cycle, scientists have a duty to comprehend the earth as an integrated whole and the effects of human activity on the environment. Overcoming the lack of commitment and cultivating the political will to safeguard people and the foundation of natural resources will be the main obstacles to achieving these goals. If action is not taken, there will be more conflicts over limited resources, more degradation of natural resources, and a wider divide between the rich and the poor. While we still have options, we must take action.

Putting sustainable energy plans into practice is one of the most significant tools available to humanity to build a sustainable world. The majority of current energy trends point to a worsening state of affairs (UNDP, 1997). Furthermore, the over-preoccupation with fossil fuels and centralized energy supply at the expense of decentralized supply, renewable energy, and energy efficiency in the current energy patterns is exacerbating this process.

9.3.3. Pollution and Environmental Degradation

Human activity linked to the industrial and automotive revolutions is what causes pollution. Pollution is a man-made issue that primarily affects wealthy nations, but it is also becoming a major issue for developing nations. Developed nations used to take advantage of the natural resources found in developing nations and export waste such as machines and materials that now contribute to pollution.

Common types of pollution include air pollution, water pollution, soil pollution, noise pollution, and radioactive pollution. These pollutants can be classified into two categories from an ecosystem perspective: non-degradable and biodegradable pollutants. Some of the main pollutants include deposited matter, carbon monoxide, sulfur dioxide, sulfuric acid droplets, hydrogen sulfide, fluoride, halogens, metals such as lead, iron, zinc, bromine, and iodine, agricultural chemicals like pesticides, insecticides, herbicides, fungicides, and fertilizers, complex organic compounds, photochemical pollutants from vehicles, radioactive materials, unwanted noise, and municipal wastes.

Since these pollutants not only cause a number of diseases but also degrade the environment, they pose a threat to humankind. As a result, quick action is needed to create an environment free from pollution. The first step in ensuring sustainable development is to address pollution at its source. Industries need to be located such that the surrounding areas cannot be impacted by their pollutants, and treatment facilities are also required. Environmental management, monitoring, education, and accountability are now necessary. Any potential damages must also be fixed. Making environmentally friendly products is a creative idea in this regard.

9.3.3.1. Problem of Waste Disposal and Its Management

The massive consumption of commodities in industrialized nations is contributing to environmental issues. These days, the issue is not just confined to developed nations; it is also becoming more and more prevalent in developing nations. Our surroundings are rapidly becoming more populated with waste and toxic substances. These materials get into the cycles of biogeochemistry. Various wastes build up in the environment of man-made ecosystems, disrupting the growth of microorganisms. A range of synthetic materials, including glass, rubber, plastic, etc., are not biodegradable, which leads to issues with the environment. Wastes include agricultural, industrial, and mining wastes as well as wastes from homes and businesses. Wastes of all kinds can be used if they are managed properly, and efforts have already been made in this regard. Gene engineering is being used to create new bacterial and microorganism strains that have the potential to break down a variety of hazardous industrial effluents into safe substances.

Additionally, scientists have created techniques and technologies to turn sewage waste into fuel, fertilizer, wood, and cattle feed. Before releasing the treated sewage into water bodies, municipal sewage treatment plants remove large objects, floating matter, and settleable solids. They also break down organic matter and eliminate harmful microbes through physico-chemical and biological processes.

Because there are inadequate facilities, untreated sewage is permitted to enter adjacent bodies of water. In addition to seriously polluting the water, untreated sewage harms the local ecology. Therefore, in order to meet all of humanity's needs, it is imperative that sewage be recycled in a way that is both environmentally safe and economically viable. Numerous methods for treating sewage and extracting water for fertilizer, fuel, fodder, and irrigation have been developed. Large-scale attempts are also being made to reclaim degraded areas, enhance agricultural lands, and build new recreational facilities using sewage. Fish farms in nations like Germany, Poland, Russia, Hungary, Israel, China, USA, Indonesia, and others frequently use sewage water for their operations. Additionally, sewage is being used to produce fuel by releasing methane gas, which is produced by bacterial growth. Thus, the creation of an effective waste management system is imperative.

9.3.3.2. Irrigation and Problem of Salinity and Water-Logging

Large-scale irrigation facilities have been installed over the past 50 years. There's no denying that this has raised global agricultural productivity. However, a large number of areas are currently experiencing salinity and waterlogging issues as a result of poor water management, farmer illiteracy, and the geology of a specific area. The microbial population and organic matter involved in decomposition gradually diminish, causing an increase in salinity and waterlogging that negatively impacts soil fertility. In areas covered by large irrigation projects, poor drainage and water management are the main causes of waterlogging and increased salinity.

9.3.3.3. Deforestation and Desertification

Deforestation has become a significant issue primarily because of the destruction of forests for development and commercial activities. Population growth, urban expansion, the need for land for agriculture, mining, infrastructure projects, and other reasons have led to widespread deforestation. This has resulted in the extinction of many valuable plant species and put thousands more at risk. Without reforestation efforts, global forest cover is decreasing rapidly. Even dense tropical forests are being heavily exploited for commercial purposes, exacerbating the problem.

Action plans should be put into place to safeguard the current forest cover as well as to undertake extensive afforestation, which would allow for the development of 30 to 33% of the land area as natural or planted forests, in order to preserve the environment and ecosystem. In order to prevent businesspeople, industrialists, and the general populace in rural areas from felling trees illegally and carelessly, it is also necessary to strictly enforce the current laws. It is important to promote agroforestry as well as social forestry, and to turn waste areas into forests. Additionally, it is important to promote public involvement at all levels. Overgrazing, the disappearance of native vegetation, and other human activities all pose a threat to desertification and the spread of desert areas. Studies conducted by UNESCO have revealed that desertification affects about 43% of the world's land area. Improved land use patterns, sand dune stabilization, the growth of native vegetation adapted to an arid climate, lessening of population pressure, and other strategies are recommended to prevent desertification.

Primarily, sustainable development ought to be predicated on preserving the delicate equilibrium between ecosystem functions and productivity functions. Our strategy for sustainable development ought to be comprehensive, encompassing all resource sectors and operating at local, regional, and global levels. We will never be able to solve the issue of environmental degradation if we do not embrace sustainable development.

9.4. CONCEPTS OF ENVIRONMENTAL MANAGEMENT

The concept of environmental management is still very new and dynamic. The management of an organization's surrounding environment falls under the purview of environmental management. It stands for the procedures, roles, organizational structure, and requirements needed to put an environmental corporate policy into action. The environment unites all inanimate objects and natural forces, including humans. Setting goals, managing information, assisting in decision-making, planning and organizing environmental management, developing environmental management programs, piloting, implementing and controlling, communicating, conducting internal and external audits, and so on are all fundamental components of effective environmental management. The current level of environmental and economic development necessitates a broader understanding of the natural environment among managers. The industrialization that is being prioritized for economic growth will lead to environmental pollution in the next year, which will be an ecological nightmare. Therefore, it is now essential to consider the environmental effects when establishing an industrial facility. Today's technology can help reduce pollution in the environment, but it also needs to be used to minimize the extent of environmental degradation and correct ecological brutality excesses. In order to achieve all of these, accurate accounting and reporting of environmental data is essential and can result in effective environmental management (Figure 9.3).

Figure 9.3. Environmental management is concerned with the management for environment encompassing a business.

Source: https://img.freepik.com/premium-photo/environmental-technology-concept-sustainable-development-goals-sdgs-green-earth-with-environment_870512-2003.jpg.

9.4.1. What Is Environmental Management?

The field of environmental management (EM) integrates socioeconomic, policy, and scientific aspects. It focuses mostly on finding solutions to real-world issues relating to resource exploitation, waste production, and human coexistence with nature. Environmental management, in a purely anthropocentric sense, revolves around addressing the fundamental problem of how to continuously innovate technology to evolve while minimizing the extent to which this process modifies the natural environment. Therefore, concerns about maintaining fair and equitable resource distribution, sustainable economic growth, and protecting natural resources for future generations are closely related to environmental management.

In light of the catastrophic effects that humans are currently having on natural ecosystems, environmental management is a response to human activity. It is reassuring to know that, in a world with fewer people and less widespread technology use, the environment might be able to recover from human misuse and abuse on its own. However, given that people place a higher value on economic growth than the preservation of natural ecosystems, it is now widely acknowledged that, in many cases, positive intervention is required for the environment to recover.

However, there is a great deal of disagreement and divergence regarding the direction that such intervention should go, leading to a multitude of approaches to environmental management. The term "deep ecology" first appeared in the 1960s alongside a wave of movements opposing technological advancement and criticizing the political underpinnings of autocracy and power. But "shallow ecology," and by extension "shallow ecologists," aimed to reach a consensus with those who maintained that the development of new technologies is the only way to address the planet's environmental issues. Therefore, environmental managers cover a wide range of positions, from technocrats to conservationists, from those who favor minimal human intervention in the environment to those who favor more intervention in order to steer natural processes in a direction that is beneficial to humanity. As a result, developers and conservationists are both represented.

It is indeed anticipated that developers and conservationists will collaborate on the need to ensure that economic growth is sustainable and does not jeopardize essential habitats and natural resources over time. The United Nations Convention on Environment and Development (a process that started at the Rio de Janeiro Earth Summit in 1992) aims to achieve this. Nevertheless, the degree to which it has altered the direction of environmental management globally has been severely constrained by underfunding and a lack of commitment at the national level.

Any business or developmental activity has a substantial impact on the environment:

- The manufacture of products involves extracting raw materials from the environment and processing them to produce saleable items. As a result of the production process, various forms of waste (solid, liquid, and gaseous) enter the environment.
- The activities surrounding the manufacturing process, such as maintenance of plant and infrastructure and the packaging and transport of goods, all have environmental impacts.

- In addition, the products that are produced will eventually be disposed of and enter the environment as waste.
- The provision of services also results in a significant environmental impact. Service companies use various products and also energy to deliver their services, both of which result in waste entering the environment. The provision of services also entails resource exploitation while rendering the environment unstable.

In other words, the environment serves as both a "source" for the raw materials used in industrial processes and a "sink" for the waste products they produce. While all businesses and organizations generate waste, not all of them also take raw materials from the environment. Only those businesses at the start of the supply chain carry out this action. As they proceed through the supply chain, these raw materials are subsequently processed in different ways. Products eventually surface and are supplied to retailers and wholesalers.

The environment gets damaged and unstable as a result of excessive raw material extraction and waste disposal. The goal of environmental management is to identify business practices that minimize or stop this deterioration. We can enjoy a better environment and ensure its preservation for future generations by taking this action.

9.4.2. Scope of Environmental Management

Through numerous global people movements in the 1970s, environmental issues and their appropriate management came to the attention of people all over the world. The Chipko movement, the Greenpeace Movement, etc., are a few of them. Scientific and technological advancements have made it possible to provide a variety of instruments and tools that are backed by statistical data in order to effectively solve environmental problems and aid in their management.

We currently have enough resources, thanks to the computer technology revolution, to apply this to environmental management. These days, highly sophisticated instruments are available to measure the water and air quality at various locations. The current method of soil testing can provide incredibly precise results regarding the composition and levels of nutrients in the soil at various locations.

The air, water, and soil are among the environmental factors that are under a great deal of stress due to rapid industrialization and population growth. A lot of forest areas have made way for factories. A significant amount of industrial waste is contaminating the nearby soil, water, and air. This contamination is not localized in any one area.

It is dispersing the population over too many areas. Environmental pollution has grown to be a worldwide problem. All societal segments must contribute to its success. Environmental health has become a major concern for people all over the world. The advancement of science and technology in all fields has left ample room for environmental management in the modern era.

The scopes in the following areas need environment management:

- Population increases and health services.

- Treatment of pollutants (air, water and solid) generated from various sources.
- Pollution level in air, water and soil.
- Development of non-polluting renewable energy sources like wind, solar, biomass, etc.
- Solid waste utilization through recycling.
- Biodiversity conservation.
- Environmental awareness in society.

9.4.3. Importance of Environmental Management

Environment considers every circumstance necessary for corporate sectors to survive. Ecological harm is severe when environmental considerations are neglected. At both the national and international levels, poverty, a lack of resources, population pressure, and global resource use inequality are creating previously unheard-of social and environmental issues. A balance between the needs of environmental process management and economic development is often struck by sustainable development. Accurate understanding of the many aspects of good environmental management is a prerequisite for sustainable development that satisfies the needs of both the current and future generations. For internal management initiatives that have a specific environmental focus, like supply chain management, cleaner production, "green" product or service design, environmentally preferable purchasing, and environmental management systems, environmental management is extremely helpful. Information of the environmental management kind is also being used more and more for external reporting needs. Among the instruments for environmental management is Environmental Management Accounting (EMA). Since environmental issues are affecting a wider range of decisions, EMA is becoming more important not only for environmental management decisions but also for all types of managerial functions, with a special emphasis on eco-efficiency and strategic position. EMA offers a broad set of principles and approaches that are necessary for the success of many other environmental management functions.

Therefore, environmental management contributes to gaining a fair assessment and helps corporations and other organizations increase public trust and confidence. The environment needs to be carefully managed because it is already deteriorating.

Through environmental management, enterprises can enjoy the following benefits:

- Pollution control being a burning subject of discussion, environmental management shows the extent to which pollution has been controlled by the corporation.
- Environmental management draws attention in another sense, which offers an idea about industrial development, a nation's economic progress, social welfare, and the fulfillment of responsibility towards society.
- Environmental management is helpful in discharging organizational accountability and increasing environmental transparency. Sustainable development is possible with the help of environmental management as it helps include the ecological ability of the enterprise.
- Negotiation between the management and society helps organizations seek to strategically manage a fresh and emerging issue with distinct users.
- Environmental management supports green reporting to

effectively combat all negative public opinions in the global economy where the existence of a strong environmental lobby against environmentally unfriendly industries is found.

- Environmental management improves performance through better management of environmental costs and thus benefits the natural and human environments.
- Environmental management forces corporate sectors to fulfill their commitments towards introduction and change, and thus appears to be responsive to new factors. Countries giving importance to the ecological aspect of activities are becoming more and more popular, particularly in Western countries.
- Environmental management reflects unsound production and consumption patterns, misuse and scanty use of resources and assets like water.
- Optimal allocation of scarce resources in the economy is possible with the help of environmental management.
- Environmental management is essential in measuring a nation's economic development, social welfare, industrial development, pollution control, and in fulfilling the needs of government. Still, the system is in its infancy and not all countries have been able to develop such a system. But with the passage of time, the system will gradually develop, research will be undertaken, and it will fulfill the requirements for which it was originated.
- Impressive decision-making through the application of environmental management reduces or eliminates many environmental problems.

It is deemed imperative to endeavor to integrate the effects of environmental resources into all business functions of a business house, given the evolutionary learning framework. The goal of environmental accounting and environmental management is to pinpoint the costs and resources that a corporation imposes on the environment. Environmental accounting is a vital component of environmental management, which in turn supports corporate development and operations for an all-encompassing environmental management system. Every corporate citizen should use environmental accounting to evaluate an environmental management system. This involves keeping track of the costs and benefits provided by the environment of a corporation and justifying these costs and benefits. The task can occasionally be very challenging due to maintaining secrecy and closely adhering to various accounting systems. However, environmental management is still very helpful for policy analysis, planning, pollution control, and managing public resources.

9.4.4. Corporate Agenda

It is impossible to dispute accountants' roles in environmental management. In order to instill an environmental culture and obtain their cooperation in enforcing it, an accountant should raise awareness of environmental issues throughout the corporate sectors, from the top down. It is imperative to raise awareness of the significance of environmental quality management among all corporate sector employees. In every aspect of corporate life, the management accountant plays a crucial role in articulating this idea or maxim.

Corporate agenda should be prepared and the following may be incorporated there:

- Corporate commitment to environmental protection should be clearly stated;
- Environmental protection is a genuine task and no stone will be left unturned in preserving and safeguarding ecological balances;
- Environmentally friendly code of conduct must be noticed at all levels and by all the people;
- Employees failing to observe environmentally friendly code of conduct must be reprimanded;
- Alterations in product design should be made as far as possible to incorporate eco-friendly materials and other inputs;
- Renewable sources of energy like solar energy, wind energy, etc., must be used and advocated;
- Maintenance, replacement, and repairs of machinery, tools, equipment, vehicles, etc., should be done properly and on time;
- Arrangements for returning or collecting back non-biodegradable packaging materials must be used to the best possible extent;
- If the operation of the corporation on any day or in any month is against environmental interests, staff doing this must apologize;
- Installation of adequate waste and sewage treatment plants before industrial waste is discharged into rivers, seas, etc., should be done;
- An adequate number of Public Relations Officers (PROs) to communicate with the external world and to clarify the corporation's commitment to the world at large must be recruited;
- Environmental engineers must be recruited to seek their opinion on environmental-related matters and to be involved in establishing a cleaner and more efficient production system, thereby minimizing waste and ensuring the optimum usage of all raw materials;
- Corporate managers should be proactive rather than reactive. Adequate measures must be initiated to protect the environment before new technology is put to use or adopted;
- Participation in activities like social forestry, community forest efforts, environmental cleaning works, etc., may be encouraged;
- Research and development activities for identifying new processes or methods of production or inputs that are sustainable from an environmental point of view should be promoted.

The management accountant should draft the aforementioned corporate environmental agenda, and top management should fully support its implementation both inside and outside the company. In terms of environmental issues, the management accountant can be thought of as the environment's "spokesperson." Frequent environmental performance reports, laws pertaining to the environment and their modifications, independent agencies conducting periodic environmental audits, etc., should be made known to all corporate stakeholders to guarantee a true environment that protects the environment.

9.4.4.1. Necessity of Accounting Standard

The urgent need of the hour is environmental accounting for environmental management. Therefore, an accounting standard is critical for tracking and reporting environmental investment. The measurement of environmental hazards created by corporations is a very challenging task, and the extent of damage varies from concern to concern based on factors such as the industry, management style, production and distribution methods, environmental pollution, employee motivation, reward and reprimand policies, etc. However, the new standard needs to take into account the costs incurred by businesses and society as a whole to mitigate negative effects, as well as the environmental risks caused by various industries. Environmental scientists, representatives of NGOs, government agencies, corporate representatives, UN representatives, and accountants foremost must be involved in this kind of national and international accounting standard setting. The standard must also make standard reporting procedures abundantly evident.

9.4.4.2. Benefits of Environmental Management Accounting

There is no one definition of environmental management accounting (EMA) that is universally recognized. EMA is the management of economic and environmental performance through the creation and application of suitable accounting procedures and systems that are related to the environment. Reporting, auditing, life-cycle costing, full-cost accounting, benefits analysis, and strategic planning for environmental management are all commonly included in EMA. EMA distinguishes, gathers, examines, and employs two categories of data for internal decision-making purposes, namely: 1. Physical data regarding the consumption, movement, and final destinations of materials, water, energy, and wastes; and 2. Financial data on earnings, savings, and expenses related to the environment. Even though EMA primarily sheds light on internal management decision-making, external stakeholders are frequently also provided with physical accounting information.

9.4.5. Guiding Principles for Environmental Management

The following is a list of international principles in common usage rewritten to apply more directly to social issues.

9.4.5.1. Precautionary Principle

The precautionary approach will be used to safeguard the environment, which is a concept that encompasses people's lifestyles and the integrity of their communities. Lack of complete knowledge about potential threats that could have a significant social impact should not be used as justification for approving a planned intervention or for avoiding the need to put strict monitoring and mitigation measures in place.

The primary aim of the precautionary principle is to guarantee that an activity or substance that poses a threat to the environment is kept from negatively impacting it, even in cases where there is insufficient scientific evidence to establish a definitive connection between that specific activity or substance and environmental harm. The terms "activity" and "substance" are the product of human invention.

Principle 15 of the Rio Declaration highlights the need to act in situations where there is a risk of significant or permanent harm. It is not acceptable to put off taking economically sensible action to stop environmental deterioration because there is not enough scientific data. As a result, the application of this principle in the production and distribution of energy resources is crucial for safeguarding both the environment and human health.

9.4.5.2. Uncertainty Principle

It must be recognized that our knowledge of the social world and of social processes is incomplete and that social knowledge can never be fully complete because the social environment and the processes affecting it are constantly changing, and vary from place to place and over time.

9.4.5.3. Intra-Generational Equity

The benefits from the range of planned interventions should address the needs of all, and the social impacts should not fall disproportionately on certain groups of the population, in particular children and women, the disabled and the socially excluded, certain generations, or certain regions.

9.4.5.4. Inter-Generational Equity

Development activities or planned interventions should be managed so that the needs of the present generation are met without compromising the ability of future generations to meet their own needs.

9.4.5.5. Recognition and Preservation of Diversity

Societies and communities are not the same. They are demographically structured by age and gender, and they are made up of diverse groups with a range of skill sets and value systems. It takes extra effort to recognize the social diversity that exists within communities and to comprehend the particular needs that may apply to special groups. It is important to take precautions to make sure that planned interventions do not cause social cohesion or diversity in a community to decline.

9.4.5.6. Internalization of Costs

No intervention should be approved or regarded as cost-effective if it achieves this by creating hidden costs to present or future generations or the environment. Instead, the full social and ecological costs of a planned intervention should be internalized through the use of economic and other instruments; these costs should be considered as part of the costs of the intervention.

9.4.5.7. The Polluter Pays Principle

The person advocating for the planned intervention should bear all expenses associated with mitigating or compensating for any social impacts. Numerous economists have proposed over the past 20 years that companies that release harmful wastewater into the environment ought to be held accountable for the amount of environmental harm they cause.

The Polluter Pays Principle (PPP) has been proposed by the OECD as a general framework for environmental policy. It states that the polluters should foot the bill if pollution-reduction measures are implemented. The so-called "Polluter Pays Principle" should be applied when allocating the costs of pollution prevention and control measures in order to promote the prudent use of finite environmental resources and prevent distortions in international trade and investment, according to the OECD Council. The primary concern of this principle is that polluters should pay for abatement without receiving any subsidies.

The Polluter Pays Principle, as defined by the Supreme Court of India, stipulates that those responsible for environmental harm must not only compensate those affected by pollution, but also cover the expenses of restoring the environment. This includes both direct costs to individuals or property and environmental costs. Restoring the environment is crucial for sustainable development, and therefore polluters are required to bear the financial burden of both compensating victims and reversing ecological damage. The application of this principle varies depending on the circumstances and cases, and has sparked controversy during the Rio Earth Summit in 1992. The South has requested increased financial assistance from the North to address environmental degradation in the South. This principle has practical implications for the distribution of economic responsibilities related to activities that harm the environment, especially in terms of liability and the utilization of economic tools.

9.4.5.8. The User Pays Principle

It is considered as a part of the PPP. The principle states that all resource users should pay for the full long-run marginal cost of the use of a resource and related services, including any associated treatment costs. It is applied when resources are being used and consumed.

9.4.5.9. The Prevention Principle

It is generally preferable and cheaper in the long run to prevent negative social impacts and ecological damage from happening than having to restore or rectify damage after the event.

9.4.5.10. The Protection and Promotion of Health and Safety

Safety and health come first. Every planned intervention should be evaluated for its potential effects on health and accident risks, particularly with regard to identifying and controlling the risks associated with potentially harmful materials, technologies, or procedures. This may involve delaying or eliminating the use of these interventions altogether. All people's physical, mental, and social well-being and safety are included in health impacts, with special attention paid to those segments of the population that are more susceptible to harm and more likely to suffer consequences, such as the economically disadvantaged, indigenous groups, women and children, the elderly, the disabled, and those most exposed to the risks associated with the planned intervention.

9.4.5.11. The Principle of Multi-Sectoral Integration

Social development requirements and the need to consider social issues should be properly integrated into all projects, policies, infrastructure programs, and other planning activities.

9.4.5.12. The Principle of Subsidiarity

Decentralized decision-making is desirable, with responsible decisions being taken as close to the individual citizen as feasible. Within the framework of SIA, this means that decisions regarding the approval of proposed interventions or the circumstances in which they might be implemented should be made in close proximity to the impacted population, with locals participating in the processes of management and approval.

9.4.5.13. Principle of Effectiveness and Efficiency

It is crucial that the employment of policy tools that provide an incentive to reduce wasteful use can also be used to achieve resource efficiency. By streamlining processes and procedures to reduce environmental costs, it also applies to various environmental governance issues.

9.4.5.14. The Principle of Responsibility

It is the responsibility of all persons, corporations, and states to maintain the ecological processes. Further, access to environmental resources carries attendant responsibilities to use them in an ecologically sustainable, economically efficient, and socially fair manner.

9.4.5.15. The Principle of Participation

Every individual has a responsibility to take part in group environmental decision-making. The use of fish, wildlife, minerals, soils, trees, and other plants for materials, food, and both consumptive and non-consumptive recreation are some areas of participation. The matter at hand pertains to solid waste, i.e., trash, building and demolition supplies, waste that poses a chemical risk, etc. Activities that produce pollution are the subject of the third participation issue.

9.4.5.16. The Principle of Proportionality

The idea of balance serves as the foundation for the proportionality principle. Maintaining equilibrium between environmental preservation and economic development is necessary. There is no denying that development is impossible without some negative impact on the environment. As a result, it is critical to balance the need to preserve the environment with people's interests. Moreover, benefits to a greater proportion of the population must be preserved, and comparative hardships must be balanced.

9.4.6. Developing Guidelines

Guidelines must be developed in the context in which they are to be applied and addressed to a specific audience because they are specific recommendations for action. They must therefore be developed in collaboration with the pertinent parties. Rather than being imposed, they must be acknowledged as the group's rules.

There are many different groups who are potentially interested in guidelines for impact assessment (IA). They include:

- Impact Assessment Practitioners – require guidelines to improve their practice;
- Regulatory Agencies – require guidelines in order to specify or audit the scope of IA activities they commission as well as the quality of IA reports they receive;
- Policy and Program Developers – require guidelines to ensure that policy and program development considers social and other relevant impacts;
- Affected People and NGOs – require guidelines to be able to participate effectively in IA processes. Local action groups (resident action groups) and NGOs often act like a regulatory agency in checking the appropriateness of IA processes.
- Developers (Proponents) and Financiers – require guidelines to be committed to good practice in environmental and social impact assessment, to adequately resource

such practice, to liaise effectively with practitioners and interested and affected parties, and with regulatory agencies.

- Development Agencies (Multilateral and Bilateral Aid Organizations) – require guidelines to ensure that the most benefit is obtained from their aid projects, that IA components are adequately resourced, and that the aid projects themselves do not have unintended environmental or social consequences.
- In addition, various sectors of the community may have special interests, and it may be appropriate for guidelines to be developed to address those special interests, such as Indigenous or Tribal People.

9.5. ENVIRONMENTAL MANAGEMENT TOOLS AND TECHNIQUES

We now concur that investing in environmental management is worthwhile given the importance of good environmental management. It is now necessary for a business or organization to understand the proper approach. The right tools are necessary for any task. A variety of environmental management tools are covered in this learning resource, which can be used by a business or organization to efficiently manage its social and environmental issues.

In the early 1990s, the International Organization for Standardization (ISO) recognized the need for standardization in the field of environmental management tools, and in 1993, it set up a committee to write standards relating to the following environmental management tools:

- Environmental policy;
- Environmental management systems;
- Environmental auditing;
- Environmental labeling;
- Life cycle assessment;
- Environmental indicators;
- Environmental policies;
- Eco-balances;
- Environmental reporting;
- Environmental charters.

9.5.1. Environmental Policy

A document that explicitly outlines a company's overall environmental goals and intentions is called an environmental policy. When a company decides to take on environmental management, they frequently start by creating an environmental policy. An environmental policy can set the stage for future environmental management initiatives by demonstrating a commitment to environmental management (Figure 9.4).

Figure 9.4. An environmental policy signals a commitment to environmental management.

Source: https://www.amsshardul.com/wp-content/uploads/2019/09/environment-02-1024x429.png.

9.5.2. Environmental Management Systems

A firm or company can manage its environmental affairs in a planned and systematic manner by using an environmental management system (EMS), which helps it identify ways to improve its environmental performance that will have the greatest positive impact on its business performance. The requirements of two EMS schemes—ISO 14001, the international standard for environmental management systems, and EMAS, the eco management and audit scheme of the European Community—are the main topic of this section.

In order to develop an EMS that meets the requirements of these schemes, a firm must take the following steps:

- First, it must produce an environmental policy that contains commitments to legislative compliance and to continually improving its environmental performance.
- Next, it must set targets relating to these commitments and devise a program for meeting these targets. Targets for improving environmental performance should be based on a comprehensive review of its environmental activities.
- It must then take the measures necessary to implement the program.
- Having implemented the program, it needs to check that it has been successful in meeting its targets. Corrective action must be taken in instances where this is not the case. The EMS must be audited periodically to check it is functioning as it should.

Finally, the firm needs to carry out a management review of the EMS, making any changes necessary in light of the audit results and changing circumstances. Having met its first set of targets, the firm must set itself a new set of targets to meet its policy commitment to continual improvement.

9.5.2.1. Environmental Auditing

An instrument for determining if a business or organization is operating ethically is environmental auditing. For example, a legislative compliance audit verifies that the company's operations covered by environmental laws (i.e., what it is doing) genuinely adhere to that law (i.e., how it ought to be operating). A company or organization can find out through an environmental audit whether its waste management procedures (i.e., what it is doing) adhere to the best practices for the industry sector that it has promised to follow (i.e., how it ought to be operating).

An essential component of an environmental management system is auditing. Similar to ISO 14001, an audit must be conducted to determine whether a company's EMAS and EMS satisfy the standards. This learning resource provides a thorough explanation of the EMS audit procedure in light of the significance of environmental management systems.

9.5.2.2. Environmental Indicators

Environmental indicators allow a firm to measure both its environmental performance and its efforts to improve its performance. Indicators can be used within an environmental management system to check that a firm has met the targets it is required to set for itself, but can equally well be used in firms that have not developed an EMS.

9.5.2.3. Eco-Balance

An organization's eco-balance keeps track of all the different raw materials, energy, resources, products, and wastes that come into, are held within, and are released from the company over a given time frame. Put another way, it offers a record of the tangible inputs, inventory, and outputs of a business. An organization can start evaluating the specific environmental effects of its inputs and outputs once it is aware of exactly what is coming in and going out. Therefore, an eco-balance enables a business to set goals for enhancing its environmental performance and carry out the thorough environmental review of its operations that is mandated by ISO 14001 and EMAS.

9.5.3. Life Cycle Assessment

A tool for determining and evaluating the numerous environmental effects connected to a specific product is life cycle assessment, or LCA. LCA adopts a "cradle to grave" perspective, examining the effects of the product at every stage of its life cycle, i.e., from the procurement of raw materials (the "cradle"), through its manufacture and application, to its ultimate disposal (the "grave"). Manufacturers can substantiate their claims about the environmental impact of their products and discover ways to reduce the environmental impact of their products over time at a reasonable cost by using life-cycle assessment (LCA) (Figure 9.5).

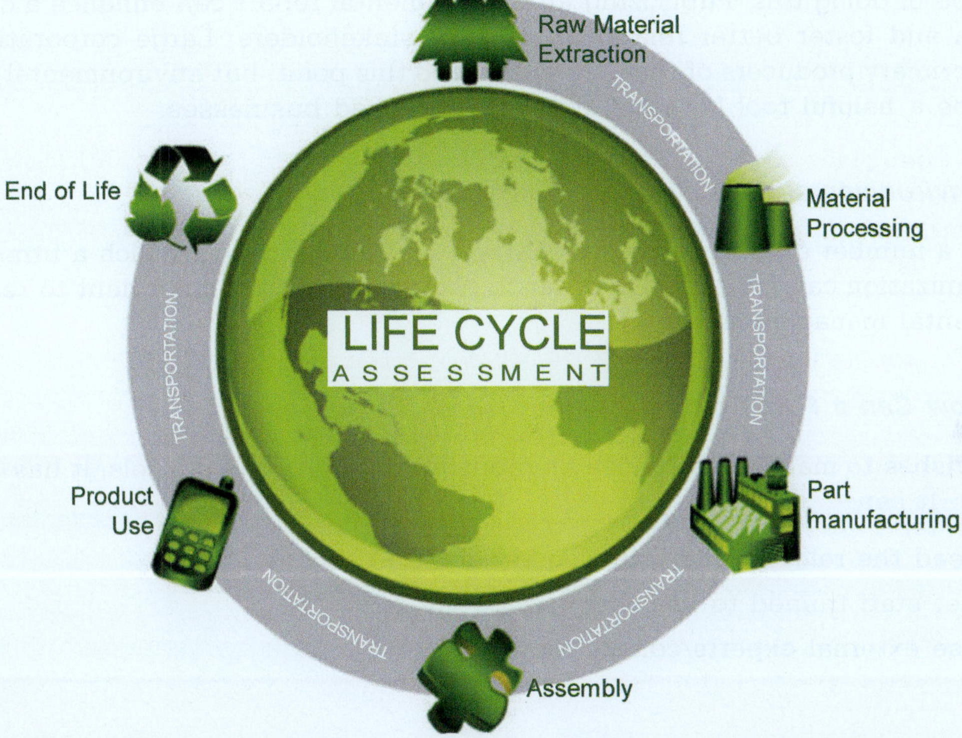

Figure 9.5. Life cycle assessment (LCA) is a tool for identifying and assessing the various environmental impacts associated with a particular product.

Source: https://3.imimg.com/data3/WK/IM/MY-9179847/life-cycle-assessment-lca.png.

9.5.3.1. Environmental Labeling

Environmental labeling schemes give environmental labels to products that are determined to be more environmentally friendly than others in the same category. Companies interested in obtaining a label for their products must submit an application to the organizer of the scheme. Products must meet specific environmental standards set by the scheme organizer, based on the product's life cycle and using Life Cycle Assessment (LCA). These criteria are designed so that only a limited percentage of products in each category can qualify for the label, typically around 20–30%. Therefore, environmental labels can serve as a valuable marketing tool, indicating that a product is among the most environmentally friendly options within its category.

9.5.3.2. Environmental Reporting

An organization or company that has implemented a variety of environmental management initiatives to enhance its environmental performance might want to share the outcomes of these efforts with the general public. An environmental report can be published as one method of doing this. Publicizing an environmental report can enhance a company's reputation and foster better relationships with stakeholders. Large corporations have been the primary producers of these reports up to this point, but environmental reporting can also be a helpful tool for small and medium-sized businesses.

9.5.3.3. Environmental Charters

There are a number of environmental charters and guidelines to which a firm/company or an organization can subscribe in order to demonstrate its commitment to responsible environmental management.

9.5.3.4. How Can a Firm/Company Make Use of These Tools?

If a firm wishes to make use of these environmental management tools, it has a number of options. It can:

- Read the relevant literature on the tools;
- Get staff trained to use them;
- Use external experts/consultancy.

CASE STUDY

SUSTAINABLE CONSTRUCTION PROJECT

Background

A multinational construction company, XYZ Construction Ltd., was awarded a contract to build a large commercial complex in an urban area. The project involved significant land excavation, construction of multiple buildings, and infrastructure development. Recognizing the environmental impact of such a project, XYZ Construction Ltd. made a commitment to integrate environmental management principles into every stage of the construction process.

- *Environmental Management Approach:*
  - **Environmental Impact Assessment (EIA):** Before commencing construction, XYZ Construction Ltd. conducted a comprehensive EIA to identify potential environmental impacts. This included assessing soil erosion, water pollution, air quality, and habitat disturbance.
  - **Stakeholder Engagement:** The company actively engaged with local communities, environmental organizations, and regulatory authorities throughout the project. Regular meetings and consultations were held to address concerns, gather feedback, and ensure transparency.
  - **Sustainable Design:** XYZ Construction Ltd. adopted sustainable design principles to minimize environmental impact. This included using eco-friendly building materials, incorporating energy-efficient technologies, and designing green spaces within the complex.
  - **Waste Management:** The company implemented a waste management plan to minimize construction waste and promote recycling. Materials such as concrete, steel, and wood were sorted and recycled wherever possible. Hazardous materials were disposed of safely and in accordance with regulations.
  - **Water Conservation:** Given the water-intensive nature of construction activities, XYZ Construction Ltd. implemented measures to conserve water. This included using water-saving equipment, minimizing runoff through sediment control, and implementing rainwater harvesting systems.
  - **Biodiversity Conservation:** Recognizing the importance of biodiversity conservation, the company took steps to protect and preserve natural habitats surrounding the construction site. Buffer zones were established to minimize disturbance to wildlife, and native vegetation was replanted in affected areas.

- **Environmental Monitoring:** Throughout the construction process, XYZ Construction Ltd. conducted regular environmental monitoring to track progress and ensure compliance with environmental standards. This included air quality monitoring, water quality testing, and noise level assessments.

Results and Impact:

1. **Reduced Environmental Footprint:** By integrating environmental management practices, XYZ Construction Ltd. significantly reduced the environmental footprint of the construction project. This included minimizing resource consumption, reducing pollution, and preserving biodiversity.
2. **Positive Stakeholder Feedback:** The company's proactive approach to environmental management garnered positive feedback from stakeholders, including local communities, environmental groups, and regulatory agencies. This helped build trust and credibility throughout the project.
3. **Cost Savings:** While implementing environmental management measures required initial investment, XYZ Construction Ltd. ultimately realized cost savings through improved resource efficiency, reduced waste disposal costs, and enhanced project efficiency.
4. **Legacy of Sustainability:** Upon completion, the commercial complex stood not only as a testament to architectural excellence but also as a model of sustainable construction practices. XYZ Construction Ltd.'s commitment to environmental management left a lasting legacy of sustainability for future generations.

Conclusion

The case study of XYZ Construction Ltd. demonstrates the importance of integrating environmental management principles into construction projects. By adopting a holistic approach that considers environmental impacts at every stage, companies can minimize their ecological footprint, build positive relationships with stakeholders, and contribute to a more sustainable future.

CLASS ACTIVITY

To reduce the environmental impact of all our activities, products, and services related to the manufacturing and distribution of welding structures such as construction machinery parts, we promote environmental management activities and aim to create harmony with the global environment based on the following policies:

- Always considering the environmental impact of our activities, products, and services, we promote the prevention of environmental pollution and continuously aim to improve our environmental management activities.
- We will observe legal and other environment-related requirements regarding our activities, products, and services.
- Out of the environmental impacts of our activities, products, and services, we focus on the following items as the high-priority themes for environmental management:
    - Power saving;
    - Reduction of loss due to defects;
    - Efficient use of vehicles;
    - Less use of specified chemical substances;
    - Awareness-raising activities such as the cleaning of the surrounding area of each factory.
- To ensure that each one of us can proactively perform environmental load reduction activities, we make this environmental declaration known to all employees and available to the public. To achieve the above policies, we set targets and regularly review them to promote the environmental management system.

SUMMARY

- Environmental management in construction is a multifaceted process aimed at minimizing the environmental impact of construction activities while promoting sustainable development.
- Waste management is another key aspect of environmental management in construction. Construction activities generate significant amounts of waste, including materials like concrete, wood, and steel.
- Sustainability is a social goal for people to coexist on Earth over a long time. Definitions of this term are disputed and have varied with literature, context, and time.
- Sustainability in construction refers to the practice of designing, constructing, operating, and demolishing buildings and infrastructure in a way that meets the needs of the present without compromising the ability of future generations to meet their own needs.
- Sustainable development is a hard topic to nail down because it consists of a wide range of things. Due to the technicality and complexity of this topic, it's best to check out its importance holistically to be able to grasp it easily.
- Crop rotation is the practice of planting different crops in the same farm to enhance soil fertility and assist in controlling diseases and insects. Crop rotation is beneficial in many ways; most importantly, it is chemical-free.
- Pollution is the result of human activities associated with industrial and automobile revolutions. Pollution is a man-made problem, mainly of affluent countries and now a serious problem in developing countries as well.
- Environmental Management is an entirely emerging and dynamic concept. Environmental Management is concerned with managing the environment within a business.
- Environmental management (EM) is a subject that combines science, policy, and socioeconomic applications. It primarily focuses on finding solutions to practical problems that people face in cohabitation with nature, resource exploitation, and waste production.
- Environmental Management Accounting (EMA) has no single univocally accepted definition. EMA is the management of environmental and economic performance with the help of development and execution of appropriate environmentally related accounting system and practices.

REVIEW QUESTIONS

1. Define environmental management and discuss its significance in modern society.
2. What are the major environmental challenges facing the world today, and how can environmental management address them?
3. Explain the concept of sustainable development and its role in environmental management.

4. What are the key components of an Environmental Management System (EMS)? Discuss their importance in achieving environmental goals.
5. Describe the process of environmental impact assessment (EIA) and its importance in decision-making for development projects.
6. How do environmental laws and regulations influence environmental management practices at the local, national, and international levels?
7. Discuss the role of corporate social responsibility (CSR) in promoting environmentally sustainable business practices.
8. Explain the concept of carbon footprint and discuss strategies for reducing it in both individual and organizational contexts.
9. What are some of the main tools and techniques used in environmental management, such as Life Cycle Assessment (LCA) or Environmental Risk Assessment (ERA)?
10. How can public participation and stakeholder engagement contribute to effective environmental management initiatives?

REFERENCES

1. Ahmed, F. (2008). An examination of the development path taken by small island developing states (PDF). Archived from the original (PDF) on October 3, 2012. (pp. 17–26).
2. Atkinson, G., Dietz, S., & Neumayer, E. (2009). Handbook of Sustainable Development. Edward Elgar Publishing. ISBN 1848444729.
3. Bakari, M. E. (2013). Globalization and sustainable development: False twins? New Global Studies, 7(3), 23–56. ISSN (Online) 1940-0004, ISSN (Print) 2194–6566. doi:10.1515/ngs-2013-021.
4. Bertelsmann Stiftung (Ed.). (2013). Winning Strategies for A Sustainable Future. Reinhard Mohn Prize 2013. Verlag Bertelsmann Stiftung. ISBN 978-3-86793-491-6.
5. Borowy, I. (2014). Defining Sustainable Development for Our Common Future: A History of The World Commission on Environment and Development (Brundtland Commission). Milton Park: Routledge.
6. Chong, H. G. (2008). The environmental management system as a competitive advantage tool for organizations. The Journal of Accounting and Finance, 22(1), 71–78.
7. Cook, S., & Dugarova, E. (2014). Rethinking social development for a post-2015 world. Development, 57(1), 30–35. doi:10.1057/dev.2014.25.
8. Danilov-Danil'yan, V. I., Losev, K. S., & Reyf, I. (2009). Sustainable Development and The Limitation of Growth: Future Prospects for World Civilization (V. Tumanov, Trans.; D. Rapp, Ed.). Springer Praxis Books. Retrieved from Google Books.
9. Edwards, A. R., & McKibben, B. (2010). Thriving Beyond Sustainability: Pathways To a Resilient Society. New Society Publishers. ISBN 0865716412.

10. Farah, P. D., & Rossi, P. (2015). Energy: Policy, legal and social-economic issues under the dimensions of sustainability and security. In World Scientific Reference on Globalization in Eurasia and The Pacific Rim. SSRN 2695701.

11. Florida, R., & Davison, D. (2001). Gaining from green management: Environmental management systems inside and outside the factory. California Management Review, 43(3), 64–85.

12. Hickel, J. (2019). The contradiction of the sustainable development goals: Growth versus ecology on a finite planet. Sustainable Development, 27(6). doi:10.1002/sd.1947.

13. Huesemann, M. H., & Huesemann, J. A. (2011). Technofix: Why Technology Won't Save Us or The Environment (Chapter 6, "Sustainability or collapse?" and Chapter 13, "The design of environmentally sustainable and socially appropriate technologies"). New Society Publishers. ISBN 0865717044.

14. Jain, T. K. (2000). Social and environmental accounting and auditing. In P. S. Bhushana Rao & P. M. Rao (Eds.), Environment Management and Audit (p. 226). Deep and Deep Publications Pvt. Ltd.

15. James, P., Magee, L., Scerri, A., & Steger, M. B. (2015). Urban Sustainability in Theory and Practice: Circles of Sustainability. Routledge.

16. James, P., Nadarajah, Y., Haive, K., & Stead, V. (2012). Sustainable Communities, Sustainable Development: Other Paths For Papua New Guinea. University of Hawaii Press.

17. Jarzombek, M. (2003). Sustainability—Architecture: Between fuzzy systems and wicked problems. Blueprints, 21(1), 6–9.

18. Mallik, A. K. (2001). Environmental accounting. In U. K. Sarkhel & J. Mallik (Eds.), Contemporary Issues in Accounting, Finance and Management (p. 21). UGC Academic Staff College, The University of Burdwan.

19. Melnyk, S. A., Sroufe, R. P., & Calantone, R. (2003). Assessing the impact of environmental management systems on corporate and environmental performance. Journal of Operations Management, 21(3), 329–351. doi:10.1016/S0272-6963(02)00109-2.

20. Patnaik, N. K. (2002). Environmental audit: A perspective. In P. S. Bhushana Rao & P. M. Rao (Eds.), Environmental Management and Audit (pp. 282–291). Deep and Deep Publications Pvt. Ltd.

21. Raudsepp-Hearne, C., Peterson, G. D., Tengö, M., Bennett, E. M., Holland, T., Benessaiah, K., MacDonald, G. M., & Pfeifer, L. (2010). Untangling the environmentalist's paradox: Why is human well-being increasing as ecosystem services degrade? BioScience, 60(8), 576–589. doi:10.1525/bio.2010.60.8.4.

22. Rogers, P., Jalal, K. F., & Boyd, J. A. (2007). An Introduction to Sustainable Development. Routledge. ISBN 1844075214.

23. Sianipar, C. P. M., Dowaki, K., Yudoko, G., & Adhiutama, A. (2013). Seven pillars of survivability: Appropriate technology with a human face. European Journal of Sustainable Development, 2(4), 1–18. ISSN 2239-5938.

24. Sroufe, R. (2003). Effects of environmental management systems on environmental management practices and operations. Production and Operations Management, 12(3), 416–431.

25. Tausch, A. (2012). Globalization, The Human Condition, And Sustainable Development in The Twenty-First Century: Cross-National Perspectives and European Implications (1st ed.). Anthem Press. ISBN 9780857284105.

26. Wallace, B. (2005). Becoming Part of The Solution: The Engineer's Guide to Sustainable Development. American Council of Engineering Companies. ISBN 978-0-910090-37-7.

CHAPTER

10

Construction Labor and Labor Management

LEARNING OBJECTIVES

After studying this chapter, you will be able to:

- Understand the types of construction labor;
- Discuss construction labor management;
- Define improve labor management in the construction industry.

KEY TERMS FROM THIS CHAPTER

Skilled labor	Unskilled labor
Labor productivity	Labor shortage
Groundskeeping	Green construction
Landscaping	Labor scheduling
Labor regulations	Labor efficiency
Labor training	Masonry assistance
Labor allocation	Job market

INTRODUCTION

Effective management of construction labor is essential for the successful executiouction project. At its core, construction labor encompasses the physical work performed by individuals skilled in various trades, ranging from carpentry and masonry to plumbing and electrical work. These laborers are the backbone of the construction industry, tasked with translating architectural designs into tangible structures. Their expertise, efficiency, and dedication directly impact the quality, timeline, and cost-effectiveness of projects.

Labor management within the construction sector involves orchestrating the diverse workforce to optimize productivity, ensure safety compliance, and mitigate potential conflicts. This entails strategic planning, recruitment, training, scheduling, and supervision of laborers throughout the project lifecycle. Effective labor management requires adept communication, problem-solving skills, and a comprehensive understanding of both the project's requirements and the capabilities of the workforce.

Moreover, labor management extends beyond mere oversight to encompass fostering a positive work environment, promoting teamwork, and addressing any issues or disputes promptly. Prioritizing the well-being and morale of construction laborers not only enhances productivity but also contributes to employee retention and overall project success. Additionally, embracing technological advancements, such as digital tools for scheduling and performance tracking, can streamline labor management processes and improve project outcomes.

10.1. TYPES OF CONSTRUCTION LABOR

Construction labor refers to the physical work performed on construction sites to build, renovate, or maintain structures such as buildings, roads, bridges, and other infrastructure. Construction laborers typically engage in tasks such as digging trenches, operating machinery, laying bricks or concrete, installing plumbing or electrical systems, and performing various other manual tasks necessary for construction projects. These workers are essential to the construction industry, as they carry out the hands-on work required to bring architectural plans to life (Mahesh & Sridevi, 2013).

Construction labor encompasses a wide range of specialized roles and skills. Here are some common types of construction labor.

10.1.1. Skilled Labor

Skilled construction labor refers to workers who have advanced training and experience in specific trades within the construction industry. These workers possess specialized knowledge and technical skills, which allow them to perform complex tasks that require precision and expertise. Unlike unskilled and semi-skilled laborers, skilled construction workers are often responsible for critical aspects of construction projects and are usually certified or licensed in their respective trades.

10.1.1.1. Historical Context

Throughout history, skilled construction labor has been crucial, from the ancient techniques used to build the Pyramids of Egypt to modern construction methods employed iancient times. From the Pyramids of Egypt and the Great Wall of China to the cathedrals of Europe and modern skyscrapers, the expertise of skilled laborers has been essential. Historically, these skills were passed down through apprenticeships within guilds, which played a crucial role in maintaining high standards of craftsmanship and trade secrets. As societies evolved, so did the methods and organization of skilled labor.

10.1.1.2. Evolution of Training and Guilds

Guilds were the primary institutions for training skilled laborers during the medieval period. These guilds regulated the quality of work, controlled the training of apprentices, and ensured that journeymen and masters met stringent standards. The Industrial Revolution brought significant changes, including the rise of trade unions and vocational schools, which formalized and expanded training opportunities.

10.1.1.3. Range of Trades in Skilled Construction Labor

Skilled construction labor encompasses a diverse array of trades, each requiring specialized knowledge and expertise. Some of the key trades include:

1. Carpentry: Carpenters are essential for constructing, installing, and repairing structures made from wood and other materials. Their tasks range from framing buildings to installing cabinets and decorative moldings.
2. Masonry: Masons work with brick, stone, and concrete to build and repair structures such as walls, walkways, and chimneys. Their craft requires precision and an understanding of various materials' properties.
3. Electrical Work: Electricians install, maintain, and repair electrical systems in residential, commercial, and industrial buildings. Their work ensures the safe and reliable operation of electrical infrastructure.
4. Plumbing: Plumbers install and repair piping systems that carry water, gas, and waste. Their work is critical for sanitation, heating, and cooling systems in buildings.
5. HVAC (Heating, Ventilation, and Air Conditioning): These technicians install and maintain systems that control the temperature, humidity, and air quality in buildings. Their expertise ensures comfort and energy efficiency.
6. Roofing: Roofers install and repair the roofs of buildings using a variety of materials, including shingles, tiles, and metal. Their work protects structures from the elements.
7. Welding: Welders join metal parts together using high heat. Their skills are essential in constructing buildings, bridges, and other infrastructure.
8. Tile Setting: Tile setters lay tiles on floors, walls, and other surfaces. Their work is both functional and decorative, requiring precision and an eye for design.
9. Painting: Painters apply paint, stain, and coatings to buildings and structures. Their work enhances aesthetics and protects surfaces from damage.

10.1.1.4. Training and Education

The path to becoming a skilled construction laborer typically involves a combination of formal education, apprenticeship programs, and on-the-job training. The specifics can vary significantly by trade and region.

10.1.1.4.1. Vocational and Technical Schools

Many skilled trades can be learned through vocational and technical schools, which offer courses and hands-on training in specific areas such as carpentry, plumbing, and electrical work. These programs often lead to certifications or associate degrees.

10.1.1.4.2. Apprenticeships

Apprenticeships are a cornerstone of training in the construction industry. These programs combine paid on-the-job training with classroom instruction, typically lasting several years. Apprentices work under the supervision of experienced journeymen, gaining practical skills and knowledge.

10.1.1.4.3. Certifications and Licensing

Many trades require workers to obtain certifications or licenses to ensure they meet industry standards. For example, electricians and plumbers often need to pass exams and complete continuing education to maintain their licenses.

10.1.1.4.4. Continuing Education

The construction industry is constantly evolving, with new technologies and techniques emerging regularly. Continuing education is crucial for skilled laborers to stay current with industry advancements and maintain their competitiveness in the job market.

10.1.1.4.5. Economic Impact

Skilled construction labor plays a vital role in the economy, contributing to job creation, economic growth, and infrastructure development.

10.1.1.4.6. Job Creation

The construction industry is a significant source of employment. According to the U.S. Bureau of Labor Statistics, construction employment is projected to grow faster than the average for all occupations. This growth is driven by the need for new infrastructure, renovation of existing buildings, and maintenance of aging infrastructure.

10.1.1.4.7. Economic Growth

Investment in construction projects stimulates economic growth by generating demand for materials, equipment, and services. This, in turn, creates a multiplier effect, as increased spending in the construction sector boosts economic activity in related industries.

10.1.1.4.8. Infrastructure Development

Skilled construction labor is essential for developing and maintaining infrastructure, including roads, bridges, schools, hospitals, and housing. High-quality infrastructure is crucial for the overall well-being of society, supporting economic activities and improving quality of life.

10.1.1.5. Challenges Faced by Skilled Construction Labor

Despite its importance, skilled construction labor faces several challenges that impact the industry's sustainability and growth.

10.1.1.5.1. Labor Shortages

One of the most significant challenges is the shortage of skilled labor. This shortage is driven by several factors, including an aging workforce, declining interest among younger generations, and the perception of construction jobs as physically demanding and low-paying.

10.1.1.5.2. Safety Concerns

Construction work is inherently hazardous, with risks including falls, equipment accidents, and exposure to harmful substances. Ensuring the safety of workers is a constant challenge, requiring stringent safety protocols, training, and compliance with regulations.

10.1.1.5.3. Technological Advancements

While technology can improve efficiency and productivity, it also poses a challenge as workers need to adapt to new tools, equipment, and methods. This requires ongoing training and investment in upskilling the workforce.

10.1.1.5.4. Economic Volatility

The construction industry is highly sensitive to economic fluctuations. Recessions can lead to reduced investment in construction projects, job losses, and financial instability for workers. Conversely, economic booms can result in labor shortages and increased pressure on wages and resources.

10.1.1.5.5. Regulatory Compliance

Compliance with building codes, environmental regulations, and labor laws is complex and can vary significantly by region. Navigating these regulations requires expertise and can be time-consuming and costly.

10.1.1.6. Future Trends in Skilled Construction Labor

The future of skilled construction labor is shaped by several emerging trends that have the potential to transform the industry.

10.1.1.6.1. Adoption of Technology

The adoption of technology in construction is accelerating. Building Information Modeling (BIM), drones, 3D printing, and robotics are among the technologies revolutionizing the industry. These advancements can improve efficiency, reduce waste, and enhance safety, but they also require workers to develop new skills.

10.1.1.6.2. Green Construction

Sustainability is becoming a critical focus in construction. Green building practices, energy-efficient designs, and the use of sustainable materials are gaining prominence. Skilled laborers need to be knowledgeable about these practices to meet the growing demand for environmentally friendly construction.

10.1.1.6.3. Modular and Prefabricated Construction

Modular and prefabricated construction methods are gaining popularity due to their potential to reduce construction time and costs. These methods involve assembling building components off-site and transporting them to the construction site. Skilled laborers must adapt to these new processes and techniques.

10.1.1.6.4. Workforce Diversity

Increasing diversity in the construction workforce is a priority. Efforts to attract more women, minorities, and young people to the industry are underway to address labor shortages and bring different perspectives and skills to the field.

10.1.1.6.5. Improved Training and Education

Innovations in training and education are helping to address the skills gap. Virtual reality (VR) and augmented reality (AR) are being used for training, providing immersive, hands-on experiences without the risks associated with on-site training. Online learning platforms and partnerships between industry and educational institutions are also enhancing access to training.

10.1.2. Semi-Skilled Labor

Semi-skilled labor refers to workers who possess more than basic skills but do not have the advanced expertise of skilled tradespeople. These laborers are proficient in using a variety of tools and equipment and can perform tasks that require some level of judgment and technical knowledge. Unlike unskilled laborers, who primarily handle basic, repetitive tasks, semi-skilled workers contribute to more complex aspects of construction projects, often under the supervision of skilled workers or site managers (Hemant, 2016).

10.1.2.1. Distinguishing Features

The distinguishing features of semi-skilled construction labor include:

1. Intermediate Skill Level: Semi-skilled workers possess more skills than unskilled laborers but less than fully trained tradespeople.
2. Specific Task Proficiency: They are proficient in specific tasks and can perform them efficiently without needing constant supervision.
3. Assistive Role: They often assist skilled tradespeople, providing support and performing preparatory or supplementary tasks.
4. Training and Experience: Semi-skilled workers typically acquire their skills through on-the-job training, short-term vocational courses, or apprenticeships.

10.1.2.2. Types of Work Involved in Semi-Skilled Construction Labor

Semi-skilled construction labor encompasses a wide range of tasks that vary depending on the type of construction project. Some common roles and tasks include:

1. Concrete Finishing: Concrete finishers are responsible for smoothing and finishing surfaces of poured concrete, such as floors, walkways, and roads. Their tasks include leveling, smoothing, and shaping concrete surfaces to ensure they meet specified standards.
2. Drywall Installation: Drywall installers, also known as hangers, are responsible for measuring, cutting, and installing drywall panels in buildings. They ensure that the panels fit correctly and are properly secured to walls and ceilings.
3. Insulation Installation: Insulation installers add insulating materials to buildings to regulate temperature and improve energy efficiency. They work with various materials, including fiberglass, foam, and cellulose, to insulate walls, ceilings, and floors.
4. Painting and Coating: Painters and coaters apply paint, stain, and other finishes to buildings and structures. Their tasks include surface preparation, color mixing, and applying coatings to achieve the desired appearance and protection.
5. Roofing Assistance: Semi-skilled laborers in roofing assist skilled roofers in installing, repairing, and maintaining roofs. They may handle materials, set up scaffolding, and perform basic installation tasks under supervision.
6. Tile and Floor Installation: Tile and floor installers lay tiles and

other flooring materials in buildings. They are responsible for surface preparation, cutting and fitting tiles, and ensuring proper alignment and adhesion.

7. Carpentry Assistance: This support skilled carpenters in various tasks, such as measuring and cutting materials, assembling structures, and installing fixtures. They perform simpler tasks while learning more complex carpentry skills.
8. Masonry Assistance: Masonry assistants help skilled masons in constructing and repairing structures made of brick, stone, and concrete. Their tasks include mixing mortar, transporting materials, and basic bricklaying.
9. Basic Electrical Work: Semi-skilled electrical workers assist electricians with installing and maintaining electrical systems. They may handle tasks such as running conduit, wiring, and installing basic fixtures under the supervision of a licensed electrician.
10. Basic Plumbing Work: Semi-skilled plumbing workers assist plumbers with installing and maintaining piping systems. Their tasks may include cutting and fitting pipes, installing fixtures, and basic repairs under supervision.

10.1.2.3. Training and Education

Training and education for semi-skilled construction laborers vary depending on the specific trade and regional requirements. The primary pathways include on-the-job training, vocational courses, and apprenticeships.

10.1.2.3.1. On-the-Job Training

On-the-job training is a common pathway for semi-skilled construction laborers. New workers learn from experienced colleagues, gradually acquiring the skills and knowledge needed to perform specific tasks. This hands-on approach allows workers to gain practical experience and develop proficiency.

10.1.2.3.2. Vocational Courses

Many community colleges and vocational schools offer short-term courses and certifications in various construction trades. These programs provide foundational knowledge and skills, often including both classroom instruction and practical training.

10.1.2.3.3. Apprenticeships

While apprenticeships are more commonly associated with skilled trades, they also provide opportunities for semi-skilled workers to gain experience and training. Apprentices work under the supervision of skilled tradespeople, learning through a combination of hands-on work and classroom instruction.

10.1.2.3.4. Certifications and Licenses

Some semi-skilled roles may require certifications or licenses, especially for tasks involving safety or specialized equipment. For example, workers handling certain insulation materials may need specific safety certifications.

10.1.2.3.5. Economic Impact

Semi-skilled construction labor significantly impacts the economy, contributing to job creation, economic growth, and infrastructure development.

10.1.2.3.6. Job Creation

The construction industry is a major source of employment, and semi-skilled labor plays a crucial role in this sector. These jobs provide opportunities for individuals who may not have advanced education but possess the necessary skills and willingness to work.

10.1.2.3.7. Economic Growth

Investment in construction projects drives economic growth by creating demand for materials, equipment, and services. Semi-skilled labor is essential for completing these projects, ensuring that construction activities continue smoothly and efficiently.

10.1.2.3.8. Infrastructure Development

Semi-skilled construction labor is vital for developing and maintaining infrastructure, including residential, commercial, and industrial buildings. High-quality infrastructure supports economic activities, improves living standards, and enhances overall societal well-being.

10.1.2.4. Challenges Faced by Semi-Skilled Construction Labor

Despite its importance, semi-skilled construction labor faces several challenges that affect the industry's sustainability and growth.

10.1.2.4.1. Labor Shortages

Labor shortages are a significant issue in the construction industry, affecting both skilled and semi-skilled workers. Factors contributing to these shortages include an aging workforce, declining interest among younger generations, and competition from other industries.

10.1.2.4.2. Safety Concerns

Construction work is inherently hazardous, and semi-skilled laborers are often exposed to risks such as falls, equipment accidents, and exposure to harmful substances. Ensuring worker safety requires stringent safety protocols, training, and adherence to regulations.

10.1.2.4.3. Economic Volatility

The construction industry is quite affected by economic changes. During recessions, there might be less investment in construction projects, leading to job losses and financial uncertainty for workers. On the other hand, economic booms can cause labor shortages and put pressure on wages and resources.

10.1.2.4.4. Technological Advancements

While technology can improve efficiency and productivity, it also poses challenges as workers need to adapt to new tools, equipment, and methods. This requires ongoing training and investment in upskilling the workforce.

10.1.2.4.5. Regulatory Compliance

Understanding and complying with building codes, environmental regulations, and labor laws can be quite intricate and may differ greatly depending on the region. Navigating these regulations demands expertise and can consume time and resources.

10.1.2.5. Future Trends in Semi-Skilled Construction Labor

The future of semi-skilled construction labor is shaped by several emerging trends that have the potential to transform the industry.

10.1.2.5.1. Adoption of Technology

The adoption of technology in construction is picking up speed. Tools like Building Information Modeling (BIM), drones, 3D printing, and robotics are transforming the industry. These advancements can boost efficiency, cut down on waste, and enhance safety, but they require workers to acquire new skills.

10.1.2.5.2. Green Construction

Sustainability is gaining importance in construction. Green building practices, energy-efficient designs, and sustainable materials are gaining popularity. Semi-skilled laborers need to grasp these practices to meet the growing demand for eco-friendly construction.

10.1.2.5.3. Modular and Prefabricated Construction

Modular and prefabricated construction methods are becoming more popular due to their potential to reduce construction time and costs. These methods involve assembling building components off-site and transporting them to the construction site. Semi-skilled laborers need to adapt to these new processes and techniques.

10.1.2.5.4. Workforce Diversity

Boosting diversity in the construction workforce is a key focus. Efforts to attract more women, minorities, and young people to the industry are underway to tackle labor shortages and bring diverse perspectives and skills to the field.

10.1.2.5.5. Improved Training and Education

Innovations in training and education are proving effective in addressing the skills gap. The use of virtual reality (VR) and augmented reality (AR) for training offers immersive, hands-on experiences while eliminating the risks associated with on-site training. Additionally, online learning platforms and partnerships between industry and educational institutions are enhancing training accessibility.

10.1.3. Unskilled Labor

Unskilled construction labor refers to workers who perform basic, manual tasks on construction sites that require minimal training or technical skills. These tasks are typically labor-intensive and crucial for the smooth operation of construction projects. Unskilled laborers often work under the supervision of skilled tradespeople and site managers, providing essential support to ensure project timelines and quality standards are met.

10.1.3.1. Distinguishing Features

The distinguishing features of unskilled construction labor include:

1. Minimal Training Required: Tasks can be learned quickly and do not require formal education or extensive training.
2. Manual and Physical Nature: The work is predominantly physical, involving lifting, carrying, digging, and cleaning.
3. Supportive Role: Unskilled laborers support skilled tradespeople by preparing sites, transporting materials, and maintaining cleanliness and order.
4. Entry-Level Positions: These roles are often entry points into the construction industry, providing opportunities for advancement through experience and additional training.

10.1.3.2. Types of Work Involved in Unskilled Construction Labor

Unskilled construction labor encompasses a wide range of tasks essential to the completion of construction projects. Some common roles and tasks include:

1. Site Preparation and Cleanup: Unskilled laborers are responsible for preparing construction sites by clearing debris, leveling ground, and setting up temporary structures such as scaffolding. They also ensure that sites are kept clean and free of hazards throughout the project.
2. Material Handling: Laborers transport materials around the construction site, including loading and unloading trucks, moving supplies to designated areas, and organizing materials for easy access by skilled tradespeople.
3. Digging and Excavation: Unskilled laborers perform manual digging and excavation tasks using shovels and other basic tools. They prepare trenches for foundations, utilities, and drainage systems.
4. Basic Demolition: Laborers assist in demolition work by tearing down structures, removing debris, and sorting recyclable materials. They use hand tools and, occasionally, small machinery to carry out these tasks.
5. Mixing and Pouring Concrete: Laborers mix and pour concrete for various construction applications, including foundations, walkways, and slabs. They help ensure the proper consistency and placement of concrete.
6. Assisting Skilled Tradespeople: Unskilled laborers support skilled tradespeople by holding materials, fetching tools, and assisting with basic tasks. This assistance allows skilled workers to focus on more complex aspects of their trades.
7. Erecting and Dismantling Scaffolding: Laborers are often tasked with erecting and dismantling scaffolding and temporary structures, ensuring that they are safe and stable for use by other workers.

8. Landscaping and Groundskeeping: Laborers perform basic landscaping and groundskeeping tasks, such as planting trees and shrubs, installing sod, and maintaining green spaces around construction sites.
9. Painting and Surface Preparation: Laborers assist with painting and surface preparation by cleaning, sanding, and masking surfaces to be painted. They may also apply primers and undercoats.
10. General Maintenance and Repairs: Laborers handle general maintenance and minor repairs on construction sites, ensuring that tools and equipment are in good working order and that the site remains functional and safe.

10.1.3.3. Economic Impact

Unskilled construction labor significantly impacts the economy, contributing to job creation, economic growth, and infrastructure development.

10.1.3.3.1. Job Creation

The construction industry is a major source of employment, and unskilled labor provides numerous job opportunities for individuals without advanced education or specialized skills. These jobs offer a starting point for workers to gain experience and potentially advance to more skilled positions.

10.1.3.3.2. Economic Growth

Investment in construction projects stimulates economic growth by generating demand for materials, equipment, and services. Unskilled labor is essential for completing these projects, ensuring that construction activities continue smoothly and efficiently.

10.1.3.3.3. Infrastructure Development

Unskilled construction labor is vital for developing and maintaining infrastructure, including residential, commercial, and industrial buildings. High-quality infrastructure supports economic activities, improves living standards, and enhances overall societal well-being.

10.1.3.3.4. Training and Education

While unskilled construction labor requires minimal formal education, training is still essential to ensure workers perform tasks safely and effectively. Training methods include on-the-job training, basic safety courses, and informal mentorship.

10.1.3.3.5. On-the-Job Training

On-the-job training is the most common method for unskilled laborers to learn their tasks. New workers learn by observing and assisting experienced colleagues, gradually acquiring the skills and knowledge needed to perform specific tasks.

10.1.3.3.6. Basic Safety Courses

Safety is a critical concern in the construction industry, and unskilled laborers must be trained in basic safety practices. Courses may cover topics such as proper lifting techniques, use of personal protective equipment (PPE), hazard recognition, and emergency procedures.

10.1.3.3.7. Informal Mentorship

Experienced workers often mentor new laborers, providing guidance and instruction on how to perform tasks correctly and efficiently. This informal mentorship helps new workers develop practical skills and confidence.

10.1.3.3.8. Certifications and Licenses

While not typically required for unskilled labor, some positions may benefit from basic certifications, such as Occupational Safety and Health Administration (OSHA) safety training or first aid certification.

10.1.3.4. Challenges Faced by Unskilled Construction Labor

Despite its importance, unskilled construction labor faces several challenges that impact the industry's sustainability and growth.

10.1.3.4.1. Labor Shortages

Labor shortages are a significant issue in the construction industry, affecting both skilled and unskilled workers. Factors contributing to these shortages include an aging workforce, declining interest among younger generations, and competition from other industries.

10.1.3.4.2. Safety Concerns

Construction work is inherently hazardous, and unskilled laborers are often exposed to risks such as falls, equipment accidents, and exposure to harmful substances. Ensuring worker safety requires stringent safety protocols, training, and adherence to regulations.

10.1.3.4.3. Economic Volatility

The construction sector is greatly impacted by economic changes. Downturns can cause decreased investment, job cuts, and financial insecurity for workers. Conversely, economic upswings can lead to labor shortages and higher wage and resource demands.

10.1.3.4.4. Job Stability and Wages

Unskilled construction laborers often face challenges related to job stability and wages. These positions are typically lower-paying, and employment can be seasonal or project-based, leading to periods of unemployment or underemployment.

10.1.3.4.5. Career Advancement

Unskilled laborers may find it challenging to advance their careers without additional training or education. Limited opportunities for upward mobility can lead to job dissatisfaction and high turnover rates.

10.1.3.5. Future Trends in Unskilled Construction Labor

The future of unskilled construction labor is shaped by several emerging trends that have the potential to transform the industry.

10.1.3.5.1. Adoption of Technology

The adoption of technology in construction is picking up speed. Tools like Building Information Modeling (BIM), drones, 3D printing, and robotics are transforming the industry. These advancements can enhance efficiency, reduce waste, and improve safety but require workers to acquire new skills.

10.1.3.5.2. Green Construction

Sustainability is becoming a crucial focus in construction. Green building practices, energy-efficient designs, and the use of sustainable materials are gaining importance. It's essential for unskilled laborers to be well-informed about these practices to meet the increasing demand for environmentally friendly construction.

10.1.3.5.3. Modular and Prefabricated Construction

Modular and prefabricated construction methods are becoming more popular because they can help reduce construction time and costs. With these methods, building components are assembled off-site and then transported to the construction site. Unskilled workers need to familiarize themselves with these new processes and techniques.

10.1.3.5.4. Workforce Diversity

Enhancing diversity in the construction workforce is a key focus. Initiatives to attract more women, minorities, and young individuals to the sector are underway to address labor shortages and introduce diverse perspectives and skills to the field.

10.1.3.5.5. Improved Training and Education

Innovations in training and education are helping to tackle the skills gap. Virtual reality (VR) and augmented reality (AR) are being used for training, offering immersive, hands-on experiences without the risks associated with on-site training. Online learning platforms and partnerships between industry and educational institutions are also improving access to training.

10.1.3.5.6. Health and Safety Enhancements

As safety remains a top concern, there is a growing focus on enhancing health and safety standards for unskilled laborers. This includes better training programs, stricter enforcement of safety regulations, and the development of new technologies to monitor and improve safety on construction sites.

10.1.3.5.7. Economic and Policy Support

Economic and policy support can play a significant role in addressing the challenges faced by unskilled construction laborers. Government initiatives and policies aimed at supporting infrastructure development, providing job training programs, and ensuring fair wages and working conditions can help stabilize and grow the workforce.

10.1.4. Technical Labor

Technical construction labor refers to workers who possess specialized skills and knowledge to perform advanced tasks within the construction industry. These tasks often involve technical precision, critical thinking, and adherence to stringent standards and regulations. Technical laborers work across various sectors, including civil engineering, electrical systems, plumbing, HVAC (heating, ventilation, and air conditioning), and more.

10.1.4.1. Distinguishing Features

The distinguishing features of technical construction labor include:

1. Advanced Training and Education: Workers undergo extensive training, often including formal education, apprenticeships, and certification.
2. Specialized Skills: They possess specialized technical skills that are crucial for performing complex tasks.
3. Precision and Accuracy: Their work requires high levels of precision and accuracy, often involving technical calculations and adherence to detailed specifications.
4. Regulatory Compliance: Technical laborers must comply with strict industry standards and regulations.
5. Problem-Solving Abilities: They must have strong problem-solving skills to address challenges that arise during construction projects.

10.1.4.2. Types of Work Involved in Technical Construction Labor

Technical construction labor spans various trades and specializations, each with its unique set of tasks and responsibilities. Some common roles and tasks include:

1. Civil Engineering Technicians: They assist civil engineers in designing, planning, and overseeing construction projects. Their tasks include:
   - Conducting site surveys and assessments;
   - Preparing and reviewing project plans and blueprints;
   - Testing construction materials;
   - Monitoring project progress and ensuring compliance with specifications.
2. Electrical Technicians: These are responsible for installing, maintaining, and repairing electrical systems in buildings and infrastructure. Their tasks include:
   - Reading and interpreting electrical blueprints and diagrams;
   - Installing wiring, outlets, and fixtures;
   - Troubleshooting electrical issues and performing repairs;
   - Ensuring compliance with electrical codes and safety standards.
3. Plumbing Technicians: They install and maintain plumbing systems, ensuring the proper functioning of water supply and waste disposal systems. Their tasks include:
   - Installing pipes, fittings, and fixtures;
   - Testing plumbing systems for leaks and other issues;
   - Performing routine maintenance and repairs;
   - Ensuring compliance with plumbing codes and regulations.
4. HVAC Technicians: They install, maintain, and repair heating, ventilation, and air conditioning systems. Their tasks include:
   - Installing HVAC systems and components;

- Conducting routine maintenance and inspections;
- Troubleshooting and repairing HVAC systems;
- Ensuring energy efficiency and compliance with environmental standards.

5. Structural Technicians: They work on the design and construction of building frameworks and structures. Their tasks include:
   - Preparing and reviewing structural plans and drawings;
   - Conducting structural assessments and inspections;
   - Assisting in the fabrication and installation of structural components;
   - Ensuring structural integrity and compliance with building codes.
6. Surveying Technicians: They assist surveyors in measuring and mapping land for construction projects. Their tasks include:
   - Conducting field surveys and measurements;
   - Using surveying equipment and technology to collect data;
   - Preparing maps, reports, and legal descriptions;
   - Assisting in the planning and layout of construction projects.
7. Geotechnical Technicians: They work on assessing soil, rock, and other earth materials to determine their suitability for construction projects. Their tasks include:
   - Conducting soil and rock sampling and testing;
   - Analyzing geotechnical data and preparing reports;
   - Assisting in the design of foundations and earthworks;
   - Monitoring site conditions and ensuring compliance with geotechnical recommendations.

10.1.4.3. Training and Education

Training and education for technical construction laborers vary depending on the specific trade and regional requirements. The primary pathways include formal education, apprenticeships, certification programs, and continuing education.

10.1.4.3.1. Formal Education

Many technical construction roles require formal education, such as an associate or bachelor's degree in a relevant field. For example:

- Civil Engineering Technicians: Typically hold an associate degree in civil engineering technology or a related field.
- Electrical Technicians: Often have formal training from a technical school or community college.
- HVAC Technicians: May complete a certificate or associate degree program in HVAC technology.

10.1.4.3.2. Apprenticeships

Apprenticeships provide a combination of on-the-job training and classroom instruction, allowing aspiring technicians to gain hands-on experience while learning theoretical concepts. Apprenticeships are common in trades such as electrical work, plumbing, and HVAC.

10.1.4.3.3. Certification Programs

Certifications validate the skills and knowledge of technical construction laborers and are often required by employers or regulatory bodies. Examples include:

- Electricians: May obtain certifications such as the Journeyman or Master Electrician license.
- HVAC Technicians: Can earn certifications such as the EPA Section 608 Certification for handling refrigerants.

10.1.4.3.4. Continuing Education

Continuing education is essential for technical construction laborers to stay current with industry advancements, new technologies, and regulatory changes. Many trades require periodic recertification or continuing education credits.

10.1.4.3.5. Economic Impact

Technical construction labor significantly impacts the economy, contributing to job creation, economic growth, and infrastructure development.

10.1.4.3.6. Job Creation

The construction industry is a major source of employment, and technical labor provides numerous job opportunities for individuals with specialized skills and training. These roles offer stable and often well-paying careers.

10.1.4.3.7. Economic Growth

Investment in construction projects helps boost economic growth by creating demand for materials, equipment, and services. Skilled labor plays a crucial role in ensuring the smooth progress of these projects.

10.1.4.3.8. Infrastructure Development

Technical construction labor is essential for the development and upkeep of infrastructure, such as residential, commercial, and industrial buildings. Well-built infrastructure not only supports economic activities but also enhances living standards and societal well-being.

10.1.4.4. Challenges Faced by Technical Construction Labor

Despite its significance, technical construction labor encounters various challenges that affect the industry's sustainability and expansion.

10.1.4.4.1. Labor Shortages

One major issue in the construction sector is the shortage of labor, impacting both skilled and unskilled workers. Reasons for these shortages include an aging workforce, decreased interest among younger individuals, and competition from other industries.

10.1.4.4.2. Safety Concerns

Construction work is inherently risky, with laborers facing significant hazards; for instance, falls accounted for 33% of all construction fatalities in 2020, and equipment accidents continue to be a leading cause of injuries on site, emphasiziidents, and exposure to harmful substances. Keeping workers safe involves following strict safety protocols, providing training, and complying with regulations.

10.1.4.4.3. Economic Volatility

Economic fluctuations impact the construction industry significantly. During recessions, there's reduced investment in projects, job cuts, and financial uncertainty for workers. Conversely, economic booms can result in labor shortages and increased pressure on wages and resources.

10.1.4.4.4. Technological Advancements

Technological advancements can enhance efficiency and productivity, but they also bring challenges as workers must adapt to new tools, equipment, and methods. This necessitates continuous training and investment in upskilling the workforce.

10.1.4.4.5. Regulatory Compliance

Compliance with building codes, environmental regulations, and labor laws can vary significantly by region. Navigating these regulations demands expertise and can be time-consuming and costly.

10.1.4.5. Future Trends in Technical Construction Labor

The future of technical construction labor is shaped by several emerging trends that have the potential to transform the industry.

10.1.4.5.1. Adoption of Advanced Technologies

The adoption of advanced technologies in construction is accelerating. Tools such as Building Information Modeling (BIM), drones, 3D printing, and robotics are revolutionizing the industry. These advancements can improve efficiency, reduce waste, and enhance safety but require workers to developnew skills.

10.1.4.5.2. Green Construction

The focus on sustainability in construction remains paramount, with green building practices and energy-efficient designs being essential for technical workers to effecte rising need for eco-friendly construction.

10.1.4.5.3. Modular and Prefabricated Construction

Modular and prefabricated construction methods are increasingly popular for their ability to cut construction time and costs. They require assembling building parts off-site and then transporting them to the construction site. Technical workers need to adjust to these new methods and techniques.

10.1.4.5.4. Workforce Diversity

Increasing diversity in the construction workforce remains a priority. Efforts to attract more women, minorities, and young people to the industry are currently underway to address labor shortages and bring different perspectives and skills to the field.

10.1.4.5.5. Improved Training and Education

Innovations in training and education have helped address the skills gap. Virtual reality (VR) and augmented reality (AR) are used for training, offering immersive, hands-on experiences without the risks associated with on-site training. Online learning platforms and partnerships between industry and educational institutions also enhance access to training.

10.1.4.5.6. Enhanced Safety Measures

As safety remains a paramount concern, there has been a growing emphasis on improving health and safety standards for technical laborers. This includes better training programs, stricter enforcement of safety regulations, and the development of new technologies to monitor and enhance safety on construction sites.

10.1.4.5.7. Economic and Policy Support

Economic and policy support can play a significant role in addressing the challenges faced by technical construction laborers. Government initiatives and policies aimed at supporting infrastructure development, providing job training programs, and ensuring fair wages and working conditions can help stabilize and grow the workforce.

10.2. CONSTRUCTION LABOR MANAGEMENT

Construction labor management is a crucial aspect of any construction project, ensuring that tasks are completed efficiently, on time, and within budget. Effective labor management involves several key components which are discussed in subsections.

10.2.1. Planning and Scheduling

Construction projects are complex endeavors involving numerous tasks, resources, and stakeholders. Effective planning and scheduling are critical for orchestrating these elements harmoniously to achieve project objectives (Abhiram & Asadi, 2016).

Importance of planning and scheduling:

1. Time Management: Efficient scheduling ensures tasks are completed in a timely manner, minimizing delays and project overruns.
2. Resource Optimization: Proper planning allocates resources effectively, preventing shortages or wastage.
3. Risk Mitigation: Anticipating and planning for potential challenges helps mitigate risks and uncertainties.
4. Quality Assurance: A well-structured schedule allows sufficient time for quality control measures, ensuring deliverables meet specifications.

10.2.1.1. Key Components of Planning and Scheduling

1. Work Breakdown Structure (WBS): It is a hierarchical decomposition of the project into smaller, manageable components. Each component represents a distinct deliverable or work package, facilitating detailed planning and resource allocation.
2. Project Scope and Objectives: Understanding the project scope and objectives is essential for developing an effective plan. Clear objectives provide direction and help prioritize tasks accordingly.
3. Task Sequencing and Dependency Analysis: Tasks are sequenced based on their logical order and dependencies. Dependency analysis identifies tasks that must be completed sequentially and those that can be performed concurrently, optimizing project duration.

4. Duration Estimation: Accurate estimation of task durations is crucial for developing a realistic schedule. Historical data, expert judgment, and mathematical models such as PERT (Program Evaluation and Review Technique) are commonly used for duration estimation.
5. Resource Allocation: Resources, including labor, equipment, and materials, are allocated based on project requirements and availability. Resource leveling ensures a balanced workload and prevents resource conflicts.
6. Critical Path Method (CPM) and Gantt Charts: The Critical Path Method (CPM) identifies the longest path of dependent tasks, determining the minimum project duration. Gantt charts visually represent project tasks and their scheduled durations, providing a timeline view of the project.
7. Contingency Planning: This involves identifying potential risks and developing strategies to mitigate their impact. Contingency reserves are allocated to address unforeseen events or delays.
8. Communication and Collaboration: Effective communication and collaboration among project stakeholders are vital for successful planning and scheduling. Regular meetings, progress reports, and collaboration tools facilitate information sharing and decision-making.

10.2.1.2. Methodologies and Techniques

1. Critical Chain Method (CCM): This extends the Critical Path Method (CPM) by considering resource constraints. It identifies the critical chain of tasks based on resource availability, optimizing resource utilization and project duration.
2. Lean Construction: These principles focus on eliminating waste and maximizing value throughout the construction process. Techniques such as Just-In-Time (JIT) delivery and Last Planner System (LPS) improve efficiency and minimize delays.
3. Building Information Modeling (BIM): It enables collaborative planning and visualization of construction projects in a virtual environment. BIM software facilitates clash detection, quantity takeoffs, and 4D scheduling, enhancing project coordination and efficiency.
4. Agile Project Management: These methodologies, adapted from software development, emphasize flexibility, iterative development, and stakeholder collaboration. Agile approaches like Scrum and Kanban can improve responsiveness to changing project requirements and market conditions.

10.2.1.3. Best Practices for Effective Planning and Scheduling

1. Start Early and Plan Thoroughly: Commence planning and scheduling activities as early as possible in the project lifecycle. Thorough planning lays the foundation for a successful project outcome.
2. Involve Key Stakeholders: Engage stakeholders, including clients, architects, engineers, contractors, and subcontractors, in the planning process. Their input and

expertise ensure a comprehensive understanding of project requirements and constraints.

3. Utilize Technology and Tools: Take advantage of construction management software, scheduling tools, and Building Information Modeling (BIM) software to streamline planning and scheduling processes. These tools facilitate collaboration, visualization, and data-driven decision-making.
4. Monitor Progress and Adapt: Regularly monitor project progress against the schedule and adjust plans as necessary. Identify deviations from the plan early and implement corrective actions to keep the project on track.
5. Prioritize Safety and Quality: Integrate safety and quality considerations into the planning and scheduling process. Allocate sufficient time and resources for safety training, inspections, and quality assurance activities.
6. Maintain Clear Communication: Ensure clear communication channels are established among project stakeholders. Regularly communicate project updates, changes, and milestones to maintain alignment and transparency.
7. Foster a Culture of Continuous Improvement: Encourage a culture of continuous improvement within the project team. Learn from past experiences, seek feedback, and implement lessons learned to enhance future planning and scheduling efforts.

10.2.2. Hiring and Training

Hiring and training are crucial aspects of construction labor management, essential for building a skilled workforce capable of executing projects efficiently and safely.

10.2.2.1. Hiring Process

1. Define Job Roles and Requirements:
   - Job Analysis: Conduct a thorough analysis of the roles and responsibilities within your construction projects.
   - Skills and Qualifications: Determine the necessary skills, qualifications, certifications, and experience required for each job role.
2. Recruitment Strategy:
   - Targeted Channels: Utilize various recruitment channels such as job boards, industry associations, social media, and referrals to attract potential candidates.
   - Networking: Build relationships with trade schools, vocational programs, and industry organizations to access a pool of skilled candidates.
3. Screening and Selection:
   - Resume Review: Screen resumes to shortlist candidates based on their qualifications, skills, and experience.
   - Interview Process: Conduct structured interviews to assess candidates' technical skills, problem-solving abilities, and fit with the company culture.
   - Reference Checks: Verify candidates' employment

history, qualifications, and performance through reference checks.

4. Onboarding:
   - Orientation: Provide new hires with an orientation session to familiarize them with company policies, safety procedures, and job expectations.
   - Introduction to Tools and Equipment: Train new hires on the tools, equipment, and machinery they will be using on the job.
   - Buddy System: Pair new hires with experienced employees ("buddies") to help them acclimate to the job and learn on-the-job skills.

10.2.2.2. Training Programs

1. Safety Training:
   - OSHA Compliance: Ensure that all workers receive Occupational Safety and Health Administration (OSHA) training compliant with safety regulations.
   - Job-Specific Safety Training: Provide job-specific safety training tailored to the tasks and hazards encountered on construction sites (e.g., fall protection, scaffolding safety, electrical safety).
   - Emergency Response Training: Train workers on emergency procedures, including first aid, CPR, and evacuation protocols.
2. Technical Skills Training:
   - Trade-Specific Training: Offer training programs focused on developing trade-specific skills such as carpentry, masonry, plumbing, electrical work, welding, and concrete work.
   - Cross-Training: Provide opportunities for cross-training to enable workers to develop skills in multiple trades, enhancing their versatility and employability.
   - Apprenticeship Programs: Establish apprenticeship programs in collaboration with trade unions or vocational schools to provide hands-on training and mentorship to aspiring construction workers.
3. Leadership and Soft Skills Training:
   - Supervisory Training: Offer leadership development programs for foremen, supervisors, and project managers to enhance their communication, leadership, and conflict resolution skills.
   - Teamwork and Communication: Provide training on effective communication, teamwork, and collaboration to improve productivity and morale among construction teams.
   - Time Management and Problem-Solving: Equip workers with time management and problem-solving skills to address challenges and adapt to changing project requirements.
4. Continuing Education:
   - Professional Development: Encourage employees to pursue continuing education and professional development opportunities

to stay updated on industry trends, technologies, and best practices.

- Certification Programs: Support employees in obtaining industry certifications relevant to their roles (e.g., LEED certification for green building, project management certification).
- In-House Training Workshops: Organize in-house training workshops, seminars, and lunch-and-learn sessions on topics relevant to construction management, safety, and sustainability.

10.2.3. Performance Monitoring

Performance monitoring is crucial in construction labor management to ensure that projects are executed efficiently, safely, and within budget. By tracking and assessing the performance of construction labor, project managers can identify areas for improvement, optimize resource allocation, and mitigate risks (Shaik & Karthikeyan, 2012).

10.2.3.1. Key Performance Indicators (KPIs)

1. Labor Productivity:
    - Output per Hour: Measure the amount of work completed by labor resources per unit of time (e.g., square feet of concrete poured per hour, linear feet of piping installed per hour).
    - Unit Cost of Labor: Calculate the cost of labor per unit of work completed, taking into account labor hours, wages, and overhead costs.
2. Schedule Performance:
    - Percentage of Tasks Completed: Track the percentage of planned tasks completed on schedule to assess project progress.
    - Schedule Variance: Compare actual progress against the planned schedule to identify delays or deviations from the schedule.
3. Safety Performance:
    - Incident Rate: Monitor the number of safety incidents, accidents, and near misses per labor hour or per project to gauge safety performance.
    - Lost Time Injury Frequency Rate (LTIFR): Measure the frequency of lost time injuries per number of hours worked to assess workplace safety.
4. Quality Performance:
    - Defect Rate: Track the number of defects, rework, or callbacks related to labor activities to assess quality performance.
    - Client Satisfaction: Gather feedback from clients or stakeholders regarding the quality of workmanship and adherence to specifications.

10.2.3.2. Methods of Performance Monitoring

1. Direct Observation:
    - Site Visits: Conduct regular site visits to observe construction activities, assess progress, and identify any issues or challenges.
    - Time and Motion Studies: Use time and motion studies

to observe and analyze the efficiency of labor tasks and workflows.

2. Work Sampling:
   - Random Sampling: Select random samples of labor activities or tasks to assess productivity, efficiency, and compliance with quality standards.
   - Activity Sampling: Observe and record the activities of labor resources at regular intervals to understand how time is allocated and utilized.
3. Performance Metrics:
   - Data Collection: Collect data on key performance metrics, such as labor hours, output, costs, incidents, and defects, using timesheets, logbooks, incident reports, and project management software.
   - Analysis: Analyze performance data to identify trends, patterns, and areas for improvement, using statistical techniques and data visualization tools.

10.2.3.3. Continuous Improvement Strategies

1. Root Cause Analysis:
   - Identify Causes: Investigate the root causes of performance issues, delays, accidents, or quality defects to address underlying problems.
   - Corrective Actions: Implement corrective actions and preventive measures to mitigate risks and improve performance in the future.
2. Training and Development:
   - Skills Enhancement: Provide targeted training and development programs to address skill gaps, improve competencies, and enhance the capabilities of labor resources.
   - Safety Training: Offer additional safety training to reinforce safe work practices, hazard awareness, and emergency procedures.
3. Performance Feedback:
   - Regular Feedback: Provide constructive feedback to labor resources on their performance, highlighting areas of strength and areas for improvement.
   - Performance Reviews: Conduct regular performance reviews or evaluations to assess individual performance, set goals, and provide opportunities for career development.
4. Incentives and Recognition:
   - Performance-based Incentives: Implement incentive programs tied to performance metrics, such as productivity targets, safety goals, or quality standards.
   - Recognition Programs: Recognize and reward exemplary performance, safety initiatives, and contributions to project success to motivate and engage labor resources.

10.2.4. Communication and Coordination

Communication and coordination are fundamental pillars of construction labor management, ensuring that tasks are executed efficiently, safely, and according to project requirements. Effective communication fosters collaboration, enhances productivity, and mitigates risks, while robust coordination ensures that labor resources are deployed optimally and project goals are achieved.

10.2.4.1. Importance of Communication and Coordination

1. Clear Instructions:
   - Task Clarity: Clear communication ensures that labor resources understand their roles, responsibilities, and tasks, reducing confusion and errors.
   - Expectation Setting: Communicating project objectives, timelines, and quality standards helps align labor resources with project goals and client expectations.
2. Problem Resolution:
   - Issue Identification: Effective communication facilitates the early identification and resolution of issues, allowing for timely corrective action.
   - Collaborative Problem-Solving: Open communication channels enable labor resources to collaborate in solving challenges and overcoming obstacles encountered on-site.
3. Safety Awareness:
   - Safety Communication: Clear communication of safety protocols, hazard identification, and emergency procedures promotes a culture of safety and reduces the risk of accidents and injuries.
   - Risk Mitigation: Communicating safety concerns and near-miss incidents enables proactive risk mitigation and the implementation of preventive measures.
4. Team Cohesion:
   - Team Building: Effective communication fosters a sense of camaraderie and teamwork among labor resources, enhancing morale, motivation, and job satisfaction.
   - Conflict Resolution: Open dialog and transparent communication facilitate the resolution of conflicts and misunderstandings that may arise among team members.

10.2.3.4. Strategies for Communication and Coordination

1. Daily Meetings:
   - Morning Briefings: Conduct daily or pre-shift meetings to discuss project objectives, tasks for the day, safety reminders, and any issues or concerns.
   - Progress Updates: Provide updates on project progress, milestones achieved, and upcoming tasks to keep labor resources informed and motivated.

2. Clear Communication Channels:
   - Two-Way Communication: Establish open channels for communication, allowing labor resources to voice concerns, provide feedback, and ask questions.
   - Hierarchy of Communication: Define a clear hierarchy of communication to ensure that information flows effectively from project managers to supervisors to labor resources.
3. Written Instructions and Documentation:
   - Work Orders: Issue written work orders or task assignments outlining specific instructions, requirements, and safety guidelines for each task.
   - Project Documentation: Maintain project documentation, including plans, specifications, and safety manuals, accessible to all labor resources for reference.
4. Technology-Enabled Communication:
   - Mobile Apps: Utilize construction management apps and mobile platforms to facilitate real-time communication, task assignment, and progress tracking.
   - Digital Collaboration Tools: Implement digital collaboration tools such as project management software, cloud-based platforms, and communication apps to streamline communication and coordination among dispersed teams.
5. Site Signage and Visual Aids:
   - Safety Signage: Install signage and visual cues throughout the construction site to communicate safety warnings, restricted areas, and emergency procedures.
   - Site Maps and Plans: Display site maps, floor plans, and task schedules in common areas to provide visual guidance and enhance understanding of project requirements.

10.2.5. Compensation and Incentives

Compensation and incentives play a critical role in construction labor management, influencing worker motivation, productivity, and retention. By offering competitive compensation packages and attractive incentives, construction firms can attract, retain, and motivate skilled labor resources, leading to improved performance and project outcomes.

10.2.5.1. Compensation Strategies

1. Competitive Wages:
   - Market Analysis: Conduct regular market surveys to benchmark wage rates and ensure that compensation packages remain competitive within the industry.
   - Skill-based Pay: Offer differential pay based on the skills, experience, certifications, and qualifications of labor resources to incentivize skill development and retention.

2. Performance-based Pay:
   - Merit Increases: Provide performance-based salary increases or bonuses to recognize and reward outstanding performance, productivity, and contribution to project success.
   - Piece-Rate Pay: Implement a piece-rate pay system, where workers are compensated based on the quantity or quality of work completed, incentivizing efficiency and output.
3. Benefits and Perks:
   - Health Insurance: Offer comprehensive health insurance coverage, including medical, dental, and vision plans, to provide financial security and peace of mind to employees and their families.
   - Retirement Plans: Provide retirement savings plans, such as 401(k) or pension plans, with employer contributions or matching contributions to help employees build long-term financial stability.
   - Paid Time Off: Offer paid vacation days, holidays, and sick leave to promote work-life balance and employee well-being.
4. Recognition and Rewards:
   - Employee of the Month: Recognize outstanding performance and dedication by naming an "Employee of the Month" and offering rewards such as gift cards, certificates, or additional time off.
   - Longevity Awards: Celebrate employee loyalty and tenure by presenting longevity awards or service milestones, accompanied by tangible rewards or tokens of appreciation.

10.2.5.2. Incentive Programs

1. Safety Incentives:
   - Safety Bonuses: Offer safety bonuses or rewards to labor resources who demonstrate exemplary safety practices, adherence to safety protocols, and the prevention of accidents or injuries.
   - Safety Contests: Organize safety contests or competitions with prizes for teams or individuals who achieve specific safety goals or milestones.
2. Performance-based Incentives:
   - Production Bonuses: Provide production bonuses or incentives based on achieving production targets, meeting project milestones, or completing tasks ahead of schedule.
   - Quality Incentives: Offer quality incentives for maintaining high-quality workmanship, minimizing defects, and ensuring adherence to specifications and standards.
3. Training and Skill Development Incentives:
   - Tuition Reimbursement: Reimburse tuition costs or provide financial assistance for

employees pursuing further education, vocational training, or certification programs related to their job roles.

- Skill Upgradation Grants: Offer grants or subsidies for employees to attend workshops, seminars, or training programs aimed at enhancing their skills, knowledge, and competencies.

4. Referral Programs:

- Employee Referral Bonuses: Encourage employees to refer qualified candidates for open positions by offering referral bonuses or rewards for successful hires.
- Retention Bonuses: Provide retention bonuses or incentives to employees who refer candidates that remain with the company for a specified period, promoting employee engagement and loyalty.

10.3. IMPROVE LABOR MANAGEMENT IN THE CONSTRUCTION INDUSTRY

Effective management is a critical requirement for delivering projects on time and within the stipulated budget in the construction industry. With over two-thirds of construction projects failing to reach the handover phase within the estimated budget, labor management in construction is mandatory.

10.3.1. Tips for Improving Labor Management in Construction

1. Collaborate with the Workers: Generally, a construction project involves several subcontractors who provide labor. Collaboration and good communication are necessary to ensure that all contractors are on the same page and there is an open line of communication with the workers.

Construction projects are usually large, and proper construction labor management is necessary for the overall efficiency and effectiveness of the project. If you can facilitate seamless collaboration between the different groups of workers, then you will successfully minimize the errors that arise due to miscommunication.

Also, any mistakes committed in construction work can be detected and corrected quickly and easily. It will prevent a domino effect that can cause project delays.

2. Prioritize the Health and Safety of Laborers: The construction industry is often marked with mishaps and accidents unless proper safety measures are taken. Labor management in construction must prioritize the health and safety of laborers.

First, laborers must have reasonable working hours to ensure they are well-rested. A fatigued workforce will cause underperformance, and too much work will also reduce their cognitive abilities, which can lead to on-site accidents.

Construction work involves operating heavy machines, handling hazardous materials, and working at dangerous heights. All of these risks pose a substantial danger to workers' lives. Therefore, it is mandatory to provide safety in the workplace. Project managers must have proper safety measures implemented at all construction sites.

Furthermore, workers must cooperate by practicing safety measures. Laborers are more likely to stay with a company that emphasizes their safety.

3. Choose a Good Workforce Management Software: Construction work needs long and tedious administrative work that involves accurately logging laborers' time on the site, overtime, daily reports, and estimates, managing a large number of workers, ensuring workers are present on-site when logging into the system, and much more.
4. Balance the Skilled Workers Where Required: Invariably, there will be gaps in the workers' skills when you hire them and what the project requires. An essential part of labor management in construction is creating an ecosystem to foster the development of the workers into skilled labor.

This will require you to optimize the workforce utilization rate. Simply put, it means putting workers in roles that will develop their skills.

5. Identify the Issues That Workers Are Facing: Construction site disputes can cause laborers to lose motivation. It can also create a hostile environment for workers, which must be avoided at all costs. The construction staff is a valuable asset for any company, and project managers must identify workers' issues.

With effective labor management techniques, it can be ensured that workers are not facing problems that are hindering their performance.

6. Give Laborers all the Tools and Resources That are Needed: Another aspect of construction labor management is providing workers with the tools and resources they need to do their jobs smoothly. When allocating a budget to acquire new construction equipment, ask project managers what tools and resources are required to speed up the progress of the projects.

CASE STUDY

STREAMLINING CONSTRUCTION LABOR MANAGEMENT:

- **Background:** ABC Construction Company is a mid-sized construction firm specializing in residential and commercial projects. With an increasing number of projects in their pipeline, they are facing challenges in effectively managing their construction labor force. The current manual processes for scheduling, tracking hours, and ensuring compliance with safety regulations are becoming inefficient and prone to errors.

Challenges:

1. **Inefficient Scheduling***:* The current scheduling process relies heavily on manual input, leading to conflicts, delays, and underutilization of resources.
2. **Labor Tracking:** There is no centralized system for tracking labor hours, making it difficult to monitor productivity and accurately allocate costs to projects.
3. **Safety Compliance:** Ensuring that all workers adhere to safety regulations is crucial, but there is no streamlined process in place for monitoring and enforcing compliance.
4. **Communication:** This between project managers, supervisors, and laborers is often fragmented, leading to misunderstandings and delays.
   - **Solution:** ABC Construction Company decides to implement a comprehensive labor management system to address these challenges and improve efficiency across their projects.

Implementation Steps:

i. Labor Management Software: The company invests in labor management software that integrates scheduling, time tracking, and payroll functionalities. This software allows project managers to create optimized schedules, assign tasks, and track labor hours in real-time.

ii. Training and Onboarding: To ensure successful adoption of the new software, ABC Construction conducts training sessions for all employees involved in labor management. This includes project managers, supervisors, and laborers. Clear guidelines are provided on how to use the software effectively.

iii. Safety Compliance Protocol: A standardized safety compliance protocol is established and integrated into the labor management system. Before starting work on any project, laborers are required to complete safety training modules and acknowledge their understanding of safety procedures. Supervisors are responsible for conducting regular safety inspections and ensuring that all safety protocols are followed.

iv. Centralized Communication Platform: A centralized communication platform, such as a project management software or mobile app, is implemented to facilitate seamless communication between all stakeholders. This allows for instant updates, task assignments, and progress reports, reducing delays and misunderstandings.

Results:

1. **Improved Efficiency:** The implementation of the labor management software streamlines scheduling and tracking processes, leading to improved resource utilization and project timelines.
2. **Enhanced Safety Compliance:** With a standardized safety protocol in place and real-time monitoring capabilities, ABC Construction achieves higher levels of safety compliance across all projects.
3. **Better Communication:** The centralized communication platform fosters collaboration and transparency among project teams, resulting in fewer communication errors and delays.
4. **Cost Savings:** By accurately tracking labor hours and costs, ABC Construction is able to allocate resources more efficiently, leading to cost savings and improved profitability.

CLASS ACTIVITY

- **Objective:** To understand the challenges and strategies involved in managing construction labor effectively.
- **Materials Needed:** Construction project scenario (fictional or based on a real project). Role-play instructions for different stakeholders (project manager, foreman, workers, client representative, etc.). Whiteboard or flip chart for notes. Timer.
- **Assessment:** Briefly explain the importance of labor management in construction projects. Introduce the scenario: describe the construction project, its scope, timeline, budget, and key stakeholders.

SUMMARY

- In the construction industry, labor management is essential because it determines the company's bottom line. The workforce is the most critical of all the resources utilized in a construction project. Laborers and skilled workers are required to keep units operational and projects on-site within budget and on schedule.
- Construction labor can be categorized into skilled and unskilled. Skilled labor typically involves specialized tasks such as carpentry, plumbing, or electrical work, while unskilled labor includes general tasks like carrying materials or cleaning.
- Effective labor management involves tasks like hiring, scheduling, training, and supervising workers. Proper management ensures that the right people are in the right place at the right time, maximizing productivity and efficiency.
- Construction sites are inherently dangerous environments, so labor management must prioritize safety. This includes providing proper training, enforcing safety protocols, and ensuring compliance with regulations like OSHA (Occupational Safety and Health Administration) standards.
- Technical construction labor refers to workers with specialized skills and knowledge to carry out advanced tasks in the construction industry. These tasks usually require technical precision, critical thinking, and compliance with strict standards and regulations.

REVIEW QUESTIONS

1. What are the different types of construction labor commonly employed in the industry?
2. Discuss the importance of skill assessment and training programs in construction labor management.
3. Explain the concept of labor forecasting and its significance in construction project planning.
4. Discuss the role of effective communication and collaboration in enhancing labor management within construction teams.
5. How to improve labor management in the construction industry? Explain it.

REFERENCES

1. Abhiram, P., Asadi, S. S., & Prasad, A. V. S. (2016). Implementation of lean methodology in Indian construction. International Journal of Civil Engineering and Technology, 7(6), 641–649.
2. Achuth Kumar, N. V., & Asadi, S. S. (2017). Assessment on factors declining labour productivity in construction projects. International Journal of Civil Engineering and Technology, 8(1), 340–348.

3. Akin, C. (2014). Ecology dimension in construction management. International Journal of Civil Engineering and Technology, 5(3), 375–382.

4. Babu, S. S., & Sudhakar, B. (2016). Construction project management during economic crisis. International Journal of Management, 7(7), 371–381.

5. Dubey, A. (2015). Earned value analysis for a construction project. International Journal of Civil Engineering and Technology, 6(6), 53–66.

6. Jeelani, S. A. K., Karthikeyan, J., & Aldosary, A. S. (2012). Performance evaluation of design-build (D-B) projects with and without agency construction management. International Journal of Civil Engineering and Technology, 3(2), 265–278.

7. Katole, H. J. (2016). A study of contract labor at a real estate and construction company. International Journal of Civil Engineering and Technology, 7(3), 128–135.

8. Kumar, J. K. (2016). A study on key factors influencing bid/no-bid decisions for different construction projects in India. International Journal of Civil Engineering and Technology, 7(6), 732–737.

9. Mahesh, C., & Sridevi, K. (2013). Implementing Eichelay formula in govt. construction projects. International Journal of Civil Engineering & Technology, 4(3), 63–72. ISSN Print: 0976–6308, ISSN Online: 0976–6316.

10. Nwaogazie, I. L., Augustine, O. O., & Henshaw, T. (2016). Multi-skilling in construction industry and dual skill labor strategy: A case of construction companies in Port Harcourt. International Journal of Civil Engineering and Technology, 7(4), 208–222.

11. Reddy, G. T. P. K., & Naik, B. H. (2016). Enhancing cost efficiency in construction using earned value management. International Journal of Civil Engineering and Technology, 7(6), 357–363.

Index

D

E

I

J

K

L

M

N

O

P

S

T

U

V

W

Printed and bound by CPI Group (UK) Ltd, Croydon, CR0 4YY

21/06/2026

14904686-0001